Advances in Spatial Science

Springer
Berlin
Heidelberg
New York
Barcelona
Budapest
Hong Kong
London
Milan
Paris
Tokyo

Titles in the Series

Cristoforo S. Bertuglia, Manfred M. Fischer and Giorgio Preto (Eds.)
Technological Change, Economic Development and Space
XVI, 354 pages. 1995

Harry Coccossis, Peter Nijkamp (Eds.)
Overcoming Isolation
Information and Transportation
Networks in Development Strategies for Peripheral Areas
VIII, 272 pages. 1995

Luc Anselin · Raymond J. G. M. Florax (Eds.)

New Directions in Spatial Econometrics

With 62 Figures

Springer

Professor Dr. Luc Anselin
West Virginia University
Regional Research Institute
511 North High Street
Morgantown, WV 26506, USA

Dr. Raymond J.G.M. Florax
Wageningen Agricultural University
Department of General Economics
P.O. Box 8130
NL-6700 EW Wageningen, The Netherlands

Die Deutsche Bibliothek - CIP-Einheitsaufnahme

New directions in spatial econometrics / Luc Anselin ; Raymond J. G. M Florax (eds.). - Berlin ; Heidelberg ; New York ; Barcelona ; Budapest ; Hong Kong ; London ; Milan ; Paris ; Tokyo : Springer, 1995
(Advances in spatial science)
ISBN-13:978-3-642-79879-5
NE: Anselin, Luc [Hrsg.]

ISBN-13: 978-3-642-79879-5 e-ISBN-13: 978-3-642-79877-1
DOI: 10.1007/978-3-642-79877-1

Softcover reprint of the hardcover 1st edition 1995

SPIN 10505604 42/2202 - 5 4 3 2 1 0 – Printed on acid-free paper

Preface

The field of spatial econometrics, which is concerned with statistical and econometric techniques to be used to handle spatial effects in multiregional models, was first touched upon in the 1950s. It was given its name in the early 70s by Jean Paelinck and has expanded since. Its development can be monitored in various monographs that have been published since, starting with the seminal work by Andrew Cliff and Keith Ord. Also, the wide array of journals in which contributions to spatial econometrics have been published, shows that the relevance of the field is not restricted to regional science, but extends to geography, spatial statistics, biology, psychology, political science and other social sciences.

This volume contains a collection of papers that were presented at special sessions on spatial econometrics organized in the context of the European and North American conferences of the Regional Science Association International, that took place in Louvain la Neuve (August 25-28, 1992) and in Houston (November 11-14, 1993), respectively. Apart from these conference papers some contributions were written especially for this volume. The central idea of this book is to communicate the state of the art of spatial econometrics and to offer a number of new directions for future research. In order to do so, the editors sought contributions of leading scholars currently active in this field. The new directions in spatial econometrics include suggestions for new specifications, tests and estimators, the inclusion of spatial effects in binary choice models as well as the development of econometric techniques for space-time models (including Bayesian approaches). The application of these new ideas in actual research practice is encouraged by including computer code.

The editorial work was greatly facilitated by a travel grant of Wageningen Agricultural University (Office for International Relations and Department of General Economics). The type-setting was financially supported by two grants from the LEB-Fonds (i.e., Fonds Landbouw Export Bureau 1916/1918) based in Wageningen, The Netherlands. Support of the U.S. National Science Foundation to the

National Center for Geographic Information and Analysis, grant SES-88-10917, is also gratefully acknowledged.

Finally, editorial work and type-setting of manuscripts as complicated as this one can only be managed successfully with assistance of those willing to do the laborious and meticulous work of putting *i*'s, *j*'s, italics, fonts, etcetera in their proper places. Given the numerous tables, graphs, mathematics and computer code, we are particularly satisfied with the result, and greatly appreciate the efforts of Lianne Hamilton and Joop Heppenstrijdt, without whom the preparation of the camera-ready copy would not have been possible. Marc Loman not only assisted in the aforementioned work but also did a great job in the proof-reading stage. Erik Schmieman lend a hand in preparing the final camera ready copy.

March 1995
Morgantown, Luc Anselin
Wageningen, Raymond Florax

Contents

PART I-B: Spatial Effects in Linear Regression Models
Spatial Data and Model Transformations

List of Figures

List of Tables

INTRODUCTION

1 New Directions in Spatial Econometrics: Introduction

Luc Anselin[1] and Raymond J.G.M. Florax[2]
[1] West Virginia University, Morgantown, WV, U.S.A.
[2] Wageningen Agricultural University, Wageningen, The Netherlands

1.1 Introduction

Since Paelinck coined the term 'spatial econometrics' in the early 1970s to refer to a set of methods that deal with the explicit treatment of space in multiregional models, the field has come a long way. The early results established in regional economics by Blommestein, Hordijk, Klaassen, Nijkamp, Paelinck and others [e.g., Hordijk and Nijkamp (1977), Hordijk (1979), Paelinck and Klaassen (1979), Blommestein (1983)], as well as the path breaking work of Cliff and Ord in geography [Cliff and Ord (1973, 1981), Ord (1975)] have grown into a broad set of models, tests and estimation techniques that incorporate space more effectively in econometric modeling [for recent overviews, see Anselin (1988, 1992a), Haining (1990), Cressie (1991)]. In spite of these important methodological developments, it would be an overstatement to suggest that spatial econometrics has become accepted practice in current empirical research in regional science and regional economics. On the positive side, the sad state of affairs reflected in the literature surveys of Anselin and Griffith (1988) and Anselin and Hudak (1992) seems to have taken a turn for the better, since there is evidence of an increased awareness of the importance of space in recent empirical work in 'mainstream' economics. For example, this is indicated by the use of spatial models in the study of fiscal spill-overs in Case et al. (1993), the analysis of the productivity effects of public sector capital in Holtz-Eakin (1994), and the assessment of land price volatility in Benirschka and Binkley (1994), among others. However, most studies that do take spatial effects into account are typically not based on the application of 'state of the art' spatial econometric methods, and, in fact, rarely go beyond the use of methods outlined in Cliff and Ord (1981). In other words, the findings published in the literature in regional science and analytical geography over the past ten to fifteen years still seem to be unfamiliar to most analysts in empirical economics. Moreover, many new results and models have been suggested that go well beyond the linear case that is emphasized in Cliff and Ord (1981). A main motivation for our effort to bring together the current collection of fourteen papers in an edited volume is to attempt to rectify this situation.

We follow Anselin's (1988, p. 7) definition of spatial econometrics and consider it as 'the collection of techniques that deal with the peculiarities caused by space in the statistical analysis of regional science models'. We refer to 'regional science models' as those specifications that incorporate regions, location and spatial interaction explicitly and/or that are based on geo-referenced (i.e., cross-sectional or space-time) data for their statistical estimation and validation. As in Anselin (1988), we focus on two broad classes of spatial effects: *spatial dependence* and *spatial heterogeneity*. The first implies not only a lack of independence, but also a *spatial structure* underlying this 'spatial' correlation (auto- or cross-correlation), in the sense that the strength and specification of the spatial dependence are determined by absolute and relative location (topology, distance). Spatial dependence is relevant in two respects. In one, referred to as *substantive* spatial dependence, the main interest is in the specification of the spatial interaction behind the variable of interest (e.g., the dependent variable in a regression model). For example, one may want to discover the spatial pattern of the adoption of technical innovations by farmers, as in Case (1992), or fiscal 'copy-catting' between neighboring jurisdictions, as in Case et al. (1993). Alternatively, the spatial dependence between the ignored variables in the model will be reflected in the error terms, which we refer to as *nuisance* dependence. For example, this may be necessary to express neighborhood effects in models of housing values, as in Dubin (1992), or due to the limitations of administrative units of observation (such as counties) to analyze agricultural land values, as in Benirschka and Binkley (1994). While substantive spatial dependence necessitates the development of spatially explicit models, nuisance dependence typically only involves adjustments to the interpretation and inferential use of existing specifications.

The second type of spatial effect, spatial heterogeneity, deals with spatial or regional differentiation. In other words, this is a special case of the more general problem of structural instability, but where the structure in question pertains to *spatial structure*, i.e., to specific locations or subregions in a data set. In practice, this issue is often encountered in the choice between fixed effects and random effects models. In the former, each location (region) would have its own functional specification (and/or coefficient), whereas in the latter, all locations are assumed to conform to the same encompassing model and region-specific characteristics are conceptualized as random deviations from the overall mean [see Anselin (1990) for further discussion].

We suggest that the importance and relevance of methods to deal with spatial effects in econometric models is on the increase. We see three major reasons for this:

- The renewed interest for the role of space and spatial interaction in *social science theory*. In mainstream economic theory, for example, this is reflected in the interest for the 'new economic geography' of Krugman (1991, 1993), among others. However, this recently rediscovered attention to space is not limited to economics, but is also reflected in other social sciences, such as the study of international relations [e.g., in the new geopolitics, Ward (1992)]. More importantly, the role of 'space' is central in the development of new interdisciplinary theoretical frameworks to deal with human-environmental interactions in the study of global change [e.g., Lambin (1994), Turner et al. (1994)].
- The increased availability of large *socio-economic data sets* with geo-referenced observations. Both in the United States and in Europe, official statistical agencies are providing large data sets with information at the regional and local scale in accessible and low cost formats. For example, in the United States, detailed information at the county level is contained in the REIS (Regional Economic Information System) CD-ROM of the U.S. Department of Commerce. A similar data set for the European Community is the REGIO databank of Eurostat. These types of data allow for the analysis of unprecedented numbers of cross-sectional and/or space-time observations, for which spatial (and/or space-time) autocorrelation is likely to be the rule rather than the exception.
- The development of an efficient and low cost computational technology to handle spatially referenced observations, in the form of geographic information systems (GIS) and spatial data analysis software. The use of a GIS allows for the efficient storage, rapid retrieval and interactive visualization of patterns in geographical data sets and provides a great opportunity for the implementation of state of the art spatial analytical techniques [e.g., Goodchild et al. (1992), Anselin and Getis (1992), Anselin et al. (1993), Fischer and Nijkamp (1993), Fotheringham and Rogerson (1993)]. Furthermore, the lack of software for spatial data analysis and spatial econometrics, which has often been cited as a reason for the lack of dissemination of the techniques [e.g., in Haining (1989)] has largely been addressed, at least as far as linear models are concerned, e.g., by the *SpaceStat* software package [Anselin (1992b, 1994a)].

Given these developments, we felt it was important and timely to collect a series of papers from active scholars in the field that reflect the state of the art in spatial econometrics and indicate promising new directions for research and applications. These new directions consist of alternative model specifications, estimators and tests for regression models in general, but, more importantly, of new perspectives on dealing with spatial effects in models with limited dependent variables and in space-time models. We have divided the volume accordingly into three parts. The chapters included range from the introduction of new model specifications and tests for spatial dependence of both substantive and nuisance form (Anselin and Florax, Kelejian and Robinson, Bolduc et al., Dubin, Griffith) as well as for spatial heterogeneity (McMillen, Ferguson and Kanaroglou, Casetti and Poon) to the discussion of practical implementation issues, such as model and data transformation (Van

Gastel and Paelinck, Haining, Getis), the choice of spatial weights (Florax and Rey) and Bayesian approaches (Duncan et al., LeSage). Each chapter is illustrated with an empirical example, either a numerical experiment, a Monte Carlo simulation study or the application of a technique to actual data sets. The latter include observations on the regional product in Dutch provinces (Van Gastel and Paelinck), trips to work among 147 traffic zones in Winnipeg, Canada (Bolduc et al.), mortality rates for 87 community medicine areas in Glasgow, Scotland (Haining), urban crime in 49 neighborhoods in Columbus, Ohio, regional disparities in 16 regional divisions of Turkey and state and local government expenditures for 48 continental U.S. states (Getis), milk yields for five years in 73 municipios of Puerto Rico (Griffith), the relationship between economic growth and exports for three time periods and 91 countries (Casetti and Poon), annual income tax collections for 40 school districts in Allegheny county, Pennsylvania over 15 years (Duncan et al.), and weekly CD rates for 21 banks and 19 thrifts in the state of Ohio over a 200 week period (LeSage). Whenever possible, software code to implement the methods covered in a given chapter is included as an appendix.

In the next three sections, we briefly review the contents of each part of the volume and provide further perspective on the specific chapters included. In the concluding section, we suggest some promising directions for future research in spatial econometrics.

1.2 Spatial Effects in Regression Models

The first part of the volume focuses on spatial effects in linear and nonlinear regression models. We have divided it into two sections, one in which the specification of spatial dependence is emphasized, the other with chapters that devote special attention to spatial data and model transformation issues.

1.2.1 Specification of Spatial Dependence

The four chapters contained in this section deal with the formal specification of spatial dependence in regression models. In the first, 'Small Sample Properties of Tests for Spatial Dependence in Regression Models: Some further results', by Anselin and Florax, the empirical size and power is compared for eight tests against spatial autocorrelation. The tests include the familiar Moran's I, four spatial Lagrange Multiplier (LM) tests (for first and second order error dependence, lag dependence, and a first order spatial autoregressive, moving average or SARMA process [see Anselin (1988, 1994b), for a review]), two robust forms of the LM test due to Bera and Yoon (1992), and the robust test of Kelejian and Robinson (1992). In a series of Monte Carlo experiments for both irregular (economic regions in the Netherlands) and regular square lattice configurations, the performance of these tests is compared against three one-directional alternatives (spatial AR error dependence, spatial MA error dependence, and spatial AR lag) and three two-directional alternatives (second order spatial AR errors; second order spatial MA errors, and a SARMA process). For all but Moran's *I*, the LM-error and LM-lag tests, which were extensively assessed in Anselin and Rey (1991), there is no earlier evidence on the finite sample performance of the tests. Also, this is the first time that tests are considered against higher order forms of spatial dependence, and against both autoregressive and moving average forms of spatial processes. The results of the experiments confirm earlier findings on the power of Moran's *I* against any form of spatial dependence. However, this power is a blessing in disguise, since the statistic provides little insight into which form of spatial dependence is the correct alternative hypothesis. Another confirmation of earlier findings is the high power of the LM-lag test, even in relatively small samples (49 observations). The LM tests provide the basis for a useful rule of thumb in terms of choosing an alternative, in that the most significant test consistently points to the proper model [see also Florax and Folmer (1992)]. They also closely approximate their asymptotic properties in the largest sample considered, with 127 observations. Given the increasing size of spatial samples under consideration in empirical work and the ease by which these tests can be implemented, even with existing econometric software packages [see, e.g., Anselin and Hudak (1992)], they should become a standard element in the set of diagnostic tools in spatial econometrics.

While Anselin and Florax consider the traditional AR and MA forms of spatial processes as alternative hypotheses, Kelejian and Robinson introduce a new error components form for this model in their chapter 'Spatial Correlation: A suggested alternative to the autoregressive model'. They start with a critical examination of the traditional interpretation of the coefficient ρ in a first order spatial autoregressive process as a 'correlation' coefficient. They focus in particular on the singularity points in the familiar Jacobian $\boldsymbol{I} - \rho\boldsymbol{M}$ (where $\boldsymbol{M}$ is a spatial weights matrix in their notation), and suggest an interpretation that allows the autoregressive coefficient to take on any value, except for a set of singularity points. In such a framework, they argue that the singularity points need to be excluded from the parameter space, which causes problems in terms of the usual large sample theory basis for the asymptotic properties of the estimates. Kelejian and Robinson complement their formal arguments with a number of simple numerical illustrations. They go on to formulate an alternative specification for spatial error dependence in which there are two (independent) stochastic shocks associated with each location (region). One of these is location-specific, in that it does not lead to spatial spill-overs, while the other does. The regression error term is then seen as consisting of the sum of the location-specific disturbance and a weighted sum of the 'spill-over' disturbance terms for neighboring regions. The latter is very similar to a spatial moving average form, except that no moving average parameter is associated with it [see Anselin (1994b), for details]. Instead, the only two parameters to be estimated are the variances of the location-specific and spill-over error terms, as in an error components model. Kelejian and Robinson illustrate the straightforward interpretation of their specification in terms of correlations between the error terms. Also, in contrast to the more familiar spatial autoregressive specification, no problems of singularity occur for the error variance matrix of their form. The model can easily be extended to include other forms of stochastic shocks in an additive manner as well. It is the basis for the General Methods of Moments (GMM) estimator for spatial processes that is outlined in Kelejian and Robinson (1993).

A nonlinear form for spatial error dependence is outlined in the chapter by Bolduc, Laferrière and Santarossa on 'Spatial Autoregressive Error Components in Travel Flow Models: An application to aggregate mode choice'. This chapter, which is an extension of the model presented in Bolduc et al. (1992), treats the error term in a spatial interaction model as consisting of three (independent) components: one associated with the origin, one with the destination and one with each flow. Each component is assumed to follow a spatial autoregressive process of the usual form. The model is nonlinear in the sense that the elements of the spatial weights matrix for each autoregressive component are themselves a function of distance, with a parameter to be estimated (in addition to the autoregressive parameter). In other words, in addition to the regression coefficients, the model contains nine error parameters, three for each component: an error variance, a spatial autoregressive coefficient and a distance decay parameter (for the elements in the weights matrix). Assuming normality, this specification can be estimated by means of maximum likelihood procedures, as a special case of the model with general nonscalar error covariance outlined in Magnus (1978). This approach is applied to an empirical study of mode choice between automobile and bus trips for 147 travel zones in Winnipeg. The model is estimated by means of ordinary least squares, standard error components and spatial autoregressive error components. The results clearly demonstrate the superiority of the latter specification in this example. In addition to using actual data, the authors also provide an illustration of the estimation procedure for a synthetic data set. Again, the superiority of the spatial autoregressive error components specification is clearly demonstrated.

Whereas in the three first chapters in this section, the spatial weights matrix is assumed to be correctly specified, the consequences of failing to do so are addressed in the chapter by Florax and Rey, 'The Impacts of Misspecified Spatial Interaction in Linear Regression Models'. This chapter gives an overview of the problems that may result from the misspecification of the (fixed) weights matrix. It is essentially the modifiable areal unit problem that is the basis for the introduction of spatial weights, in the sense that a specific level of spatial aggregation has to be chosen as well as a spatial arrangement in terms of patterns of contiguity. The latter, usually defined as a (linear) transformation of the observations on contiguous spatial units or spatial units at a fixed distance, is needed for reasons of identification. In series of Monte Carlo experiments, the effect of a wrong choice for the (fixed) spatial weights matrix is assessed on the power of three tests for spatial dependence (Moran's I, and the Lagrange Multiplier tests for the spatial error and the spatial lag) and on the bias and efficiency of three estimators (ordinary least squares, estimated generalized least squares and maximum likelihood). Particular attention is paid to pre-test aspects as in Florax and Folmer (1992). The experimental results indicate that overspecification of the weights matrix leads to reduced power of the tests, whereas underspecification has a less serious impact. Further, misspecification of the weights matrix is shown to have an impact on the sampling properties of alternative estimators for spatial process models, but the effects do not alter the relative performance of the alternative specification approaches that would occur when the weights matrix is specified correctly. The development of new tests for weights matrix misspecification is considered to be an area where additional research is required.

1.2.2 Spatial Data and Model Transformations

The second section of this first part contains three chapters that focus on issues related to the transformation of spatial models and spatial data. Van Gastel and Paelinck extend the use of the Box-Cox transformation to spatial econometric applications in 'Computation of Box-Cox Transform Parameters: A new method and its application to spatial econometrics'. Their point of departure is the expression of the nonlinear Box-Cox model for a single explanatory variable in the form of partial elasticities, which are subsequently approximated by linear expansion. In a series of numerical illustrations, the authors investigate the extent of linearization, discretization and specification bias. A spatial econometric form of the transformation is introduced for a bivariate model of per capita income in Dutch provinces and a suggestion is formulated for its extension to multiple explanatory variables.

In contrast to van Gastel and Paelinck, Haining does not emphasize spatial models per se, but focuses on the characteristics of spatial data and their effect on the traditional assessment of goodness-of-fit in a linear regression model in his chapter on 'Data Problems in Spatial Econometric Modeling'. Haining starts with a typology of spatial data types, based on insights from data modeling as viewed in the literature on geographic information systems. From this he develops a classification of generic spatial data problems that result from the discrete representation of a continuous reality, in which he differentiates between problems with attributes, problems with the areal framework and area/attribute interactions. He emphasizes that in contrast to the failure to adhere to statistical assumptions (the object of specification tests), data problems may give rise to effects that are difficult to identify. He goes on to focus on a particular type of data problem, i.e., the detection of influential observations, and outlines the difficulties encountered when implementing standard diagnostics (such as the use of studentized residuals, interpretation of the hat matrix, and Cook's distance) in a context where the regression error terms are spatially correlated [see also Haining (1994)]. Haining illustrates these 'adjusted' diagnostics in an empirical example of a bivariate regression of cancer mortality rates on a deprivation index, for 87 community medicine areas in Glasgow, Scotland. A clear difference between the non-spatial diagnostics (based on ordinary least squares regression) and the spatially adjusted ones (based on a maximum likelihood estimation of a spatial error model) is indicated.

In Getis's chapter on 'Spatial Filtering in a Regression Framework: Examples using data on urban crime, regional inequality and government expenditures', an attention to the spatial characteristics of the data is the basis for a particular type of transformation, referred to as spatial filtering [see also Getis(1990)]. Getis' rationale for the filter is that the fit and interpretation of regression models will improve after spatial autocorrelation is removed from both dependent and explanatory variables. The filter is based on the $G_i(d)$ statistic of Getis and Ord (1992), which is a local (i.e., observation-specific) indicator of spatial association. For a given distance d, chosen in function of the observed $G_i(d)$ statistics, a filter is applied to the original observations to remove spatial autocorrelation. The difference between the original variable and the filtered one is posited to represent the spatial effects embedded in the original. Getis outlines four criteria to assess the effectiveness of the filter: 1) the lack of spatial autocorrelation in the filtered variable; 2) spatial autocorrelation in the difference between the original and the filtered variable; 3) no residual spatial autocorrelation in a regression with filtered variables; and 4) improved significance of the regression variables after filtering. He goes on to illustrate the filter for three empirical examples, replicating published studies where no filter was used. These examples are a regression for urban crime in 49 Columbus, Ohio neighborhoods, a study of regional disparities among 16 regional divisions of Turkey, and the relation of state and local government expenditures to per capita federal grants, population and per capita income for 48 U.S. states. In all three instances, the filtering approach satisfies the four criteria.

1.3 Spatial Effects in Limited Dependent Variable Models

The second part of the volume is devoted to the treatment of spatial effects in models with limited dependent variables. Unlike the linear regression case, for which a large body of results already exists, the literature on spatial logit and probit models is very recent and still limited in scope. Only in Case (1991, 1992) and McMillen (1992) are attempts formulated to incorporate a form of spatial dependence or spatial heterogeneity in probit models, but neither is of general applicability. The three chapters in this part extend these findings and add new perspectives to this issue.

In the chapter by McMillen, on 'Spatial Effects in Probit Models: A Monte Carlo investigation', a series of Monte Carlo simulations are carried out to investigate the effects of both spatial autocorrelation and spatial heterogeneity on estimation and specification testing of probit models. McMillen is particularly interested in the influence of potential spatial autocorrelation on the properties of the heteroscedastic probit. The latter can be a special case of Casetti's spatial expansion model, as in

McMillen's earlier work [McMillen (1992)], but in this chapter it is of a general form. In order to distinguish between the heteroscedasticity and spatial autocorrelation, he specifies the latter in a form that does not induce heteroscedasticity (as is the case for the more traditional spatial autoregressive error process), as a direct function of the distance between two locations (observations). By varying two parameters of a heteroscedastic function and the coefficient that indicates spatial dependence, he is able to assess the relative influence of both factors in a series of Monte Carlo simulations for 200, 500 and 800 observations. In terms of estimation, the standard probit turns out to be more affected by heteroscedasticity than by spatial autocorrelation. For moderate amounts of heteroscedasticity, the probit performs acceptably, but this deteriorates as heteroscedasticity increases. Also, the influence is more pronounced when attention is focused on the marginal effects, rather than on the coefficients themselves. The standard probit estimates are still consistent in the presence of spatial autocorrelation (of the form that does not induce heteroscedasticity), though the standard error estimates are biased. The effect of spatial autocorrelation is also investigated with respect to the power of Lagrange Multiplier (LM) and Likelihood Ratio (LR) tests against heteroscedasticity. The LM test turns out to have low power in the smaller samples, but approaches the power of LR as the sample size increases. Neither test is affected by the presence of spatial autocorrelation, which contrasts with the case in the standard linear regression model, where there is a strong relation between the two, due to the induced heteroscedasticity of spatial AR or MA error processes. The extensive set of simulation results in McMillen's chapter is the first published small sample evidence on the effect of spatial autocorrelation in probit models.

In her chapter on 'Estimating Logit Models with Spatial Dependence', Dubin introduces a specification for a logit model with space-time dependence. The model is motivated in the context of innovation diffusion, where the decision to adopt at a given point in time is posited to depend (in addition to other explanatory variables) on the decision by 'neighbors' to adopt in a previous point in time. In other words, this model is the expression of the classic spatial innovation diffusion process as a logit model. As in her earlier work [e.g., Dubin (1992)], she uses an exponential distance decay form to specify the spatial interaction. Parenthetically, this is also the specification used by McMillen in the previous chapter. Dubin formulates the log likelihood function for this model and derives its first and second partial derivatives, needed to carry out a nonlinear optimization and to obtain the information matrix. She illustrates this approach with a synthetic example in which the spatial (more precisely, space-time) dependence is simulated. This allows her to compare the results of the standard logit with those from the spatial logit. Due to the misspecification of the standard logit in the presence of spatially autocorrelated errors (caused by the ignored space-time dependence), its coefficient estimates are inconsistent. The spatial logit performs well and obtains a superior fit. The model suggested by Dubin is a useful addition to the literature and clearly appropriate in situations where there are observations on past occurrences of a limited dependent variable, but only contemporaneous observations are available for the other explanatory variables in the model. For example, this would be the case when the explanatory variables do not vary over time, which frequently occurs in empirical work.

Ferguson and Kanaroglou focus on the underlying causes for the presence of spatial heterogeneity and dependence in the context of spatial choice models in 'Utility Variability within Aggregate Spatial Units and its Relevance to Discrete Models of Destination Choice'. The point of departure is an explicit treatment of aggregation problems in situations where there is a mismatch between the scale of the spatial choice units observed in the data and the scale that is likely to be evaluated by decision makers. The authors argue that this necessitates an explicit treatment of the utility variability within aggregates. They introduce corrective size and heterogeneity terms to modify the utility associated with aggregate choice units in a heterogeneous choice model and compare it to the standard multinomial logit form. The crucial correction term is a ratio of variances in a random utility model that corresponds to an overall measure of correlation of elemental units within aggregate spatial alternatives, i.e., some form of spatial autocorrelation. This effect can vary between alternatives and among decision makers, reflecting a form of spatial heterogeneity as well. Using a generalized logit framework, Ferguson and Kanaroglou express the likelihood function for the heterogeneous model and derive the expressions for the first and second order partial derivatives, needed to carry out nonlinear optimization. The properties of this model are assessed in a number of simulation studies for both aggregate data and individual micro data. The results indicate the bias in the parameters of the ordinary multinomial logit model when the spatial utility variability is ignored and the superior fit and inference yielded by the heterogeneous model. This chapter constitutes and important step forward in an effort to integrate spatial choice analysis and spatial econometrics, a promising area of research deserving of more attention.

1.4 Heterogeneity and Dependence in Space-Time Models

The third part of the volume contains four chapters that deal with various aspects related to the treatment of spatial effects in models that combine cross sectional and time series data. The first two chapters fit in a classical statistical framework, while the other two represent Bayesian approaches to spatial econometrics. The Bayesian perspective has been underutilized in spatial econometrics and the two chapters included in the this part illustrate its potential to lead to important contributions, especially in the context of space-time forecasting.

The chapter by Griffith on 'The General Linear Model and Spatial Autoregressive Models' considers the introduction of a spatial autoregressive element into multivariate linear models. The discussion is set in the context of an interesting data set, consisting of milk production yields for 73 municipios in Puerto Rico, for which five time points are considered with data separately collected by two agencies, the Commonwealth of Puerto Rico and the U.S. Department of Agriculture. In addition to its unique character in terms of identifying spatial data quality issues pertaining to the two agencies, this data set also allows for the treatment of a number of other important data issues, such as dealing with missing values, data transformations and the elimination of spatial autocorrelation. In the specific context of a canonical correlation model, Griffith introduces a procedure to incorporate a spatial autoregressive process. This also yields an approach to filter the data for spatial autocorrelation, based on the estimates of that process, which provides an alternative to the spatial filtering procedure advocated in the chapter by Getis.

Casetti and Poon introduce a diagnostic to assess the contextual drift of a model in their chapter on 'Econometric Models and Spatial Parametric Instability: Relevant concepts and an instability index'. The instability index is outlined within Casetti's (1972) spatial expansion framework, though it has applicability to many other models that allow for parameter variation. In the expansion approach, the coefficients of an initial model are expressed in function of additional variables (in so-called expansion functions) to explain contextual drift. This yields a so-called terminal model. As shown in Casetti (1986), each terminal model generated from a linear initial model has a second (dual) initial model that yields the same terminal model. Accordingly, two instability indexes are suggested, one based on the primal, the other on the dual expansion equations. In each, the parametric instability is measured by the variance of the predicted values in the terminal model where the explanatory variables of the initial model are replaced by their means. The resulting index is made dimensionless by dividing it by the variance in the predicted values in the usual expression (with the explanatory variables taking their observed values). The difference between the two indices lies in the use of the primal versus the dual expansion equations, which allows an assessment of the relative importance of each source of contextual drift. The indices can be computed over the whole data set, as well as over any subset of the data. Casetti and Poon suggest that this approach can also be applied to investigate contextual drift for any theoretical model for which empirical results have been published and a data set is available to compute the expansions. They illustrate the approach with a study of the relationship between economic growth and export growth for 91 countries over three periods in time. The primal expansion is in function of development level and yields strong evidence of a contextual effect. Moreover, they find that the relation between growth of product per capita and export growth is more unstable (the primal expansion) than the relationship between rates and levels of product per capita (the dual expansion). The instability indices provide a useful addition to the set of diagnostic tools available in spatial econometrics.

In the first of the Bayesian chapters, Duncan, Gorr and Szczypula provide further evidence on 'Bayesian Hierarchical Forecasts for Dynamic Systems: Case study on backcasting school district income tax revenues'. This chapter extends their earlier work on Bayesian forecasting [Duncan et al. (1993)] with a more general discussion and an empirical application to backcasting government revenues for small subcounty spatial units. The approach consists of the combination of a multi-state Kalman filter and a conditionally independent hierarchical model into a cross-sectional multi-state Kalman filter. More specifically, the Kalman filter time-varying parameter model is updated by means of a Bayesian shrinkage procedure that combines individual location parameters with cross-sectional averages, thereby 'borrowing strength' through pooling across space. This is particularly useful when the time series are short and contain considerable short term variation and structural shocks. In general terms, the procedure is described as a sequential and recursive estimation system with four modular components: 1) the transformation of time series data to remove cross sectional variation due to scale differences; 2) clustering of observations into independent reference groups, based on similarity and exchangeability; 3) Kalman filter time series adaptive estimation; and 4) cross-sectional pooling to borrow strength and increase the reliability of the estimate. The cross sectional multi-state Kalman filter is compared to the standard multi-state Kalman filter in a backcasting exercise for income tax collection over a 15 year period, in 40 school districts of Allegheny county, Pennsylvania. In this illustration, the cross sectional approach turned out to be more accurate than the univariate multi-state Kalman filter, and more so as the time series were shorter, the sensitivity to economic cycles increased and the backcasting horizon was longer. This provides a strong argument for further consideration of ways in which the performance of time series forecasts may be improved by means of incorporating spatial information.

A different form of incorporating information from spatially contiguous observations from a Bayesian perspective is presented in the chapter by LeSage on 'A Multiprocess Mixture Model to Estimate Space-Time Dimensions of Weekly Pricing of Certificates of Deposit'. LeSage starts with a dynamic model to describe optimal CD rate setting by banks and thrifts in function of the desired level of deposits, adjustment costs and the rates offered by competing institutions. He is particularly interested in discovering the extent of spatial competition in rate setting. The empirical expression of the theoretical model takes the form of a dynamic linear model with spatial time-varying parameters, following the general Bayesian framework developed by West and Harrison (1989). Its distinguishing characteristic is the use of a competitive influence parameter, which follows a transition pattern that reflects a linear combination of a traditional Markov scheme and a spatial weighting of contiguous observations. An estimable parameter determines the extent to which each factor contributes towards the variation of the competitive coefficient over time. As in the chapter by Duncan et al., the type of parameter transition employed by LeSage uses the mean of a parameter for the spatially contiguous observations as Bayesian prior information. Estimation of the model is based on a set of Kalman filter recursions. The model is implemented for a sample of 21 banks and 19 thrifts in Cincinnati, Cleveland and Toledo, Ohio, over a 200 week period which includes the 1985 Ohio Deposit Insurance crisis. Banks in the same metropolitan area are considered to be spatially contiguous and so are thrifts, resulting in a block-wise spatial weights matrix which precludes competition between banks and thrifts and between institutions in different areas. The spatial model is compared to a traditional Markov transition scheme to specify the varying parameters and found to be superior. An interesting side result of the model is the indication how the spatial effects varied over time, and, in particular, how the spatial competitive influence increased after the deposit crisis. In addition, LeSage points out the importance of specifying the extent of spatial influence, i.e., the choice of the spatial weights matrix, an issue also addressed in the chapter by Florax and Rey. The detailed insight into the variability of parameters over time and across space is one of the distinguishing characteristics of the Bayesian approach, and one with great scope for application in policy modeling. Despite the fact that its use in spatial econometrics is only starting, it constitutes a very promising avenue for future research.

1.5 Future Directions

The issues touched upon in this volume are many and provide a glimpse of the ongoing research in spatial econometrics. Three themes in particular indicate some potentially fruitful directions for future research. One pertains to the familiar issue of the specification of the extent and intensity of spatial interaction in the form of spatial weights. Recent developments in computation intensive approaches to estimation and inference may lead to new ways of attacking this issue. One example is the semi-parametric approach outlined in a recent paper by Brett et al. (1994) in which the weights are approximated by a very general functional form. A second theme is the need to more effectively incorporate spatial effects in nonlinear models in general, and limited dependent variable models, such as probit and tobit, in particular. Here again, recent advances in computation intensive approaches, such as econometric inference using simulation techniques [e.g., Brown et al. (1993)] may provide a means to avoid the analytical intractability that follows from the two-dimensional and multi-directional nature of spatial dependence. A third theme is the importance of dealing with spatial heterogeneity and spatial structural change, particularly in models that combine observations over time and across space. The development of a Bayesian spatial econometric perspective, while still in its infancy, seems to be particularly promising in this respect. An example is the application of the Gibbs sampler to a number of spatial modeling situations, e.g., as illustrated in recent papers by LeSage (1994a, 1994b).

We hope that the collection of papers in this volume will provide a useful background as well as a stimulus to pursue these future directions.

References

Anselin, L., *Spatial Econometrics: Methods and Models*, Dordrecht: Kluwer Academic, 1988.

Anselin, L., Spatial Dependence and Spatial Structural Instability in Applied Regression Analysis, *Journal of Regional Science*, 30, 185-207, 1990.

Anselin, L. (ed.), Space and Applied Econometrics (Special Issue), *Regional Science and Urban Economics*, 22, 1992a.

Anselin, L., *SpaceStat: A Program for the Analysis of Spatial Data*, Santa Barbara, CA: National Center for Geographic Information and Analysis, University of California, 1992b.

Anselin, L., *SpaceStat Version 1.50: Revision Notes*, Morgantown, WV: West Virginia University, Regional Research Institute, Research Paper 9428, 1994a.

Anselin, L., *Testing for Spatial Dependence in Linear Regression Models: A Review*, Morgantown, WV: West Virginia University, Regional Research Institute, Research Paper 9416, 1994b.

Anselin, L., R. Dodson and S. Hudak, Linking GIS and Spatial Data Analysis in Practice, *Geographical Systems*, 1, 3-23, 1993.

Anselin, L. and A. Getis, Spatial Statistical Analysis and Geographic Information Systems, *The Annals of Regional Science*, 26, 19-33, 1992.

Anselin, L. and D.A. Griffith, Do Spatial Effects Really Matter in Regression Analysis?, *Papers of the Regional Science Association*, 65, 11-34, 1988.

Anselin, L. and S. Hudak, Spatial Econometrics in Practice, A Review of Software Options, *Regional Science and Urban Economics*, 22, 509-536, 1992.

Anselin, L. and S. Rey, Properties of Tests for Spatial Dependence in Linear Regression Models, *Geographical Analysis*, 23, 112-131, 1991.

Benirschka, M. and J.K. Binkley, Land Price Volatility in a Geographically Dispersed Market, *American Journal of Agricultural Economics*, 76, 185-195, 1994.

Bera, A.K. and M.J. Yoon, *Simple Diagnostic Tests for Spatial Dependence*, Champaign, IL: Department of Economics, University of Illinois, 1992.

Blommestein, H., Specification and Estimation of Spatial Econometric Models: A Discussion of Alternative Strategies for Spatial Economic Modelling, *Regional Science and Urban Economics*, 13, 250-271, 1983.

Bolduc, D., R. Laferrière and G. Santarossa, Spatial Autoregressive Error Components in Travel Flow Models, *Regional Science and Urban Economics*, 22, 371-385, 1992.

Brett, C., J. Pinkse and M. E. Slade, *Spatial Price Competition, A Semiparametric Approach*, Paper Presented at the 41st North American Meetings of the Regional Science Association International, Niagara Falls, ON, 1994.

Brown, B.W., A. Monfort and H. K. van Dijk (eds.), Econometric Inference Using Simulation Techniques (Special Issue), *Journal of Applied Econometrics*, 8, 1993.

Case, A., Spatial Patterns in Household Demand, *Econometrica*, 59, 953-965, 1991.

Case, A., Neighborhood Influence and Technological Change, *Regional Science and Urban Economics*, 22, 491-508, 1992.

Case, A.C., H.S. Rosen and J.R. Hines, Budget Spillovers and Fiscal Policy Interdependence: Evidence from the States, *Journal of Public Economics*, 52, 285-307, 1993.

Casetti, E., Generating Models by the Expansion Method: Applications to Geographical Research, *Geographical Analysis*, 4, 81-91, 1972.

Casetti, E., The Dual Expansion Method: An Application for Evaluating the Effects of Population Growth on Development, *IEEE Transactions on Systems, Man and Cybernetics*, SMC-16, 29-39, 1986.

Cliff, A. and J.K Ord, *Spatial Autocorrelation*, London: Pion, 1973.

Cliff, A. and J.K. Ord, *Spatial Processes: Models and Applications*, London: Pion, 1981.

Cressie, N., *Statistics for Spatial Data*, New York: Wiley, 1991.

Dubin, R., Spatial Autocorrelation and Neighborhood Quality, *Regional Science and Urban Economics*, 22, 433-452, 1992.

Duncan, G., W. Gorr and J. Szczypula, Bayesian Forecasting for Seemingly Unrelated Time Series: Applications to Local Government Revenue Forecasting, *Management Science*, 39, 275-293, 1993.

Fischer, M. and P. Nijkamp, *Geographic Information Systems, Spatial Modelling and Policy Evaluation*, Berlin: Springer Verlag, 1993.

Florax, R. and H. Folmer, Specification and Estimation of Spatial Linear Regression Models: Monte Carlo Evaluation of Pre-Test Estimators, *Regional Science and Urban Economics*, 22, 405-432, 1992.

Fotheringham, A.S. and P. Rogerson, GIS and Spatial Analytical Problems, *International Journal of Geographical Information Systems*, 7, 3-19, 1993.

Getis, A., Screening for Spatial Dependence in Regression Analysis, *Papers of the Regional Science Association*, 69, 69-81, 1990.

Getis, A. and K. Ord, The Analysis of Spatial Association by Use of Distance Statistics, *Geographical Analysis*, 24, 189-206, 1992.

Goodchild, M.F., R.P. Haining, S. Wise, et al., Integrating GIS and Spatial Analysis - Problems and Possibilities, *International Journal of Geographical Information Systems*, 6, 407-423, 1992.

Haining, R., Geography and Spatial Statistics: Current Positions, Future Developments, in: B. Macmillan (ed.), *Remodelling Geography*, Oxford: Basil Blackwell, 1989.

Haining, R., *Spatial Data Analysis in the Social and Environmental Sciences*, Cambridge: Cambridge University Press, 1990.

Haining, R., Diagnostics for Regression Modeling in Spatial Econometrics, *Journal of Regional Science*, 34, 325-341, 1994.

Holtz-Eakin, D., Public-Sector Capital and the Productivity Puzzle, *Review of Economics and Statistics*, 76, 12-21, 1994.

Hordijk, L., Problems in Estimating Econometric Relations in Space, *Papers, Regional Science Association*, 42, 99-115, 1979.

Hordijk, L. and P. Nijkamp, Dynamic Models of Spatial Autocorrelation, *Environment and Planning A*, 9, 505-519, 1977.

Kelejian, H.H. and D.P. Robinson, Spatial Autocorrelation: A New Computationally Simple Test with an Application to Per Capita County Policy Expenditures, *Regional Science and Urban Economics*, 22, 317-331, 1992.

Kelejian, H.H. and D.P. Robinson, A Suggested Method of Estimation for Spatial Interdependent Models with Autocorrelated Errors, and an Application to a County Expenditure Model, *Papers in Regional Science*, 72, 297-312, 1993.

Krugman, P., Increasing Returns and Economic Geography, *Journal of Political Economy*, 99, 483-499, 1991.

Krugman, P., First Nature, Second Nature, and Metropolitan Location, *Journal of Regional Science*, 33, 129-144, 1993.

Lambin, E., *Modelling Deforestation Processes: A Review*, Trees Series B, Research Report no. 1, Luxembourg: Office for Official Publications of the European Community, 1994.

LeSage, J.P. and W.A. Muraco, *Spatial Modeling of Gypsy Moth Movements*, Toledo, OH: Department of Economics Working Paper, University of Toledo, 1994a.

LeSage, J.P. and J.J. Stevens, *A Bayesian Approach to Spatial Expansion*, Paper Presented at the 41st North American Meetings of the Regional Science Association International, Niagara Falls, ON, 1994b.

Magnus, J., Maximum Likelihood Estimation of the GLS Model with Unknown Parameters in the Disturbance Covariance Matrix, *Journal of Econometrics*, 7, 281-312, 1978 (Corrigenda, *Journal of Econometrics*, 10, 261).

McMillen, D.P., Probit with Spatial Autocorrelation, *Journal of Regional Science*, 32, 335-348, 1992.

Ord, J.K., Estimation Methods for Models of Spatial Interaction, *Journal of the American Statistical Association*, 70, 120-126, 1975.

Paelinck, J. and L. Klaassen, *Spatial Econometrics*, Farnborough: Saxon House, 1979.

Turner, B.L., W.B. Meyer and D.L. Skole, Global Land-use/land-cover Change: Towards an Integrated Study, *Ambio*, 23, 91-95, 1994.

Ward, M.D., *The New Geopolitics*, London: Gordon and Breach, 1992.

West, M. and J. Harrison, *Bayesian Forecasting and Dynamic Models*, New York: Springer Verlag, 1989.

PART I-A

Spatial Effects in Linear Regression Models

Specification of Spatial Dependence

2 Small Sample Properties of Tests for Spatial Dependence in Regression Models: Some Further Results

Luc Anselin[1] and Raymond J.G.M. Florax[2]
[1] West Virginia University, Morgantown, WV, U.S.A.
[2] Wageningen Agricultural University, Wageningen, The Netherlands

2.1 Introduction

It has now been more than two decades since Cliff and Ord (1972) and Hordijk (1974) applied the principle of Moran's *I* test for spatial autocorrelation to the residuals of regression models for cross-sectional data. To date, Moran's *I* statistic is still the most widely applied diagnostic for spatial dependence in regression models [e.g., Johnston (1984), King (1987), Case (1991)]. However, in spite of the well known consequences of ignoring spatial dependence for inference and estimation [for a review, see Anselin (1988a)], testing for this type of misspecification remains rare in applied empirical work, as illustrated in the surveys of Anselin and Griffith (1988) and Anselin and Hudak (1992). In part, this may be due to the rather complex expressions for the moments of Moran's *I*, and the difficulties encountered in implementing them in econometric software [for detailed discussion, see Cliff and Ord (1981), Anselin (1992), Tiefelsdorf and Boots (1994)]. Recently, a number of alternatives to Moran's *I* have been developed, such as the tests of Burridge (1980) and Anselin (1988b, 1994), which are based on the Lagrange Multiplier (LM) principle, and the robust tests of Bera and Yoon (1992) and Kelejian and Robinson (1992). These tests are all asymptotic and distributed as χ^2 variates. Since they do not require the computation of specific moments of the statistic, they are easy to implement and straightforward to interpret. However, they are all large sample tests and evidence on their finite sample properties is still limited.

In general, the study of finite sample properties of tests for spatial dependence in regression models has not received much attention in the literature. In Bartels and Hordijk (1977) and Brandsma and Ketellapper (1979), the power of Moran's *I* was compared for a number of different estimates for regression residuals. In both simulation studies, the conclusion was that standard ordinary least squares residuals yielded the highest power. In Anselin and Rey (1991), an extensive set of simulation experiments was carried out, comparing Moran's I to two Lagrange Multiplier tests for a wide range of regular lattice sizes and error distributions, and for both error and lag forms of spatial dependence. Their conclusion confirmed the theoretical findings on the power of Moran's I [King (1981)], but also indicated a tendency for this test to have power against several types of alternatives, including non-normality, heteroscedasticity and different forms of spatial dependence. In contrast, the use of Lagrange Multiplier tests seemed to provide a better basis for the indication of the proper alternative hypothesis, while being fairly close in power to Moran's I for spatial error dependence and superior for spatial lag dependence.

In this paper, we provide some further evidence on the finite sample properties of tests for spatial dependence in linear regression models. We extend the existing studies in two important respects. First, we study the properties of some tests for which no results on small sample performance exist to date. This includes the robust Kelejian-Robinson and Bera-Yoon tests, and the Lagrange Multiplier tests for higher order forms of spatial dependence, i.e., second order spatial error dependence and spatial autoregressive moving average (SARMA) processes. Secondly, we compare the power of the tests for both autoregressive and moving average forms of spatial dependence. The latter has not yet been considered.

In all, we consider eight tests for spatial dependence that are based on the results of a classical ordinary least squares regression: Cliff and Ord's (1972) Moran's I (Moran); the Lagrange Multiplier test for spatial error dependence (LM-ERR) due to Burridge (1980); the Kelejian and Robinson (1992) robust test for spatial error dependence (K-R); the test for spatial error dependence robust to the presence of spatial lag dependence (LM-EL) of Bera and Yoon (1992); the LM test for second order spatial error dependence (LM-ERR(2)) of Anselin (1994); the LM test for a spatial autoregressive moving average process (SARMA) of Anselin (1994); the LM test for spatial lag dependence (LM-LAG) of Anselin (1988b); and the test for a spatial lag, robust to the presence of spatial error dependence (LM-LE) of Bera and Yoon (1992). The Monte Carlo experiments are carried out for two irregular and two regular lattice configurations, for both normal and lognormal error terms, and for six alternative hypotheses: three one-dimensional alternatives, i.e., spatial autoregressive error dependence, spatial moving average error dependence and spatial autoregressive lag dependence; and three two-dimensional alternatives, second order spatial autoregressive error dependence, second order spatial moving average error dependence, and a SARMA process.

In the remainder of this chapter paper, we first outline the various tests in more formal terms. This is followed by a description of the experimental design. The results of the Monte Carlo simulations are discussed next. We close with some concluding remarks and recommendations on strategies to carry out specification testing for spatial dependence in practice.

2.2 Tests for Spatial Dependence

2.2.1 Null and Alternative Hypotheses

All tests considered in this chapter are based on estimation under the null hypothesis of no spatial dependence, i.e., by means of ordinary least squares. The specification of this linear regression equation is, for $r = 1, ..., R$ spatial units:

$$y = X\beta + \varepsilon \tag{1}$$

with y as a R by 1 vector of observations on the dependent variable, X as a R by K matrix of observations on the explanatory variables, ε as a R by 1 vector of uncorrelated (independent) error terms with zero mean and variance matrix $\sigma^2 I$, and β as a K by 1 vector of regression coefficients. In this chapter, we will assume the error terms to be homoscedastic. An extension to heteroscedastic errors can be obtained in a straightforward manner [see Anselin (1988a)].

The most general alternative of spatial dependence is a spatial autoregressive moving average or SARMA process [Huang (1984)]. The first order form of such a SARMA process, or, a SARMA(1,1) process can be expressed as:

$$y = \rho W_1 y + X\beta + \theta_1 W_1 \mu + \mu \tag{2}$$

where y, X and β are as before, W_1 is an R by R spatial weights matrix (typically standardized such that each row sums to one), and μ is an R by 1 vector of error terms.[1] The first term in equation (2), $\rho W_1 y$ is a spatial lag (or, spatially lagged dependent variable), with associated autoregressive parameter ρ. The next to last term, $\theta_1 W_1 \mu$, represents the lag in a spatial moving average, with associated parameter θ_1. A test on the absence of any spatial dependence in (2) then becomes a test on the joint null hypothesis H_0: $\rho = 0$ and $\theta_1 = 0$. Typically, only tests for one type of dependence are carried out, and the other type is assumed to be absent. In other words, the null hypothesis in a test for a spatial autoregressive process is H_0: $\rho = 0$, conditional upon $\theta_1 = 0$. Similarly, the null hypothesis in a test for a spatial moving average error process is H_0: $\theta_1 = 0$, conditional upon $\rho = 0$. When these conditions are not satisfied, i.e., when spatial dependence of the other form is present, a test can no longer be based on OLS regression results, but must be computed from the maximum likelihood estimates of the appropriate spatial model, such as a spatial autoregressive lag model [see Anselin (1988a)]. Alternatively, tests must be constructed that are robust to the presence of the other form of dependence, as in the approach suggested by Bera and Yoon (1992).

---

[1] A recently suggested alternative is a form of spatial error components, in which the spatial dependence in the error term is as $\varepsilon = W\nu + \mu$, where ν and μ are uncorrelated error terms [see Kelejian and Robinson (1993) for further details].

The SARMA(1, 1) process can easily be extended to higher orders of spatial dependence. For example, a second order moving average process for the error terms can be specified as:

$$\varepsilon = \theta_1 W_1 \mu + \theta_2 W_2 \mu + \mu \tag{3}$$

where W_2 is a second spatial weights matrix, different from W_1 (e.g., reflecting second order contiguity), with associated parameter θ_2. A test for a second order spatial error MA process is then a test on the joint null hypothesis H_0: $\theta_1 = 0$ and $\theta_2 = 0$. If one of these parameters is assumed to be nonzero, a test for the other one can still be constructed, but must be based on estimates in the appropriate spatial model (typically estimated by maximum likelihood) and no longer on OLS results. Again, robust forms may be constructed based on the principles outlined in Bera and Yoon (1992) [for details, see Anselin (1994)].

Tests for spatial error dependence are typically couched in terms of an alternative hypothesis that is a spatial autoregressive rather than a moving average process. Such a SAR error process is specified as:

$$\varepsilon = \lambda_1 W_1 \varepsilon + \mu \tag{4}$$

for a first order process, with λ_1 as the autoregressive parameter, or as:

$$\varepsilon = \lambda_1 W_1 \varepsilon + \lambda_2 W_2 \varepsilon + \mu \tag{5}$$

for a second order process, with λ_2 as the autoregressive parameter associated with the second weights matrix, W_2. Tests for spatial error dependence are then tests on the null hypothesis H_0: $\lambda_1 = 0$, or on the joint null hypothesis H_0: $\lambda_1 = 0$ and $\lambda_2 = 0$. In practice, the distinction between autoregressive and moving average error processes is mostly irrelevant, since tests for either form that are based on OLS estimation results are identical.[2] However, the distinction is important with respect to the comprehensive model (2) in the sense that a spatial autoregressive process with spatial autoregressive error terms is not identified, unless the weights matrices for the lag dependence and the error dependence are different. Typically, these weights will be the same, so that the SARMA specification is the only allowable one that combines the two forms of spatial dependence [see Anselin (1988a), for details].

[2] Note that this property is not specific to spatial dependence, but a general property of Lagrange Multiplier tests [see Bera and Ullah (1991) for a review].

2.2.2 Tests for Spatial Error Dependence

The tests for spatial error dependence considered in this chapter are summarized in the first five rows of Table 2.1. The most familiar among these is Moran's I (Moran in the sequel), which, for row-standardized weights is defined as:[3]

$$I = \frac{e'W_1 e}{e'e} \tag{6}$$

where e is a R by 1 vector of regression residuals from OLS estimation of equation (1). Inference for this test is carried out on the basis of an asymptotically normal standardized z-value, obtained by subtracting the expected value and dividing by the standard deviation. The detailed moments are derived and discussed at length in Cliff and Ord (1972) and are not repeated here [see also Anselin (1988a), Anselin and Rey (1991), Anselin and Hudak (1992)]. It is important to note that, in contrast to the tests based on the Lagrange Multiplier principle, Moran's does not have a direct correspondence with a particular alternative hypothesis.

The second test, LM-ERR, is based on the Lagrange Multiplier principle and was originally suggested in Burridge (1980). The test is identical for spatial autoregressive and spatial moving average errors. It is defined as:

$$\text{LM-ERR} = \frac{\left(e'W_1 e/s^2\right)^2}{T_1} \tag{7}$$

where $s^2 = e'e/R$, and $T_1 = \text{tr}(W_1'W_1 + W_1^2)$, with tr as the matrix trace operator. This statistic is distributed as χ^2 with one degree of freedom.

The third test, K-R, is the robust large sample test suggested by Kelejian and Robinson (1992). This test does not assume normality, nor linearity, and is derived from an auxiliary regression using cross products of residuals of observations that are potentially spatially correlated, and cross products of the corresponding explanatory variables. Specifically, the dependent variable in the auxiliary regression is:

$$C_h = e_i e_j \tag{8}$$

where h is an index for each cross product, e is a residual term and i, j are contiguous observations. The explanatory variables in the auxiliary regression, Z_h, are formed as cross products of X_i and X_j. With γ as the coefficient vector obtained from OLS estimation in a regression of C on Z, and α as the associated vector of residuals, the K-R statistic results as:

---

[3] Since all simulations are carried out for row-standardized weights, only the expressions of the tests for this case are listed. The general form of Moran's I is given in a number of sources, most notably Cliff and Ord (1972).

$$\text{K-R} = \frac{\gamma' Z' Z \gamma}{\alpha'\alpha / h_R} \tag{9}$$

where h_R is the number of observations in the auxiliary vector (8). The statistic is distributed as χ^2 with K degrees of freedom, where K is the number of explanatory variables in $\boldsymbol{Z}$.

The fourth test, LM-EL, is the adjusted Lagrange Multiplier test of Bera and Yoon (1992) that is robust to local misspecification in the form of a spatial lag term. The test is computed as:

$$\text{LM-EL} = \frac{\left(e'W_1e/s^2 - T_1(R\tilde{J}_{\rho\text{-}\beta})^{-1}(e'W_1y/s^2)\right)^2}{\left(T_1 - T_1^2(R\tilde{J}_{\rho\text{-}\beta})^{-1}\right)} \tag{10}$$

with:

$$(R\tilde{J}_{\rho\text{-}\beta})^{-1} = \left[T_1 + (W_1X\beta)'M(W_1X\beta)/s^2\right]^{-1} \tag{11}$$

where $W_1X\beta$ is a spatial lag of the predicted values from an OLS regression of (1), $M = I - X(X'X)^{-1}X'$ is the familiar projection matrix, and the other notation is as before. Just like its LM-ERR counterpart, this statistic is also distributed as χ^2 with one degree of freedom.

The final test for spatial error dependence is a Lagrange Multiplier test for a second order spatial dependence, LM-ERR(2). In general, as shown in Anselin (1994), tests for higher order error dependence are simply the sum of the corresponding one-directional tests, distributed as χ^2 with degrees of freedom equal to the number of terms in the sum. The second order test is thus:

$$\text{LM-ERR(2)} = \frac{\left(e'W_1e/s^2\right)^2}{T_1} + \frac{\left(e'W_2/s^2\right)^2}{T_2} \tag{12}$$

with $T_2 = \mathrm{tr}(W_2'W_2 + W_2^2)$, and the rest of the notation as before.

2.2.3 Tests for Spatial Lag Dependence

The bottom three rows in Table 2.1 give a summary of three tests for spatial lag dependence. The first, SARMA, is a Lagrange Multiplier test for a joint spatial lag and spatial moving average error, as in equation (2). As shown in Anselin (1994), it is identical to the LM test for a joint spatial lag and spatial autoregressive error of Anselin (1988b), except that the restriction on the weights matrices is relaxed (a process with identical weights matrices for the lag and the error is not identified).

Table 2.1: Tests for spatial dependence in regression models (with notation as defined in main text).

Test	Formulation	Distribution	Source
Moran	$\dfrac{e'\mathbf{W}_1 e}{e'e}$	$N(0,1)$	Cliff and Ord (1972)
LM-ERR	$\dfrac{\left(e'\mathbf{W}_1 e/s^2\right)^2}{T_1}$	$\chi^2(1)$	Burridge (1980)
K-R	$\dfrac{\gamma'\mathbf{Z}'\mathbf{Z}\gamma}{\alpha'\alpha/h_R}$	$\chi^2(K)$	Kelejian and Robinson (1992)
LM-EL	$\dfrac{\left(e'\mathbf{W}_1 e/s^2 - T_1(R\tilde{J}_{\rho-\beta})^{-1}(e'\mathbf{W}_1 y/s^2)\right)^2}{\left(T_1 - T_1^2(R\tilde{J}_{\rho-\beta})^{-1}\right)}$	$\chi^2(1)$	Bera and Yoon (1992)
LM-ERR(2)	$\dfrac{\left(e'\mathbf{W}_1 e/s^2\right)^2}{T_1} + \dfrac{\left(e'\mathbf{W}_2/s^2\right)^2}{T_2}$	$\chi^2(2)$	Anselin (1994)
SARMA	$\dfrac{\left(e'\mathbf{W}_1 y/s^2 - e'\mathbf{W}_1 e/s^2\right)^2}{\left(R\tilde{J}_{\rho-\beta} - T_1\right)} + \dfrac{\left(e'\mathbf{W}_1 e/s^2\right)^2}{T_1}$	$\chi^2(2)$	Anselin (1988b, 1994)
LM-LAG	$\dfrac{\left(e'\mathbf{W}_1 y/s^2\right)^2}{\left(R\tilde{J}_{\rho-\beta}\right)}$	$\chi^2(1)$	Anselin (1988b)
LM-LE	$\dfrac{\left(e'\mathbf{W}_1 y/s^2 - e'\mathbf{W}_1 e/s^2\right)^2}{\left(R\tilde{J}_{\rho-\beta} - T_1\right)}$	$\chi^2(1)$	Bera and Yoon (1992)

The corresponding statistic is:

$$\text{SARMA} = \frac{\left(e'\mathbf{W}_1 y/s^2 - e'\mathbf{W}_1 e/s^2\right)^2}{\left(R\tilde{J}_{\rho\text{-}\beta} - T_1\right)} + \frac{\left(e'\mathbf{W}_1 e/s^2\right)^2}{T_1} \tag{13}$$

in the same notation as before. This statistic is distributed as χ^2 with two degrees of freedom.

The second test, LM-LAG, is the Lagrange Multiplier test for spatial lag dependence of Anselin (1988b):

$$\text{LM-LAG} = \frac{\left(e'\mathbf{W}_1 y/s^2\right)^2}{\left(R\tilde{J}_{\rho\text{-}\beta}\right)} \tag{14}$$

distributed as χ^2 with one degree of freedom. Note that the SARMA test (13) is not simply the sum of the two one-directional tests (7) and (14), which distinguishes this case from the results in Jarque and Bera (1980) and Bera and Jarque (1982).

The final test considered, LM-LE, is the counterpart of LM-EL, i.e., a test for a spatial lag robust to local misspecification in the form of a spatial moving average error process by Bera and Yoon (1992). This test is defined as:[4]

$$\text{LM-LE} = \frac{\left(e'\mathbf{W}_1 y/s^2 - e'\mathbf{W}_1 e/s^2\right)^2}{\left(R\tilde{J}_{\rho\text{-}\beta} - T_1\right)} \tag{15}$$

distributed as χ^2 with one degree of freedom. Note that this is exactly the first term in the sum (13). In other words, LM-LE (15) and LM-ERR (7) add up to SARMA (13).

2.3 Experimental Design

The experimental design for the Monte Carlo simulations is based on a format extensively used in earlier studies, e.g., in Anselin and Griffith (1988), Anselin and Rey (1991), and Florax and Folmer (1992, 1994). The model under the null hypothesis of no spatial dependence is the classical regression model, as in equation (1). Using standard practice for Monte Carlo simulations, the R observations on the

---

[4] Note that the original formulation by Bera and Yoon (1992) is for spatial autoregressive error dependence. However, as shown in Anselin (1994), the results are identical for spatial MA errors. Given the identification problems with spatial AR errors in a spatial AR process, MA errors only are considered here.

dependent variables are generated from a vector of standard normal random variates ε and an R by 3 matrix of explanatory variables $\boldsymbol{X}$, consisting of a constant term and two variates drawn from a uniform (0,10) distribution. In addition to a normal error, a lognormal error term is generated as well, with mean and variance equal to that of the normal variates.[5] In each experiment, 5000 data sets are generated. The tests are evaluated at their theoretical (asymptotic) critical values for $\alpha = 0.05$ and the proportion of rejections (i.e., the proportion of times the computed test statistic exceeded its asymptotic critical value) is reported. For a Type I error of 0.05, the 5000 replications yield a sample standard deviation of 0.0031, which is judged sufficiently precise for our purposes.[6]

The configurations used to generate spatial dependence are formally expressed in four weights matrices. These correspond to sample sizes 40, 81 and 127. The weights matrices of size 40 and 127 are for two actual irregularly shaped regionalizations of the Netherlands, illustrated in Figures 2.1 and 2.2. The weights matrices for R = 81 correspond to a regular square 9 by 9 grid, with contiguity defined by both the rook (4 neighbors) and the queen (8 neighbors) criterion. The relative merits of using regular versus irregular spatial configurations are discussed at length in Haining (1986) and Anselin (1986). In this experiment, we included both types, the irregular ones to reflect the types of economic regions often encountered in empirical work, and the regular ones to focus on the effect of the characteristics of the connectivity structure on the properties of the tests. In contrast to what holds for time series applications, in space the sample size (R) is not the only variable important in achieving convergence to asymptotic properties of tests and estimators. As shown in Anselin (1988a), the degree of interconnectedness between observations (locations) is also an important factor in determining the extent to which the central limit theorems on dependent spatial processes hold.

A number of different criteria have been suggested to quantify the connectedness structure in a spatial weights matrix, three of which are illustrated in Table 2.2 for the configurations considered here. The percent zero cells gives and indication of the sparseness of the weights matrix: as R increases, everything else being the same, this percentage should decrease, as it does for the sequence R = 40, R = 81 (rook) and R = 127. Note the qualitative difference between the rook and queen weights for R = 81, the latter being much less sparse and more interconnected. The average number of links per observation should not increase with R.

---

[5] The lognormal variates e_i are obtained from the standard normal variates u_i by means of the transformation $e_i = \exp(0.694u_i - 1.272)$.

[6] The sample standard deviation is obtained as $\sqrt{p(1-p)/R}$. However, this is based on an assumption of independence between the rejection frequencies. Since all tests are computed for the same generated data, the actual sample standard deviation will be smaller, due to the positive covariance between the statistics [for further discussion of this issue, see, e.g., Davidson and MacKinnon (1993, pp. 738-755)].

Figure 2.1: Regionalisation of the Netherlands in 40 nodal regions.

Figure 2.2: Regionalisation of the Netherlands in 127 economic geographical regions (the islands are discarded).

Again, for the $R = 40$, $R = 81$ (rook) and $R = 127$ sequence this variable is around 4 to 5, while for the queen contiguity it is much higher, at 6.72. The difference between the averages for the rook and queen weights and their theoretical values of 4 and 8 reflects the influence of border and corner cells (with fewer neighboring grid cells). A final characteristic is represented by the largest root of the (unstandardized) weights matrix.[7] As shown in the third row of Table2.2, both irregular lattice structures achieve a higher value than the rook criterion for $R = 81$, but are lower than the queen criterion. It is often argued that this reflects a stronger overall spatial interconnectedness for the queen grid cells, and thus a greater potential for nonzero spatial covariances.[8] In our simulation experiments, all weights matrices are used in row-standardized form.

Table 2.2: Characteristics of weights matrices.

	$R = 40$	$R = 81$ (queen)	$R = 81$ (rook)	$R = 127$
% nonzero cells	10.77	8.40	4.44	3.76
average links	4.20	6.72	3.56	4.73
largest root	4.90	7.42	3.80	6.25

In total, we considered six alternative hypotheses of spatial dependence. Three of these are one-directional, i.e., a function of a single spatial parameter, and three are two-directional, i.e., a function of two spatial parameters. The spatially dependent observations are generated by means of an appropriate spatial transformation applied to a vector of errors or 'observations' of uncorrelated values, as follows:

- Spatial autoregressive error:

$$\varepsilon = (\boldsymbol{I} - \lambda_1 \boldsymbol{W}_1)^{-1}\mu$$

where μ is a vector of standard normal variates, and the other notation is as before. The resulting vector of spatially autocorrelated errors ε is added to the $\boldsymbol{X}\beta$ vector to generate a vector of observations on the dependent variable y.

[7] For row-standardized weights matrices, the largest root is always 1 and thus not informative in this respect.

[8] For a more extensive discussion of the interpretation of the maximum eigenvalue as a characteristic of a spatial network, see, e.g., Boots (1984) and Boots and Royle (1991).

- Spatial moving average error:

$$\varepsilon = (\boldsymbol{I} + \theta_1 \boldsymbol{W}_1)\mu$$

with the spatially autocorrelated errors ε added to the explanatory variables in the same way as for (a).

- Spatial autoregressive lag:

$$y = (\boldsymbol{I} - \rho \boldsymbol{W}_1)^{-1} (\boldsymbol{X}\beta + \varepsilon)$$

where ε is a vector of standard normal variates.

- Second order spatial autoregressive error:

$$\varepsilon = (\boldsymbol{I} - \lambda_1 \boldsymbol{W}_1 - \lambda_2 \boldsymbol{W}_2)^{-1}\mu$$

and proceeding in the same way as for (a).

- Second order spatial moving average error:

$$\varepsilon = (\boldsymbol{I} + \theta_1 \boldsymbol{W}_1 + \theta_2 \boldsymbol{W}_2)\mu$$

and proceeding in the same way as for (a).

- SARMA process:

$$y = (\boldsymbol{I} - \rho \boldsymbol{W}_1)^{-1} [\boldsymbol{X}\beta + (\boldsymbol{I} + \theta_1 \boldsymbol{W}_1)\mu]$$

For the one-directional alternative hypotheses, the spatial parameters take on values from 0.1 to 0.9. For ease of interpretation, negative parameter values are excluded [see Anselin and Rey (1991) for a discussion of the complications caused by negative parameter values]. The maximum value of 0.9 reflects the constraint on the Jacobian term $|\boldsymbol{I} - \zeta \boldsymbol{W}|$ for the autoregressive processes and $|\boldsymbol{I} + \zeta \boldsymbol{W}|$ for the moving average processes, where ζ represents a spatial parameter ρ, λ_1, or θ_1. As is well known, the Jacobian term simplifies to an expression in the roots of the weights matrix, as shown in Ord (1975) for autoregressive processes:

$$\ln|\boldsymbol{I} - \zeta \boldsymbol{W}| = \sum_i \ln(1 - \zeta\omega_i) \tag{16}$$

where the ω_i are the eigenvalues of the weights matrix. Consequently, the restriction on the parameter is of the form $\zeta < 1/\omega_i, \forall i$. The resulting acceptable parameter space for autoregressive processes is:[9]

$$1/\omega_{min} < \zeta < 1/\omega_{max} \qquad (17)$$

where the subscripts indicate the minimum and maximum eigenvalues in real terms [see Anselin (1982)]. For moving average processes, ζ should be replaced by $-\zeta$ in expression (17). For row-standardized weights, the largest eigenvalue is 1, and $1/\omega_{min} \leq -1$, which effectively constrains the positive parameter values to $\zeta < 1$. The combinations of parameter value, spatial configuration and error distribution yield a total of 216 cases for the one-directional alternatives.

For the two-directional alternatives, the constraints on the parameter space are slightly more complex. In a SARMA process the constraint (17) holds separately for the autoregressive and the moving average parameter, yielding 81 parameter combinations (positive values only). For the second order autoregressive and moving average processes, the Jacobian terms are, respectively, $|\boldsymbol{I} - \lambda_1 \boldsymbol{W}_1 - \lambda_2 \boldsymbol{W}_2|$ and $|\boldsymbol{I} + \theta_1 \boldsymbol{W}_1 + \theta_2 \boldsymbol{W}_2|$, yielding constraints on the parameters of the form $\lambda_1 \omega_{1,i} + \lambda_2 \omega_{2,i} < 1$ for the AR parameters, and $-(\theta_1 \omega_{1,i} + \theta_2 \omega_{2,i}) < 1$ for the MA parameters. For positive parameter values, this implies that the sum of λ_1 and λ_2 should be less than 1, yielding 36 allowable parameter combinations. For spatial moving average errors, all 81 combinations are allowed. In total, this yields 1,584 cases for the two-directional alternatives.

2.4 Results of Monte Carlo Experiments

The Monte Carlo experiments were designed to assess a number of properties of the various tests for spatial dependence in different circumstances. In addition to providing insight into the bias and power of the five tests for which no prior finite sample results exist, i.e., KR, LM-EL, LM-LE, LM-ERR(2) and SARMA, the focus of the investigation was on three issues of a more general concern: (a) a comparison of the relative power of the tests against error dependence between spatial autoregressive and spatial moving average errors; (b) a comparison of the relative power of the tests against higher order forms of dependence, and the effect of higher order dependence on the power of one-directional tests; and (c) the power trade-offs involved in using the robust LM-EL and LM-LE tests, both in the presence and in the absence of local misspecification. These issues form the common theme in the discussion of detailed results that follows.

[9] For a different perspective, see Kelejian and Robinson (1995), where the parameter space is defined over the entire range of real values, with the exception of R singularity points.

2.4.1 Empirical Size of the Tests

The proportion of rejections of the null hypothesis of no spatial dependence, when none is present, is given in Table 2.3, for each test, four spatial weights, and for both normal and lognormal error terms. Since the specified critical values were for $\alpha = 0.05$, a significant deviation from this rejection proportion would indicate a bias of the tests in finite samples. For 5000 replications and under a normal approximation to the binomial, a 95% confidence interval centered on $p = 0.05$ would include rejection frequencies between 0.044 and 0.056. It is encouraging to note that for R = 127, with normal error terms, all eight tests yield rejection frequencies within this range. Also, even for R = 40, this is the case for all but the LM-ERR(2) test. This indicates a correct size for even moderately sized and small data sets. Moreover, four tests, LM-LAG, LM-ERR, LM-EL and LM-LE yield rejection frequencies within the 95% confidence interval in all four samples. This is in general agreement with the results for LM-LAG and LM-ERR in Anselin and Rey (1991). The LM-ERR(2) test significantly under-rejects the null hypothesis for R = 40 and the queen case, but not for the others.

The poorest performance results for the rook case (relative to the queen configuration for the same number of observations), where Moran, SARMA and K-R all over-reject the null hypothesis, only marginally for the former two, but significantly so for K-R. A similar result occurred in Anselin and Rey (1991), where differences in empirical size were also found when different weights matrices were used for the same number of observations. It is not clear why the rook case would stand out in this respect. The only indication as to how it differs from the other layouts is that it yields the smallest maximum eigenvalue of the four configurations (but its rejection frequencies are always higher). To some extent, this influence of the choice of the weights matrix is counterintuitive, since there is no spatial dependence present. It further highlights the difference between the two-dimensional spatial dependence and serial dependence in time series analysis, which is one-dimensional (and one-directional). In one dimension, first order dependence (first order autocorrelation) is defined unambiguously, while this is not the case in two dimensions. As shown in Anselin and Rey (1991, Table 4), this is particularly an issue in small samples and is much less pronounced as the number of observations increases (in the limit, the size of R dominates the effect of the connectedness structure).

A misspecification in the form of a lognormal error term seems to affect the size of the tests more for the error tests than for the lag tests, as was the case in Anselin and Rey (1991). For LM-LAG and LM-LE, the rejection frequency remains in the 95% interval for the four cases, while the SARMA test significantly under-rejects for R = 40 and the queen case. Of the error tests, LM-ERR and LM-EL significantly under-reject in three configurations (for R = 40 and 127, and for the queen case), while LM-ERR(2) under-rejects in all four cases, and Moran only for R = 40.

Table 2.3: Empirical size of the tests.[a]

Test	$R = 40$	$R = 81$ (queen)	$R = 81$ (rook)	$R = 127$
		Normal distribution		
Moran	0.051	0.054	0.057	0.051
LM-ERR	0.046	0.046	0.056	0.049
K-R	0.049	0.045	0.065	0.051
LM-EL	0.046	0.049	0.053	0.051
LM-ERR(2)	0.036	0.039	0.048	0.048
SARMA	0.051	0.045	0.057	0.048
LM-LAG	0.052	0.054	0.054	0.051
LM-LE	0.055	0.052	0.055	0.054
		Lognormal distribution		
Moran	0.041	0.044	0.049	0.044
LM-ERR	0.033	0.034	0.047	0.041
K-R	0.073	0.063	0.079	0.062
LM-EL	0.038	0.041	0.046	0.043
LM-ERR(2)	0.028	0.033	0.039	0.042
SARMA	0.042	0.043	0.047	0.050
LM-LAG	0.049	0.051	0.052	0.048
LM-LE	0.050	0.052	0.053	0.056

[a] A 95% confidence interval for $p = 0.05$ with 5000 replications is $0.044 < p < 0.056$.

In practice, an under-rejection of the null hypothesis when no spatial dependence is present does not have any consequences, since the standard estimation results are interpreted as they should be (without taking spatial effects into account). On the other hand, an over-rejection would tend to result in unnecessary estimation of spatial models and cause problems with pre-testing [see Florax and Folmer (1992)]. This is the case for the K-R test with lognormal errors, which significantly over-rejects the null hypothesis in all four configurations. There are two possible explanations for the apparent bias of this test. One is that the test may not be robust to lognormal errors. Another, and more likely explanation is that the K-R test does not yet achieve its 'large sample' robust properties for the sample sizes considered in the experiments.

2.4.2 Power of Tests Against First Order Spatial Error Dependence

The rejection frequencies of the tests against an alternative of spatial autoregressive errors are listed in Table 2.A1 for normal disturbance terms.[10] To more specifically illustrate the relative performance of the five error tests, their power functions are shown in Figure 2.3, for $R = 40$, and in Figure 2.4, for $R = 127$. Of all the tests, Moran always achieves highest power, with LM-ERR as a close second, becoming virtually indistinguishable for $R = 127$ and $\lambda_1 \geq 0.5$. This confirms earlier findings on the superiority of Moran in Anselin and Rey (1991), but now relative to a wider range of competitors. Of the other error tests, the power of K-R is always considerably lower, particularly for small values of λ_1, but even in the largest sample. LM-EL entails a small loss of power relative to LM-ERR, and is slightly inferior to LM-ERR(2) in the smaller samples. However, for $R = 127$, the power functions of LM-EL and LM-ERR(2) are virtually identical (and cross at $\lambda_1 = 0.5$) and approach the one for LM-ERR, as Figure 2.4 illustrates. In other words, even in moderately sized samples, the penalty for the correction for a potential lag in LM-EL is almost negligible when no lag is actually present. On the other hand, LM-ERR(2) has high power, even when no second order error dependence is present, though it is always less than LM-ERR. The poor performance of K-R may be in part due to its higher degrees of freedom (3 in the current experiments). Note that all error tests perform rather poorly in the smallest sample, achieving the 95% rejection mark only for $\lambda > 0.7$ for Moran, LM-ERR and LM-ERR(2), and for $\lambda > 0.8$ for LM-EL and K-R.

Two of the three lag tests, SARMA and LM-LAG, achieve considerable power against AR error dependence, the former at levels comparable to LM-EL and LM-ERR(2), but always less than LM-ERR. In other words, SARMA has higher power against AR error dependence, compared to LM-LAG, which makes it slightly less suitable in a specification search aimed at distinguishing lag from error dependence. On the other hand, a comparison of LM-ERR and LM-LAG, as suggested in Anselin and Rey (1991), provides an indication of the proper alternative.[11]

---

[10] In order to increase readability the tables (for normal errors) have been grouped in Appendix 1. Tables with results for lognormal errors are given in the working paper version of this chapter [Anselin and Florax (1994)].

[11] Note that a specification search for a spatial regression model often occurs in contexts of multiple comparisons and pre-testing, as shown in Florax and Folmer (1992, 1994). Nevertheless, they found that the simple decision rule formulated in Anselin and Rey (1991) performs well.

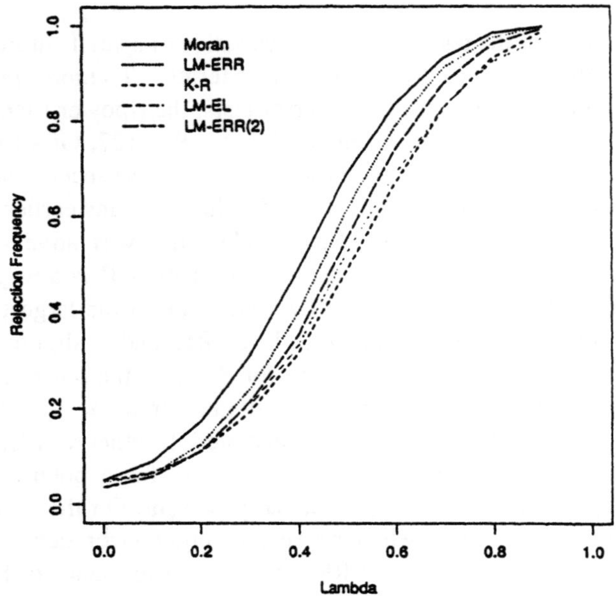

Figure 2.3: Power against spatial AR error, R = 40.

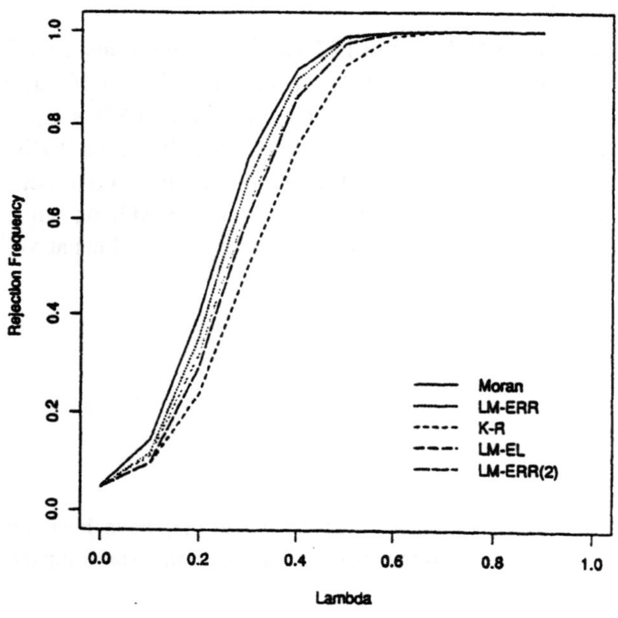

Figure 2.4: Power against spatial AR error, R = 127.

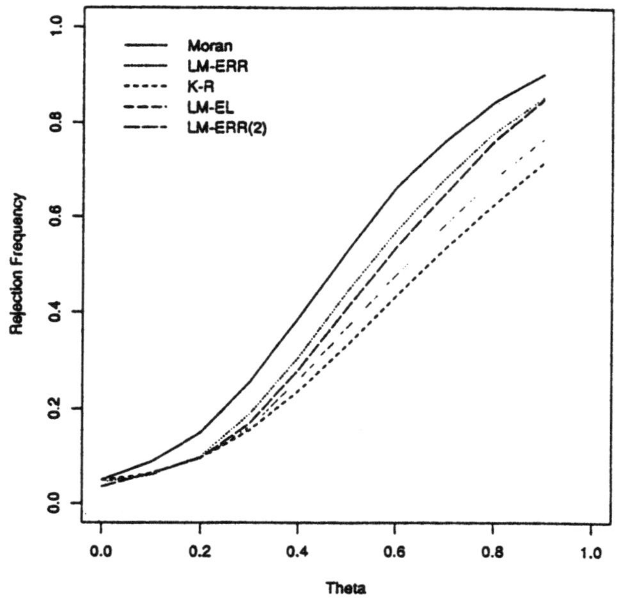

Figure 2.5: Power against spatial MA error, $R = 40$.

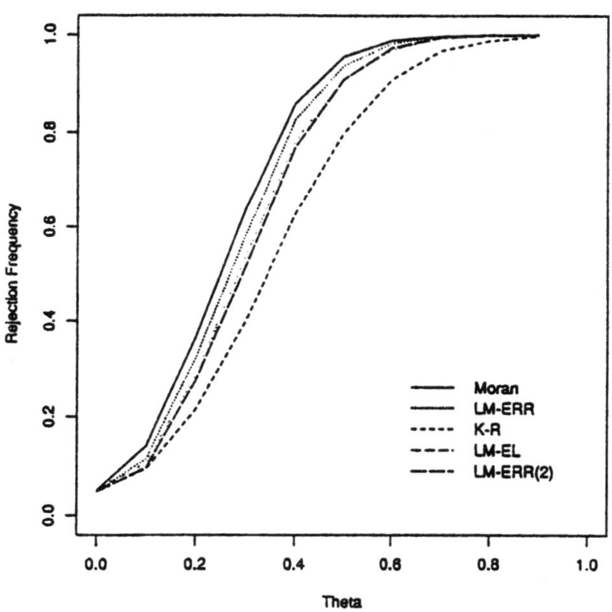

Figure 2.6: Power against spatial MA error, $R = 127$.

The robust LM-LE test performs remarkably well, yielding low rejection frequencies even for $\lambda_1 = 0.9$ (e.g., 25% for $R = 127$). The correction for error dependence in LM-LE thus seems to work in the right direction when no lag dependence is present, especially for small values of λ_1.

The effect of lognormal errors on the power of the tests is small. The power of the error tests is slightly less than for the normal case for small values of λ_1, but equal to or slightly larger for $\lambda_1 > 0.5$. The relative position of the performance of the tests is not affected, and K-R in particular is not distinguished by a higher degree of 'robustness'.

The rejection frequencies of the eight tests against spatial moving average errors are reported in Table 2.A2 for normal errors. The power functions for the five specific error tests are illustrated for $R = 40$ in Figure 2.5, and for $R = 127$ in Figure 2.6.

A striking feature of the results is that, for $\theta_1 > 0.1$, and for similar parameter values, the power of all tests is considerably lower than that against AR errors.[12] This is clearly illustrated by Figure 2.5, where even Moran never reaches a rejection frequency higher than 90%. This raises serious doubt about the usefulness of these tests against spatial MA errors in small samples such as the $R = 40$ used here. The power is more acceptable for $R = 127$, but still remains considerably lower than against AR errors for similar parameter values. On the other hand, the relative ranking of tests is not affected, and K-R continues to perform poorly, even in the largest sample, as illustrated in Figure 2.6. As in the AR case, there is a loss of power involved in using the robust LM-EL test compared to the LM-ERR tests, particularly in the smallest configuration, but this is much less the case for $R = 127$. Note that LM-EL moves from well below the power function for LM-ERR in Figure 2.5, to slightly above the one for LM-ERR(2), and much closer to LM-ERR in Figure 2.6.

Another interesting aspect of the results in Table 2.A2 is that LM-LAG no longer has very good power against error dependence. For example, for $R = 40$, the null hypothesis is rejected by LM-LAG in only 35% of the cases for $\theta_1 = 0.9$, compared to 90% for $\lambda_1 = 0.9$ in Table 2.A1. Even in the largest configuration, the rejection frequency is only about 62% for LM-LAG (with $\theta_1 = 0.9$). To some extent, this is to be expected, given the much greater similarity between a lag process and a spatial AR error process (e.g., in the form of a spatial Durbin model), while such a similarity does not exist with MA errors. On the other hand, the clear superiority of LM-ERR compared to LM-LAG in this context would tend to strengthen the decision rule of Anselin and Rey (1991).

---

[12] In a strict sense, the parameter values for an AR error process and a MA error process are not equivalent, since each process implies a different range for the spatial interaction between observations. For an AR process, all observations interact, while for a MA process, only the first and second order neighbors interact, as shown in Anselin (1994). In other words, the same parameter values imply stronger interaction for an AR process than for a MA process.

Of the other two lag tests, SARMA has good power against MA errors, less than LM-ERR (except in one instance, for $\theta_1 = 0.1$ and $R = 40$), but comparable to LM-EL. Again, in practice a SARMA test should be compared to the one-directional LM-ERR test in order to aid in identifying the proper alternative hypothesis. Finally, the robust LM-LE test performs even better (i.e., has lower power) than in the AR case. For example, for the rook configuration with $R = 81$, the rejection frequencies of LM-LE are practically within the 95% confidence interval around 0.05 for all values of θ_1.

The effect of lognormal errors is similar to that in the AR error case.

2.4.3 Power of Tests Against Spatial Autoregressive Lag Dependence

The rejection frequencies of the eight tests against a first order spatial autoregressive lag are reported in Table 2.A3 for normal disturbances. The power functions of the three lag tests are illustrated in Figure 2.7, for $R = 40$.

The LM-LAG test is clearly the most powerful test against this alternative, achieving a 95% rejection level for $\rho > 0.3$ in the smallest sample, and for $\rho > 0.1$ in the rook ($R = 81$) and $R = 127$ configurations. The two other lag tests have only slightly less power and are almost indistinguishable in the largest data set. In other words, the penalty in terms of power for the robustness against error dependence in LM-LE, when none is present, is almost negligible. On the other hand, there is hardly any power difference between this test and the SARMA test that explicitly takes error dependence into account.

Overall, the power functions of the three lag tests compare very favorably to the ones for tests against error dependence, as illustrated by the much steeper slope in Figure 2.7, even relative to the slopes for $R = 127$ in Figures 2.4 and 2.6. This reliability of the lag tests is encouraging, since the consequences of ignoring a spatial lag (as an omitted variable) when one should be included (i.e., inconsistent estimates) are much more serious than the ones of ignoring spatially correlated errors (less efficient estimates).

Four of the five error tests also have power against a spatial lag (Moran the most), but much less than the lag tests. The behavior of LM-EL is very interesting. Except somewhat for $\rho = 0.7$ and 0.8 in the queen case ($R = 81$), this test has no power against lag dependence, as it should. Moreover, its power function tends to decrease with increasing values of ρ. For small values of ρ, the rejection frequency of LM-EL is very close to its expected value of 0.05, but for large values, it becomes almost negligible (except for the queen case). Since the LM-EL test is robust to 'local' misspecification, this is not surprising, although it may be a bit disconcerting in practice. On the other hand, a clear discrepancy between the indication of LM-ERR and LM-EL, while both LM-LAG and LM-LE are significant would provide strong evidence for lag dependence as opposed to error dependence. The extent to which such a decision rule would hold in a specification search characterized by pre-testing remains to be further investigated.

Relatively speaking, the effect of a misspecification in the form of lognormal errors is much less on the lag tests than in the case of error dependence. The robustness of LM-LAG which was found in Anselin and Rey (1991) is thus extended to SARMA and LM-LE as well. Interestingly, for small values of ρ, the power of the tests is slightly higher in the presence of lognormal errors, while for larger values it is lower (although the latter effect is marginal, given the strong power of the tests for large values of ρ).

2.4.4 Power of Tests Against Second Order Spatial Error Dependence

The results on the rejection frequencies of the five tests for error autocorrelation used against higher order alternatives are reported in Tables 2.A4 to 2.A7 for second order autoregressive errors, and in Tables 2.A8 to 2.A11 for second order moving average errors. These results are for normal disturbances only. To conserve space, results for lognormal errors are not reported.[13] The power function for the LM-ERR(2) test is further illustrated in Figures 2.8 and 2.9, for the rook configuration ($R = 81$). Since two parameters vary, these graphs show a three dimensional power surface.[14]

The LM-ERR(2) test turns out to have acceptable power against both autoregressive and moving average errors, particularly whenever $\lambda_2 \geq \lambda_1$ and $\theta_2 \geq \theta_1$, but not in the smallest data set. As in the tests against one-directional alternatives, the power is higher for AR than for MA errors, for the same parameter values, as illustrated in Figures 2.8 and 2.9. Note that the power curve for $\lambda_1 = 0$ in Figure 2.8 is actually the power of LM-ERR(2) against first order AR error dependence (as in Figures 2.3 and 2.4), but for the weights matrix $\boldsymbol{W}_2$, i.e., the second order contiguity matrix. Similarly, the power curve for $\theta_1 = 0$ in Figure 2.9 illustrates the power of LM-ERR(2) against first order MA error dependence for the second order contiguity matrix (similar to Figures 2.5 and 2.6). In both Figures 2.8 and 2.9, the power functions for values of the first order coefficient of 0.0 and 0.1 are almost identical. In those instances, the LM-ERR(2) test is inferior to Moran and LM-ERR.

A striking feature of these results is the lack of symmetry between the power functions for the first and second order parameters. For example, in Table 2.A7, the rejection frequency for LM-ERR(2) with $\lambda_1 = 0.1$ and $\lambda_2 = 0.5$ is only 0.868, while for the reverse case ($\lambda_1 = 0.5$ and $\lambda_2 = 0.1$) it is 0.989. Again, this is due to

---

[13] The effect of lognormal errors is in general fairly small and qualitatively similar to their effect for the one-directional alternatives. The detailed results are available from the authors.

[14] One should note that there are certain disadvantages in presenting these three dimensional graphs. In particular, the power surface shows some irregularities (especially at the edges of the parameter space) which are due to the extrapolation and smoothing routine of the computer program that produces the graphs rather than to the actual behavior of the test. Two dimensional cross-sections are presented in Anselin and Florax (1994).

the different degree of connectedness implied by the second order weights matrix W_2, compared to the first order contiguity matrix W_1.

The power functions also show a qualitatively different pattern between Figures 2.8 and 2.9, similar to what was found for first order error dependence. For AR errors, power clearly increases with higher values of both parameters. However, this is not the case for MA errors, where for high values of θ_1, power first decreases with θ_2 and then increases, but only very slowly. The differential pattern between AR and MA errors is also found for the other tests. Whereas for AR errors, all five tests have power increasing with both λ_1 and λ_2, for MA errors the power for all but the LM-ERR(2) test first increases with θ_2 for low values of θ_1, but then actually decreases with θ_2 for high values of θ_1. This coincides with a clear power superiority of both Moran and LM-ERR whenever $\theta_1 > \theta_2$, and a superiority of LM-ERR(2) in the reverse case. K-R and LM-EL always have lowest power (but very similar), though LM-EL achieves higher rejection frequencies than LM-ERR(2) in some cases, similar to LM-ERR. Clearly, the power functions of the one-directional tests are driven by the value of the first order parameter, while LM-ERR(2) has good relative power when the second parameter is large.

In practice, the similarity between the powers of the one-directional Moran and LM-ERR tests on the one hand, and the two-directional LM-ERR(2) test on the other hand complicates the specification search for the proper spatial lag length.[15] In this respect, it would be interesting to assess the properties of a Bera-Yoon like correction to the LM-ERR(2) test, as outlined in Anselin (1994). As we found for the one-directional alternatives, the comparison of the uncorrected to the corrected LM tests may provide insight into the presence of the potential misspecification.

2.4.5 Power of Tests Against a SARMA (1, 1) Process

The rejection frequencies against a SARMA alternative are listed in Tables 2.A12 to 2.A15. We focus in particular on the properties of SARMA, LM-EL and LM-LE, for which the power functions are illustrated in Figures 2.10 to 2.12, for the queen ($R = 81$) configuration, and using the same format as for the second order processes. The four other one-directional tests are included in the tables for the sake of completeness.[16]

---

[15] In many ways, the identification of the proper lag length is similar to the selection of the proper weights matrix. Often, higher order spatial dependence will be implemented when the first order weights are incapable of capturing sufficient interaction.

[16] Results for lognormal errors are very similar and not reported here (they are available from the authors). Also, the results for LM-ERR(2) are not reported, since this test is inappropriate for a mixed process, whereas the included one-directional tests attempt to capture one aspect of the mixed process.

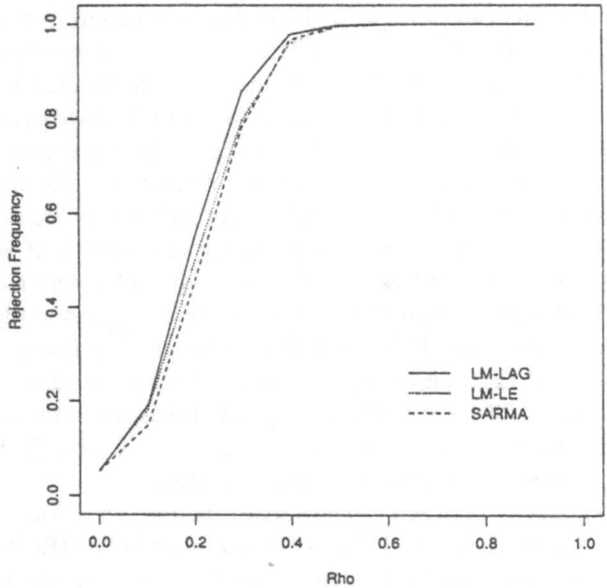

Figure 2.7: Power against spatial AR lag, R = 40.

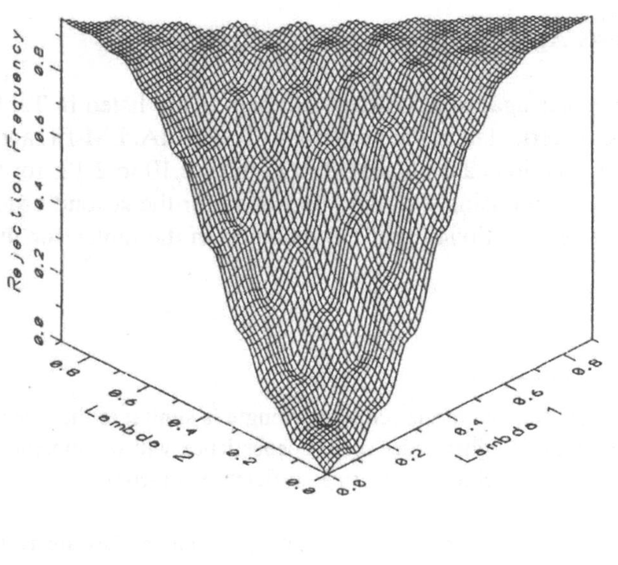

Figure 2.8: Power of LM-ERR(2) against second order AR errors, R = 81 (rook).

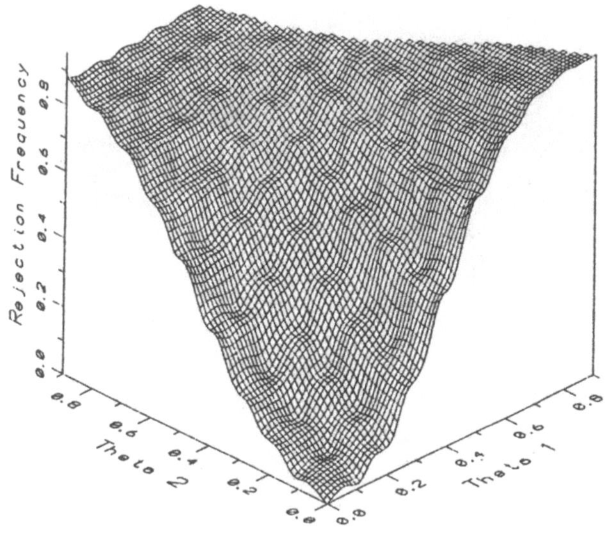

Figure 2.9: Power of LM-ERR(2) against second order MA errors, R = 81 (rook).

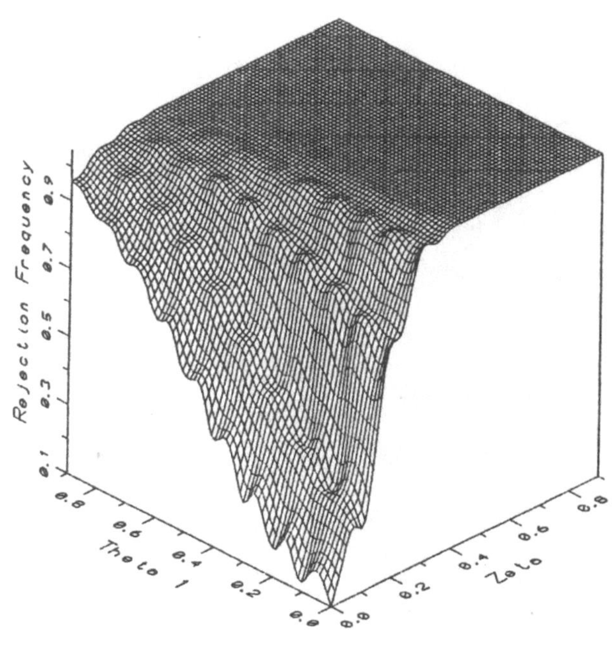

Figure 2.10: Power of SARMA against SARMA(1,1), R = 81 (queen).

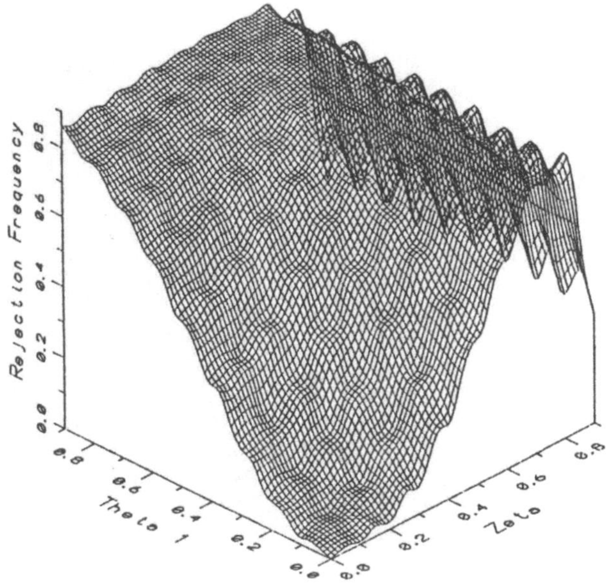

Figure 2.11: Power of LM-EL against SARMA(1,1), R = 81 (queen).

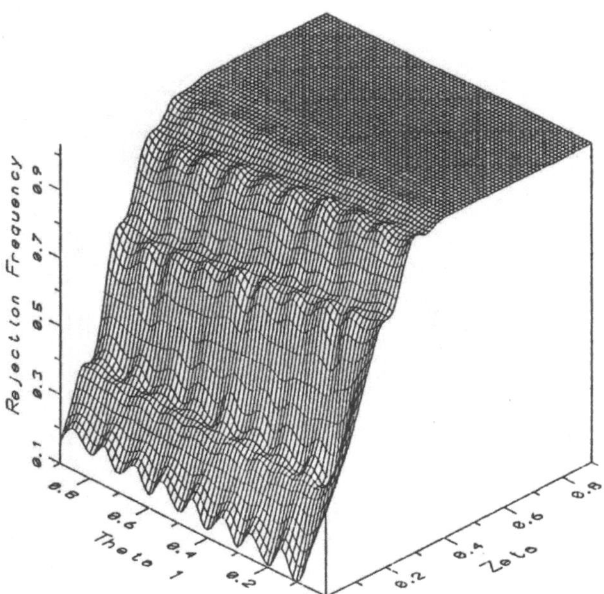

Figure 2.12: Power of LM-LE against SARMA(1,1), R = 81 (queen).

The SARMA test has excellent power against this alternative, achieving 95% rejection frequencies for $\rho \geq 0.4$, $\forall \theta_1$ even in the smallest sample. For R = 127, a 95% rejection is obtained for all but three parameter combinations ($\rho = 0.1$ and $\theta_1 \leq 0.3$). This is also illustrated by the steep slope of the power surface with increasing values of ρ, and the horizontal power surface with θ_1 for $\rho \geq 0.3$ in Figure 2.10. Except for a few instances, SARMA also has the highest power of all tests. Interestingly, in the three smaller configurations and for small values of ρ, Moran and LM-LAG achieve slightly superior power, but this pattern weakens with sample size and disappears by R = 127.

The power surface in Figure 2.11 illustrates the extent to which the LM-EL test is robust to the presence of lag dependence. For small values of ρ, the power function mimics that of the tests against first order moving average error dependence (e.g., as in Figure 2.5), with slowly increasing power with values of θ_1. The curves for $\rho = 0.0$ and $\rho = 0.1$ in particular are very close, confirming the proper behavior of this test against 'local' misspecification. For higher values of ρ, this pattern is not maintained however. The power surface also illustrates the strange behavior of LM-EL with increasing values of ρ. As shown in the discussion of one-directional alternatives, the power function is almost horizontal for $\theta_1 = 0.0$ and small values of ρ, as it should, but this is not the case for higher values of ρ, where power first goes up and then decreases. For R = 127 in particular, power becomes negligible for high ρ, even for high values of θ_1.

In contrast, the power functions of LM-LE seem almost unaffected by the value of θ_1. In Figure 2.12, the power curves for different values of θ_1 are virtually identical and illustrate the good power of this test against spatial autoregressive alternatives. Similarly, the power curves are more or less horizontal with θ_1, even for high values. The difference in properties between lag and error tests is thus maintained for their robust forms, in the presence of misspecification. Overall, LM-LE comes across as much more reliable than LM-EL, and very similar in power to SARMA. The robust tests thus seem more appropriate to test for lag dependence in the presence of error correlation than for the reverse case.

2.5 Conclusions

As in any Monte Carlo experiment, the generality of the results is limited in scope by the design. However, by using irregular lattices for two sample sizes and two different weights for the same regular lattice size, we feel we captured some of the important determinants of the properties of tests for spatial dependence.

Some of our results confirm the earlier findings of Anselin and Rey (1991), in particular with respect to the power of Moran as a test against any form of spatial dependence, and regarding the good properties of the LM tests against lag dependence. We also provided more insight into the role of the connectedness structure of the spatial weights matrix as a determinant of the properties of the tests, highlighted by the differences between the rook and queen cases for $R = 81$. For all tests, power was higher in the rook case. A new finding was the distinction between alternatives with autoregressive error dependence and moving average dependence. Clearly, tests against the former achieve higher power faster than tests against the latter. However, it should be kept in mind that the two forms of dependence are not mirror images, as illustrated by the different acceptable range for the parameters and the difference in the extent of the implied spatial interaction. In practice, it is not possible to distinguish between AR and MA error dependence based solely on the tests considered here. Instead, such a specification search should be based on non-nested tests. The extent to which the differences in power of the tests affect the properties of the resulting pre-test estimators also must be further investigated.

Another useful confirmation of earlier findings in Anselin and Rey (1991) was the higher power of lag tests in smaller samples, relative to tests against error dependence. Given the more serious consequences of ignoring a spatial lag for estimation and inference, this is a reassuring property.

Of the new tests considered, the power of K-R was rather disappointing in the samples considered. Since this is a large sample test (i.e., based on some large sample approximations), the configurations considered here may not be sufficiently large to achieve its theoretical properties. Similar conclusions may be drawn for the robustness of K-R to lognormal errors, which was not impressive in the samples considered. It is possible that the power of this test depends on the degree of spatial autocorrelation in the explanatory variables X, for which the cross-products Z are used in equation (9). In the experimental design used here, these observations are independent, while in most realistic contexts they may be spatially correlated as well [see e.g., Florax and Folmer (1992)]. This remains to be investigated. In practice, it would seem that the usefulness of K-R as an additional test (in addition to the LM tests) would be rather limited and restricted to large samples. The performance of K-R in larger samples and for other forms of error misspecification remains to be further investigated. This is also the case for the other tests, since the effect of lognormality on their power was rather marginal.

The robust LM-EL and LM-LE tests performed remarkably well against one-directional alternatives, which suggests that they may be usefully combined with the LM-ERR and LM-LAG tests to indicate which of the two forms of dependence (lag or error) is the proper alternative. In other words, this may result in an augmented decision rule, similar to the one formulated in Anselin and Rey (1991): when LM-LAG is more significant than LM-ERR and LM-LE is significant while LM-EL is not, a lag dependence is the likely alternative; and when LM-ERR is more significant than LM-LAG and LM-EL is significant while LM-LE is not, an error dependence is the likely alternative. While the simple rule was found to perform well in an actual specification search [Florax and Folmer (1992, 1994)], the extent to which this is the case for the augmented rule remains to be assessed.

However, matters are a bit more complex against SARMA alternatives. Here, there is a clear distinction between LM-LE, which performs well, and LM-EL, which is only reliable for truly 'local' misspecification.

The results on the tests against higher order forms of spatial dependence were less reassuring. While the new LM-ERR(2) and SARMA tests had good power properties, so did the one-directional tests, including Moran. While this is a useful finding for misspecification testing, in the sense that problems with the original model are clearly indicated, it is a less insightful guide for an effective specification search. In this respect, the application of a Bera-Yoon like correction to the tests should be further investigated. In practice, it would seem that a one-directional test against a spatial lag (LM-LAG or LM-LE) should usefully preceed a test against higher order dependence. The rejection of lag dependence allows the proper use of LM-ERR(2) to assess higher order dependence. On the other hand, if lag dependence is not rejected, SARMA is the proper test to check for remaining error dependence, given the strange performance of LM-EL for non-local alternatives. For these higher order alternatives, a remaining issue is the extent to which the tests based on the null hypothesis (e.g., the SARMA) compare to tests carried out for a one-directional alternative, such as a test for error dependence in a first order spatial lag model. Similarly, the effect of using multiple tests on the properties of the resulting pre-test estimators needs to be considered.

Overall, while some important issues remain to be investigated, our study reaffirmed the usefulness of Monte Carlo experimentation to assess the properties of tests against spatial dependence. As these tests are becoming more used in empirical practice, it is necessary to gain insight into their properties for typical sample sizes, such as the U.S. states (48) and the number of counties in a state (around 100). In this respect, the scant evidence for the spatial case pales in comparison to the wealth of studies on serial dependence in the time domain. While the rather poor performance of the error tests in the smallest sample (R = 40) was a bit disconcerting, it was encouraging to see the asymptotic properties of the Lagrange Multiplier tests being closely approximated for the largest sample (R = 127). Given the ease by which these tests can be implemented and interpreted, they should become a standard element in the set of diagnostic tools used by empirical researchers who apply regression models to cross-sectional data.

Acknowledgments

Anselin's research was supported in part by grant SES 88-10917 to the National Center for Geographic Information and Analysis (NCGIA) from the U.S. National Science Foundation. Programming assistence from Marc Loman is greatly appreciated.

References

Anselin, L., A Note on Small Sample Properties of Estimators in a First-Order Spatial Autoregressive Model, *Environment and Planning A*, 14, 1023-1030, 1982.

Anselin, L., Some Further Notes on Spatial Models and Regional Science, *Journal of Regional Science*, 26, 799-802, 1986.

Anselin, L., *Spatial Econometrics: Methods and Models*, Dordrecht: Kluwer Academic Publishers, 1988a.

Anselin, L., Lagrange Multiplier Test Diagnostics for Spatial Dependence and Spatial Heterogeneity, *Geographical Analysis*, 20, 1-17, 1988b.

Anselin, L., *SpaceStat: A Program for the Analysis of Spatial Data*, Santa Barbara: National Center for Geographic Information and Analysis, University of California, 1992.

Anselin, L., *Testing for Spatial Dependence in Linear Regression Models: A Review*, Morgantown: West Virginia University, Regional Research Institute Research Paper, 94-16, 1994.

Anselin, L. and R.J.G.M. Florax, *Small Sample Properties of Tests for Spatial Dependence in Regression Models: Some Further Results*, Morgantown: West Virginia University, Regional Research University Research Paper, 94-14, 1994.

Anselin, L. and D.A. Griffith, Do Spatial Effects Really Matter in Regression Analysis?, *Papers of the Regional Science Association*, 65, 11-34, 1988.

Anselin, L. and S. Hudak, Spatial Econometrics in Practice, a Review of Software Options, *Regional Science and Urban Economics*, 22, 509-536, 1992.

Anselin, L. and S. Rey, Properties of Tests for Spatial Dependence in Linear Regression Models, *Geographical Analysis*, 23, 112-131, 1991.

Bartels, C.P.A. and L. Hordijk, On the Power of the Generalized Moran Contiguity Coefficient in Testing for Spatial Autocorrelation Among Regression Disturbances, *Regional Science and Urban Economics*, 7, 83-101, 1977.

Bera, A.K. and C.M. Jarque, Model Specification Tests, A Simultaneous Approach, *Journal of Econometrics*, 20, 59-82, 1982.

Bera, A.K. and A. Ullah, Rao's Score Test in Econometrics, *Journal of Quantitative Economics*, 7, 189-220, 1991.

Bera, A.K. and M.J. Yoon, *Simple Diagnostic Tests for Spatial Dependence*, Champaign: University of Illinois, Department of Economics, 1992 (mimeo).

Boots, B., Evaluating Principal Eigenvalues as Measures of Network Structure, *Geographical Analysis*, 16, 270-275, 1984.

Boots, B.N. and G.F. Royle, A Conjecture on the Maximum Value of the Principal Eigenvalue of a Planar Graph, *Geographical Analysis*, 23, 276-282, 1991.

Brandsma, A.S. and R.H. Ketellapper, Further Evidence on Alternative Procedures for Testing of Spatial Autocorrelation Among Regression Disturbances, in: C.P.A. Bartels and R.H. Ketellapper (eds.), *Exploratory and Explanatory Statistical Analysis of Spatial Data*, Boston: Martinus Nijhoff, 1979.

Burridge, P., On the Cliff-Ord Test for Spatial Autocorrelation, *Journal of the Royal Statistical Society B*, 42, 107-108, 1980.

Case, A., Spatial Patterns in Household Demand, *Econometrica*, 59, 953-965, 1991.

Cliff, A. and J.K. Ord, Testing for Spatial Autocorrelation Among Regression Residuals, *Geographical Analysis*, 4, 267-284, 1972.

Cliff, A. and J.K. Ord, *Spatial Processes: Models and Applications*, London: Pion, 1981.

Davidson, R. and J.G. MacKinnon, *Estimation and Inference in Econometrics*, New York: Oxford University Press, 1993.

Florax, R. and H. Folmer, Specification and Estimation of Spatial Linear Regression Models: Monte Carlo Evaluation of Pre-Test Estimators, *Regional Science and Urban Economics*, 22, 405-432, 1992.

Florax, R. and H. Folmer, *The Relevance of Hendry's Econometric Methodology in Linear Spatial Process Modeling: Experimental Simulation Results for ML and IV estimators*, Working Paper, Department of General Economics, Wageningen Agricultural University, 1994.

Haining, R., Spatial Models and Regional Science: A Comment on Anselin's Paper and Research Directions, *Journal of Regional Science*, 26, 793-798, 1986.

Hordijk, L., Spatial Correlation in the Disturbances of a Linear Interregional Model, *Regional and Urban Economics*, 4, 117-140, 1974.

Huang, J.S., The Autoregressive Moving Average Model for Spatial Analysis, *Australian Journal of Statistics*, 26, 169-178, 1984.

Jarque, C.M. and A.K. Bera, Efficient Tests for Normality, Homoscedasticity and Serial Independence in Regression Residuals, *Economics Letters*, 6, 255-259, 1980.

Johnston, J., *Econometric Methods*, New York: McGraw-Hill, 1984.

Kelejian, H.H. and D.P. Robinson, Spatial Autocorrelation: A New Computationally Simple Test with an Application to Per Capita County Policy Expenditures, *Regional Science and Urban Economics*, 22, 317-331, 1992.

Kelejian, H.H. and D.P. Robinson, A Suggested Method of Estimation for Spatial Interdependent Models with Autocorrelated Errors, and an Application to a County Expenditure Model, *Papers in Regional Science*, 72, 297-312, 1993.

Kelejian, H.H. and D.P. Robinson, Spatial Correlation: The Cliff and Ord Model and a Suggested Alternative, 1995 (this issue).

King, M.L., A Small Sample Property of the Cliff-Ord Test for Spatial Correlation, *Journal of the Royal Statistical Society B*, 43, 263-264, 1981.

King, M.L., Testing for Autocorrelation in Linear Regression Models: A Survey, in: M. King and D. Giles (eds.), *Specification Analysis in the Linear Model*, London: Routledge and Kegan Paul, 1987.

Ord, J.K., Estimation Methods for Models of Spatial Interaction, *Journal of the American Statistical Association*, 70, 120-126, 1975.

Tiefelsdorf, M. and B. Boots, The Exact Distribution of Moran's *I*, *Environment and Planning A*, 1994.

Appendix 1: Tables

Table 2.A1: Power of tests against first order spatial autoregressive errors, normal distribution.

R	λ_1	Moran	LM-ERR	K-R	LM-EL	LM-ERR(2)	SAR-MA	LM-LAG	LM-LE
	0.1	0.090	0.064	0.067	0.066	0.058	0.071	0.067	0.067
	0.2	0.173	0.125	0.109	0.109	0.111	0.122	0.089	0.076
	0.3	0.313	0.242	0.194	0.207	0.215	0.222	0.125	0.081
	0.4	0.491	0.401	0.318	0.333	0.355	0.365	0.180	0.096
40	0.5	0.687	0.612	0.487	0.524	0.552	0.564	0.253	0.122
	0.6	0.835	0.790	0.670	0.689	0.742	0.753	0.379	0.141
	0.7	0.931	0.910	0.828	0.830	0.878	0.885	0.540	0.154
	0.8	0.984	0.974	0.931	0.923	0.961	0.962	0.724	0.166
	0.9	0.997	0.996	0.987	0.972	0.993	0.994	0.899	0.171
	0.1	0.100	0.066	0.071	0.065	0.066	0.072	0.063	0.063
	0.2	0.222	0.161	0.137	0.146	0.139	0.158	0.091	0.075
	0.3	0.403	0.312	0.258	0.285	0.279	0.276	0.119	0.082
81	0.4	0.612	0.533	0.445	0.490	0.489	0.494	0.174	0.105
(queen)	0.5	0.811	0.758	0.658	0.697	0.706	0.707	0.279	0.122
	0.6	0.928	0.898	0.833	0.866	0.873	0.871	0.399	0.159
	0.7	0.980	0.973	0.949	0.958	0.961	0.965	0.579	0.199
	0.8	0.998	0.995	0.989	0.991	0.991	0.992	0.767	0.264
	0.9	1.000	1.000	0.999	0.999	0.999	0.999	0.933	0.356
	0.1	0.098	0.072	0.065	0.070	0.060	0.063	0.052	0.048
	0.2	0.260	0.208	0.152	0.179	0.151	0.164	0.079	0.056
	0.3	0.508	0.431	0.307	0.389	0.326	0.350	0.107	0.057
81	0.4	0.755	0.691	0.541	0.645	0.587	0.603	0.153	0.063
(rook)	0.5	0.918	0.889	0.773	0.839	0.816	0.831	0.271	0.066
	0.6	0.981	0.974	0.922	0.952	0.948	0.953	0.402	0.083
	0.7	0.998	0.997	0.984	0.991	0.992	0.993	0.595	0.093
	0.8	1.000	1.000	0.998	0.999	0.999	0.999	0.811	0.118
	0.9	1.000	1.000	1.000	1.000	1.000	1.000	0.965	0.159
	0.1	0.145	0.118	0.097	0.112	0.099	0.108	0.068	0.058
	0.2	0.403	0.353	0.237	0.318	0.292	0.300	0.107	0.073
	0.3	0.728	0.683	0.496	0.628	0.605	0.609	0.184	0.084
	0.4	0.922	0.900	0.755	0.869	0.861	0.867	0.301	0.103
127	0.5	0.990	0.986	0.931	0.975	0.975	0.978	0.464	0.123
	0.6	0.999	0.998	0.990	0.996	0.997	0.998	0.651	0.157
	0.7	1.000	1.000	0.999	1.000	1.000	1.000	0.844	0.187
	0.8	1.000	1.000	1.000	1.000	1.000	1.000	0.964	0.221
	0.9	1.000	1.000	0.999	1.000	1.000	1.000	0.999	0.247

Table 2.A2: Power of tests against first order spatial moving average errors, normal distribution.

R	θ_1	Moran	LM-ERR	K-R	LM-EL	LM-ERR(2)	SAR-MA	LM-LAG	LM-LE
	0.1	0.090	0.061	0.066	0.063	0.065	0.072	0.063	0.058
	0.2	0.152	0.103	0.098	0.097	0.099	0.104	0.086	0.075
	0.3	0.256	0.190	0.157	0.163	0.171	0.178	0.098	0.078
40	0.4	0.389	0.307	0.237	0.259	0.281	0.279	0.135	0.085
	0.5	0.529	0.445	0.333	0.370	0.411	0.389	0.178	0.101
	0.6	0.663	0.570	0.435	0.479	0.536	0.522	0.219	0.113
	0.7	0.761	0.682	0.533	0.582	0.649	0.630	0.266	0.118
	0.8	0.844	0.779	0.627	0.682	0.762	0.729	0.313	0.143
	0.9	0.902	0.854	0.716	0.766	0.850	0.816	0.353	0.142
	0.1	0.094	0.062	0.073	0.063	0.064	0.071	0.061	0.061
	0.2	0.192	0.134	0.126	0.122	0.120	0.127	0.078	0.067
	0.3	0.329	0.250	0.206	0.226	0.227	0.227	0.106	0.085
81	0.4	0.490	0.401	0.317	0.370	0.366	0.360	0.138	0.085
(queen)	0.5	0.633	0.550	0.448	0.502	0.511	0.496	0.179	0.096
	0.6	0.754	0.686	0.566	0.638	0.649	0.630	0.223	0.107
	0.7	0.852	0.795	0.684	0.749	0.769	0.749	0.272	0.117
	0.8	0.917	0.875	0.773	0.839	0.864	0.836	0.312	0.131
	0.9	0.960	0.937	0.859	0.905	0.931	0.909	0.379	0.133
	0.1	0.099	0.071	0.072	0.069	0.056	0.065	0.055	0.050
	0.2	0.233	0.176	0.131	0.161	0.118	0.140	0.069	0.047
	0.3	0.478	0.400	0.272	0.362	0.292	0.303	0.092	0.045
81	0.4	0.712	0.646	0.452	0.582	0.517	0.531	0.131	0.051
(rook)	0.5	0.868	0.823	0.642	0.771	0.741	0.730	0.177	0.051
	0.6	0.950	0.924	0.798	0.889	0.885	0.864	0.234	0.053
	0.7	0.986	0.973	0.898	0.954	0.968	0.944	0.290	0.055
	0.8	0.997	0.995	0.957	0.984	0.996	0.984	0.342	0.057
	0.9	0.999	0.997	0.981	0.994	0.999	0.994	0.411	0.058
	0.1	0.141	0.114	0.094	0.106	0.097	0.103	0.066	0.067
	0.2	0.366	0.322	0.215	0.285	0.275	0.267	0.104	0.069
	0.3	0.639	0.583	0.401	0.538	0.517	0.509	0.154	0.074
127	0.4	0.858	0.825	0.627	0.781	0.768	0.771	0.217	0.080
	0.5	0.956	0.937	0.797	0.910	0.909	0.903	0.293	0.103
	0.6	0.989	0.983	0.909	0.970	0.974	0.968	0.385	0.111
	0.7	0.998	0.997	0.968	0.993	0.995	0.992	0.466	0.113
	0.8	1.000	1.000	0.988	0.998	0.999	0.999	0.532	0.124
	0.9	1.000	1.000	0.997	1.000	1.000	1.000	0.618	0.133

Table 2.A3: Power of tests against first order spatial autoregressive lag, normal distribution.

R	ρ	Moran	LM-ERR	K-R	LM-EL	LM-ERR(2)	SAR-MA	LM-LAG	LM-LE
	0.1	0.092	0.067	0.071	0.048	0.060	0.150	0.193	0.183
	0.2	0.206	0.147	0.132	0.040	0.123	0.458	0.554	0.501
	0.3	0.423	0.331	0.259	0.033	0.263	0.783	0.858	0.797
40	0.4	0.697	0.609	0.479	0.026	0.506	0.966	0.978	0.959
	0.5	0.877	0.822	0.703	0.018	0.746	0.996	0.998	0.994
	0.6	0.975	0.956	0.887	0.010	0.914	1.000	1.000	0.999
	0.7	0.997	0.993	0.976	0.002	0.984	1.000	1.000	1.000
	0.8	1.000	1.000	0.997	0.001	0.998	1.000	1.000	1.000
	0.9	1.000	1.000	1.000	0.006	1.000	1.000	1.000	1.000
	0.1	0.119	0.084	0.089	0.052	0.076	0.234	0.299	0.276
	0.2	0.342	0.260	0.241	0.062	0.232	0.734	0.810	0.777
	0.3	0.692	0.610	0.551	0.118	0.557	0.980	0.992	0.984
81	0.4	0.936	0.904	0.865	0.248	0.868	1.000	1.000	1.000
(queen)	0.5	0.993	0.990	0.983	0.441	0.986	1.000	1.000	1.000
	0.6	1.000	1.000	1.000	0.632	0.999	1.000	1.000	1.000
	0.7	1.000	1.000	1.000	0.802	1.000	1.000	1.000	1.000
	0.8	1.000	1.000	1.000	0.860	1.000	1.000	1.000	1.000
	0.9	1.000	1.000	1.000	0.351	1.000	1.000	1.000	1.000
	0.1	0.096	0.073	0.081	0.057	0.063	0.372	0.463	0.458
	0.2	0.210	0.162	0.164	0.063	0.174	0.931	0.967	0.955
	0.3	0.395	0.321	0.320	0.061	0.438	1.000	1.000	1.000
81	0.4	0.630	0.557	0.566	0.061	0.766	1.000	1.000	1.000
(rook)	0.5	0.858	0.806	0.805	0.051	0.953	1.000	1.000	1.000
	0.6	0.981	0.969	0.964	0.027	0.998	1.000	1.000	1.000
	0.7	1.000	0.999	0.997	0.010	1.000	1.000	1.000	1.000
	0.8	1.000	1.000	1.000	0.001	1.000	1.000	1.000	1.000
	0.9	1.000	1.000	1.000	0.000	1.000	1.000	1.000	1.000
	0.1	0.177	0.142	0.123	0.045	0.125	0.549	0.653	0.607
	0.2	0.601	0.547	0.460	0.053	0.507	0.992	0.996	0.993
	0.3	0.943	0.927	0.870	0.092	0.925	1.000	1.000	1.000
127	0.4	0.999	0.998	0.991	0.116	0.998	1.000	1.000	1.000
	0.5	1.000	1.000	1.000	0.077	1.000	1.000	1.000	1.000
	0.6	1.000	1.000	1.000	0.027	1.000	1.000	1.000	1.000
	0.7	1.000	1.000	1.000	0.002	1.000	1.000	1.000	1.000
	0.8	1.000	1.000	1.000	0.000	1.000	1.000	1.000	1.000
	0.9	1.000	1.000	1.000	0.000	1.000	1.000	1.000	1.000

Table 2.A4: Power of tests against second order spatial autoregressive errors, $R = 40$, normal distribution.

λ_1	λ_2	Moran	LM-ERR	K-R	LM-EL	LM-ERR(2)
	0.1	0.097	0.071	0.081	0.071	0.067
	0.2	0.117	0.087	0.097	0.085	0.100
	0.3	0.142	0.107	0.117	0.099	0.150
0.1	0.4	0.183	0.151	0.159	0.137	0.234
	0.5	0.233	0.194	0.202	0.182	0.341
	0.6	0.299	0.258	0.267	0.227	0.469
	0.7	0.411	0.373	0.368	0.330	0.626
	0.8	0.532	0.491	0.487	0.441	0.759
	0.1	0.194	0.141	0.130	0.129	0.136
	0.2	0.224	0.170	0.159	0.147	0.172
	0.3	0.247	0.201	0.191	0.186	0.238
0.2	0.4	0.343	0.283	0.269	0.249	0.357
	0.5	0.406	0.353	0.343	0.315	0.471
	0.6	0.516	0.465	0.449	0.422	0.614
	0.7	0.632	0.592	0.570	0.534	0.762
	0.1	0.339	0.268	0.217	0.232	0.237
	0.2	0.386	0.316	0.280	0.280	0.308
0.3	0.3	0.461	0.393	0.339	0.341	0.401
	0.4	0.524	0.462	0.423	0.412	0.510
	0.5	0.618	0.569	0.529	0.514	0.638
	0.6	0.720	0.680	0.648	0.628	0.763
	0.1	0.552	0.473	0.377	0.402	0.425
	0.2	0.595	0.531	0.446	0.461	0.494
0.4	0.3	0.656	0.596	0.522	0.524	0.587
	0.4	0.741	0.690	0.633	0.627	0.703
	0.5	0.821	0.787	0.756	0.739	0.819
	0.1	0.733	0.672	0.558	0.581	0.611
0.5	0.2	0.774	0.721	0.634	0.648	0.685
	0.3	0.830	0.790	0.729	0.726	0.776
	0.4	0.892	0.869	0.826	0.820	0.862
	0.1	0.867	0.825	0.737	0.742	0.780
0.6	0.2	0.900	0.871	0.805	0.799	0.843
	0.3	0.942	0.924	0.884	0.876	0.913
0.7	0.1	0.949	0.930	0.874	0.875	0.908
	0.2	0.974	0.964	0.930	0.935	0.952
0.8	0.1	0.990	0.986	0.964	0.961	0.978

Table 2.A5: Power of tests against second order spatial autoregressive errors, $R = 81$ (queen), normal distribution.

λ_1	λ_2	Moran	LM-ERR	K-R	LM-EL	LM-ERR(2)
	0.1	0.111	0.079	0.086	0.080	0.080
	0.2	0.131	0.094	0.097	0.095	0.114
	0.3	0.165	0.128	0.130	0.119	0.186
0.1	0.4	0.208	0.165	0.169	0.157	0.300
	0.5	0.279	0.235	0.227	0.220	0.448
	0.6	0.356	0.310	0.308	0.284	0.586
	0.7	0.486	0.441	0.429	0.410	0.756
	0.8	0.617	0.575	0.559	0.552	0.874
	0.1	0.237	0.181	0.166	0.165	0.162
	0.2	0.289	0.224	0.203	0.206	0.236
	0.3	0.320	0.259	0.247	0.247	0.308
0.2	0.4	0.407	0.345	0.324	0.317	0.449
	0.5	0.498	0.447	0.410	0.412	0.587
	0.6	0.606	0.559	0.527	0.531	0.739
	0.7	0.722	0.681	0.667	0.659	0.859
	0.1	0.436	0.358	0.303	0.328	0.331
	0.2	0.498	0.421	0.362	0.381	0.407
0.3	0.3	0.561	0.501	0.442	0.460	0.517
	0.4	0.649	0.591	0.541	0.556	0.645
	0.5	0.730	0.681	0.643	0.648	0.762
	0.6	0.825	0.792	0.760	0.759	0.873
	0.1	0.649	0.574	0.495	0.530	0.522
	0.2	0.716	0.656	0.574	0.610	0.629
0.4	0.3	0.776	0.724	0.667	0.686	0.727
	0.4	0.843	0.805	0.751	0.771	0.823
	0.5	0.900	0.875	0.845	0.858	0.903
	0.1	0.842	0.793	0.723	0.749	0.758
0.5	0.2	0.873	0.835	0.773	0.804	0.810
	0.3	0.919	0.896	0.852	0.871	0.882
	0.4	0.962	0.949	0.921	0.931	0.944
	0.1	0.949	0.927	0.881	0.902	0.906
0.6	0.2	0.968	0.954	0.926	0.936	0.939
	0.3	0.980	0.972	0.961	0.967	0.968
0.7	0.1	0.990	0.984	0.964	0.976	0.978
	0.2	0.994	0.992	0.985	0.988	0.988
0.8	0.1	0.999	0.998	0.995	0.996	0.997

Table 2.A6: Power of tests against second order spatial autoregressive errors, $R = 81$ (rook), normal distribution

λ_1	λ_2	Moran	LM-ERR	K-R	LM-EL	LM-ERR(2)
	0.1	0.124	0.095	0.089	0.097	0.099
	0.2	0.167	0.137	0.127	0.127	0.196
	0.3	0.225	0.192	0.181	0.179	0.366
0.1	0.4	0.282	0.247	0.245	0.224	0.556
	0.5	0.336	0.303	0.318	0.281	0.757
	0.6	0.424	0.393	0.418	0.358	0.905
	0.7	0.534	0.509	0.538	0.472	0.974
	0.8	0.648	0.631	0.674	0.588	0.997
	0.1	0.317	0.261	0.208	0.241	0.222
	0.2	0.388	0.324	0.265	0.300	0.340
	0.3	0.456	0.395	0.335	0.361	0.505
0.2	0.4	0.559	0.516	0.464	0.481	0.703
	0.5	0.663	0.618	0.571	0.578	0.861
	0.6	0.762	0.731	0.702	0.704	0.954
	0.7	0.846	0.828	0.820	0.798	0.992
	0.1	0.588	0.520	0.400	0.465	0.445
	0.2	0.674	0.617	0.504	0.572	0.592
0.3	0.3	0.763	0.719	0.623	0.671	0.748
	0.4	0.838	0.803	0.735	0.766	0.874
	0.5	0.905	0.887	0.846	0.854	0.951
	0.6	0.952	0.944	0.927	0.931	0.989
	0.1	0.824	0.782	0.651	0.723	0.705
	0.2	0.885	0.858	0.750	0.809	0.812
0.4	0.3	0.932	0.916	0.856	0.886	0.913
	0.4	0.963	0.951	0.918	0.938	0.966
	0.5	0.987	0.985	0.974	0.977	0.991
	0.1	0.955	0.938	0.863	0.907	0.897
0.5	0.2	0.972	0.961	0.918	0.942	0.944
	0.3	0.992	0.988	0.970	0.978	0.981
	0.4	0.997	0.996	0.992	0.994	0.995
	0.1	0.993	0.988	0.964	0.976	0.976
0.6	0.2	0.997	0.995	0.982	0.988	0.990
	0.3	0.999	0.999	0.998	0.998	0.999
0.7	0.1	0.999	0.999	0.995	0.998	0.998
	0.2	1.000	1.000	0.999	1.000	1.000
0.8	0.1	1.000	1.000	1.000	1.000	1.000

Table 2.A7: Power of tests against second order spatial autoregressive errors, $R = 127$, normal distribution.

λ_1	λ_2	Moran	LM-ERR	K-R	LM-EL	LM-ERR(2)
	0.1	0.169	0.136	0.110	0.134	0.143
	0.2	0.221	0.192	0.152	0.174	0.287
	0.3	0.288	0.251	0.211	0.229	0.482
0.1	0.4	0.357	0.323	0.273	0.305	0.706
	0.5	0.499	0.467	0.396	0.440	0.868
	0.6	0.638	0.607	0.550	0.579	0.963
	0.7	0.816	0.793	0.742	0.765	0.993
	0.8	0.938	0.934	0.912	0.915	1.000
	0.1	0.450	0.402	0.281	0.368	0.363
	0.2	0.537	0.489	0.365	0.453	0.535
	0.3	0.621	0.581	0.462	0.546	0.708
0.2	0.4	0.724	0.686	0.571	0.652	0.857
	0.5	0.842	0.822	0.726	0.793	0.953
	0.6	0.921	0.908	0.860	0.891	0.992
	0.7	0.979	0.976	0.958	0.969	0.999
	0.1	0.771	0.731	0.564	0.676	0.679
	0.2	0.826	0.799	0.644	0.752	0.804
0.3	0.3	0.888	0.867	0.751	0.835	0.902
	0.4	0.942	0.931	0.845	0.905	0.966
	0.5	0.981	0.977	0.937	0.966	0.995
	0.6	0.994	0.994	0.982	0.991	0.999
	0.1	0.944	0.931	0.817	0.900	0.907
	0.2	0.971	0.962	0.890	0.946	0.954
0.4	0.3	0.984	0.980	0.937	0.969	0.983
	0.4	0.995	0.994	0.981	0.990	0.997
	0.5	0.999	0.999	0.993	0.998	1.000
	0.1	0.994	0.992	0.960	0.987	0.989
0.5	0.2	0.996	0.996	0.981	0.991	0.993
	0.3	0.999	0.998	0.993	0.998	0.998
	0.4	1.000	1.000	0.999	0.999	1.000
	0.1	1.000	1.000	0.996	0.999	1.000
0.6	0.2	1.000	1.000	1.000	1.000	1.000
	0.3	1.000	1.000	1.000	1.000	1.000
0.7	0.1	1.000	1.000	1.000	1.000	1.000
	0.2	1.000	1.000	1.000	1.000	1.000
0.8	0.1	1.000	1.000	1.000	1.000	1.000

Table 2.A8: Power of tests against second order spatial moving average errors, R = 40, normal distribution.

θ_1	θ_2	Moran	LM-ERR	K-R	LM-EL	LM-ERR(2)
	0.1	0.095	0.070	0.078	0.072	0.065
	0.2	0.101	0.076	0.089	0.079	0.078
	0.3	0.111	0.087	0.100	0.084	0.109
	0.4	0.125	0.106	0.117	0.100	0.150
0.1	0.5	0.137	0.119	0.130	0.108	0.209
	0.6	0.152	0.131	0.144	0.124	0.281
	0.7	0.176	0.154	0.164	0.137	0.364
	0.8	0.187	0.170	0.185	0.161	0.448
	0.9	0.199	0.187	0.183	0.166	0.517
	0.1	0.157	0.112	0.112	0.108	0.103
	0.2	0.169	0.125	0.119	0.118	0.118
	0.3	0.186	0.142	0.136	0.129	0.151
	0.4	0.187	0.147	0.150	0.133	0.187
0.2	0.5	0.184	0.152	0.159	0.135	0.232
	0.6	0.212	0.171	0.178	0.159	0.299
	0.7	0.225	0.188	0.193	0.172	0.366
	0.8	0.233	0.200	0.206	0.180	0.438
	0.9	0.253	0.221	0.221	0.198	0.526
	0.1	0.262	0.190	0.164	0.171	0.167
	0.2	0.264	0.200	0.175	0.182	0.176
	0.3	0.274	0.220	0.188	0.198	0.216
	0.4	0.293	0.233	0.209	0.212	0.258
0.3	0.5	0.296	0.234	0.210	0.209	0.293
	0.6	0.305	0.258	0.234	0.218	0.351
	0.7	0.308	0.260	0.241	0.237	0.402
	0.8	0.313	0.263	0.257	0.242	0.473
	0.9	0.324	0.281	0.271	0.246	0.532
	0.1	0.383	0.310	0.250	0.262	0.266
	0.2	0.393	0.313	0.255	0.275	0.277
	0.3	0.399	0.325	0.273	0.292	0.294
	0.4	0.392	0.321	0.265	0.289	0.322
0.4	0.5	0.381	0.317	0.280	0.276	0.350
	0.6	0.396	0.334	0.294	0.294	0.393
	0.7	0.398	0.340	0.312	0.299	0.446
	0.8	0.404	0.344	0.324	0.305	0.496
	0.9	0.412	0.356	0.331	0.318	0.557
	0.1	0.533	0.448	0.338	0.379	0.385
	0.2	0.531	0.449	0.344	0.379	0.383
0.5	0.3	0.520	0.442	0.356	0.380	0.389
	0.4	0.528	0.455	0.367	0.392	0.424
	0.5	0.516	0.443	0.373	0.389	0.443

Table 2.A8: continued.

θ_1	θ_2	Moran	LM-ERR	K-R	LM-EL	LM-ERR(2)
	0.6	0.509	0.436	0.374	0.390	0.466
0.5	0.7	0.501	0.437	0.380	0.398	0.501
	0.8	0.501	0.440	0.386	0.387	0.549
	0.9	0.496	0.440	0.396	0.392	0.601
	0.1	0.659	0.578	0.439	0.491	0.517
	0.2	0.644	0.564	0.448	0.493	0.504
	0.3	0.647	0.574	0.458	0.488	0.508
	0.4	0.631	0.561	0.454	0.486	0.509
0.6	0.5	0.626	0.555	0.455	0.480	0.532
	0.6	0.617	0.543	0.465	0.489	0.548
	0.7	0.627	0.559	0.473	0.503	0.583
	0.8	0.608	0.544	0.474	0.485	0.613
	0.9	0.603	0.537	0.477	0.487	0.649
	0.1	0.765	0.691	0.543	0.600	0.643
	0.2	0.749	0.679	0.549	0.592	0.621
	0.3	0.746	0.677	0.551	0.595	0.612
	0.4	0.752	0.682	0.562	0.602	0.621
0.7	0.5	0.714	0.651	0.543	0.579	0.607
	0.6	0.708	0.642	0.547	0.578	0.620
	0.7	0.708	0.644	0.552	0.582	0.648
	0.8	0.693	0.634	0.547	0.572	0.666
	0.9	0.689	0.625	0.551	0.565	0.690
	0.1	0.849	0.788	0.643	0.692	0.756
	0.2	0.838	0.776	0.630	0.693	0.727
	0.3	0.830	0.767	0.641	0.684	0.712
	0.4	0.823	0.769	0.647	0.682	0.710
0.8	0.5	0.810	0.756	0.649	0.691	0.706
	0.6	0.799	0.745	0.633	0.676	0.706
	0.7	0.792	0.738	0.650	0.666	0.722
	0.8	0.783	0.732	0.637	0.657	0.733
	0.9	0.777	0.728	0.635	0.657	0.751
	0.1	0.901	0.849	0.706	0.770	0.827
	0.2	0.896	0.854	0.726	0.766	0.814
	0.3	0.896	0.852	0.723	0.778	0.805
	0.4	0.884	0.845	0.724	0.766	0.800
0.9	0.5	0.882	0.834	0.717	0.763	0.787
	0.6	0.870	0.828	0.717	0.755	0.782
	0.7	0.860	0.819	0.719	0.750	0.782
	0.8	0.856	0.815	0.709	0.742	0.793
	0.9	0.846	0.802	0.718	0.739	0.802

Table 2.A9: Power of tests against second order spatial moving average errors, R = 81 (queen), normal distribution.

θ_1	θ_2	Moran	LM-ERR	K-R	LM-EL	LM-ERR(2)
	0.1	0.100	0.076	0.078	0.071	0.071
	0.2	0.114	0.088	0.095	0.081	0.097
	0.3	0.130	0.096	0.106	0.101	0.151
	0.4	0.146	0.116	0.123	0.111	0.205
0.1	0.5	0.153	0.131	0.127	0.125	0.285
	0.6	0.172	0.142	0.138	0.144	0.368
	0.7	0.192	0.166	0.167	0.162	0.466
	0.8	0.211	0.188	0.186	0.175	0.560
	0.9	0.224	0.208	0.210	0.191	0.650
	0.1	0.204	0.146	0.132	0.139	0.139
	0.2	0.199	0.144	0.142	0.139	0.146
	0.3	0.221	0.173	0.158	0.154	0.196
	0.4	0.226	0.180	0.170	0.165	0.252
0.2	0.5	0.222	0.176	0.171	0.167	0.312
	0.6	0.252	0.205	0.205	0.194	0.399
	0.7	0.266	0.222	0.214	0.199	0.475
	0.8	0.281	0.241	0.244	0.223	0.571
	0.9	0.292	0.249	0.255	0.239	0.652
	0.1	0.332	0.259	0.221	0.248	0.226
	0.2	0.331	0.257	0.230	0.233	0.238
	0.3	0.335	0.266	0.241	0.248	0.269
	0.4	0.343	0.281	0.259	0.252	0.314
0.3	0.5	0.354	0.286	0.271	0.272	0.368
	0.6	0.375	0.315	0.288	0.290	0.457
	0.7	0.371	0.316	0.300	0.291	0.519
	0.8	0.379	0.321	0.307	0.300	0.588
	0.9	0.398	0.345	0.332	0.316	0.670
	0.1	0.468	0.386	0.316	0.347	0.346
	0.2	0.481	0.400	0.329	0.367	0.355
	0.3	0.473	0.391	0.341	0.361	0.371
	0.4	0.487	0.410	0.351	0.377	0.411
0.4	0.5	0.493	0.425	0.366	0.392	0.470
	0.6	0.487	0.417	0.376	0.388	0.508
	0.7	0.483	0.418	0.387	0.394	0.567
	0.8	0.491	0.435	0.409	0.393	0.632
	0.9	0.506	0.447	0.412	0.417	0.693
	0.1	0.604	0.523	0.430	0.484	0.480
	0.2	0.617	0.538	0.454	0.502	0.481
0.5	0.3	0.625	0.547	0.469	0.508	0.505
	0.4	0.619	0.548	0.478	0.506	0.528
	0.5	0.620	0.545	0.480	0.508	0.556

Table 2.A9: continued.

θ_1	θ_2	Moran	LM-ERR	K-R	LM-EL	LM-ERR(2)
	0.6	0.614	0.546	0.493	0.510	0.596
0.5	0.7	0.612	0.546	0.494	0.508	0.635
	0.8	0.609	0.547	0.493	0.510	0.684
	0.9	0.616	0.554	0.509	0.525	0.740
	0.1	0.746	0.668	0.573	0.622	0.624
	0.2	0.755	0.678	0.568	0.630	0.620
	0.3	0.729	0.665	0.570	0.619	0.610
	0.4	0.732	0.661	0.577	0.619	0.627
0.6	0.5	0.733	0.668	0.588	0.627	0.647
	0.6	0.730	0.667	0.593	0.628	0.679
	0.7	0.726	0.664	0.595	0.626	0.706
	0.8	0.716	0.663	0.604	0.624	0.738
	0.9	0.705	0.650	0.598	0.616	0.778
	0.1	0.853	0.793	0.688	0.749	0.753
	0.2	0.843	0.785	0.686	0.742	0.731
	0.3	0.828	0.773	0.690	0.740	0.722
	0.4	0.819	0.764	0.683	0.730	0.725
0.7	0.5	0.823	0.769	0.685	0.735	0.736
	0.6	0.814	0.763	0.691	0.719	0.744
	0.7	0.814	0.770	0.695	0.729	0.775
	0.8	0.806	0.758	0.688	0.729	0.787
	0.9	0.806	0.761	0.697	0.730	0.820
	0.1	0.907	0.870	0.787	0.831	0.844
	0.2	0.906	0.870	0.788	0.835	0.834
	0.3	0.907	0.870	0.785	0.833	0.824
	0.4	0.897	0.859	0.789	0.830	0.821
0.8	0.5	0.893	0.857	0.780	0.826	0.820
	0.6	0.888	0.853	0.786	0.826	0.833
	0.7	0.872	0.835	0.766	0.808	0.826
	0.8	0.873	0.840	0.786	0.804	0.849
	0.9	0.865	0.825	0.762	0.798	0.854
	0.1	0.952	0.925	0.863	0.900	0.914
	0.2	0.944	0.915	0.845	0.887	0.894
	0.3	0.944	0.920	0.860	0.898	0.897
	0.4	0.947	0.920	0.860	0.895	0.890
0.9	0.5	0.928	0.902	0.845	0.876	0.877
	0.6	0.933	0.908	0.848	0.882	0.883
	0.7	0.928	0.901	0.845	0.877	0.884
	0.8	0.924	0.897	0.842	0.875	0.890
	0.9	0.911	0.883	0.833	0.862	0.893

Table 2.A10: Power of tests against second order spatial moving average errors, R = 81 (rook), normal distribution.

θ_1	θ_2	Moran	LM-ERR	K-R	LM-EL	LM-ERR(2)
	0.1	0.122	0.096	0.093	0.091	0.095
	0.2	0.136	0.108	0.108	0.102	0.155
	0.3	0.148	0.128	0.136	0.118	0.249
	0.4	0.189	0.160	0.166	0.154	0.392
0.1	0.5	0.218	0.189	0.197	0.177	0.537
	0.6	0.239	0.212	0.224	0.206	0.689
	0.7	0.272	0.248	0.282	0.236	0.793
	0.8	0.288	0.268	0.293	0.253	0.873
	0.9	0.313	0.298	0.322	0.281	0.926
	0.1	0.275	0.218	0.158	0.194	0.174
	0.2	0.288	0.238	0.193	0.213	0.247
	0.3	0.302	0.250	0.217	0.229	0.335
	0.4	0.322	0.272	0.240	0.249	0.463
0.2	0.5	0.343	0.302	0.275	0.280	0.587
	0.6	0.335	0.295	0.291	0.274	0.697
	0.7	0.373	0.333	0.327	0.300	0.802
	0.8	0.368	0.331	0.350	0.308	0.872
	0.9	0.394	0.361	0.375	0.340	0.932
	0.1	0.492	0.423	0.305	0.382	0.324
	0.2	0.502	0.435	0.327	0.394	0.382
	0.3	0.525	0.463	0.360	0.420	0.464
	0.4	0.524	0.465	0.381	0.431	0.557
0.3	0.5	0.533	0.475	0.404	0.442	0.664
	0.6	0.524	0.467	0.421	0.437	0.749
	0.7	0.524	0.473	0.446	0.443	0.835
	0.8	0.521	0.483	0.452	0.452	0.895
	0.9	0.538	0.496	0.472	0.456	0.940
	0.1	0.708	0.641	0.478	0.586	0.519
	0.2	0.724	0.663	0.517	0.613	0.578
	0.3	0.723	0.667	0.528	0.618	0.623
	0.4	0.706	0.662	0.547	0.604	0.682
0.4	0.5	0.710	0.655	0.563	0.619	0.750
	0.6	0.701	0.650	0.558	0.609	0.813
	0.7	0.699	0.653	0.592	0.613	0.871
	0.8	0.678	0.632	0.581	0.600	0.918
	0.9	0.674	0.635	0.607	0.601	0.946
	0.1	0.865	0.825	0.661	0.774	0.730
	0.2	0.863	0.825	0.686	0.783	0.738
0.5	0.3	0.867	0.827	0.696	0.779	0.764
	0.4	0.859	0.823	0.714	0.782	0.798
	0.5	0.864	0.829	0.729	0.788	0.841

Table 2.A10: continued.

θ_1	θ_2	Moran	LM-ERR	K-R	LM-EL	LM-ERR(2)
	0.6	0.844	0.813	0.731	0.777	0.875
0.5	0.7	0.832	0.797	0.729	0.767	0.910
	0.8	0.817	0.781	0.729	0.749	0.942
	0.9	0.803	0.772	0.727	0.737	0.962
	0.1	0.953	0.927	0.807	0.892	0.871
	0.2	0.952	0.931	0.824	0.905	0.875
	0.3	0.952	0.932	0.844	0.904	0.881
	0.4	0.949	0.933	0.853	0.902	0.899
0.6	0.5	0.944	0.922	0.847	0.902	0.911
	0.6	0.936	0.918	0.848	0.887	0.927
	0.7	0.918	0.896	0.845	0.874	0.944
	0.8	0.912	0.891	0.839	0.864	0.962
	0.9	0.903	0.880	0.835	0.848	0.975
	0.1	0.990	0.980	0.917	0.966	0.965
	0.2	0.988	0.980	0.928	0.964	0.958
	0.3	0.985	0.977	0.934	0.961	0.956
	0.4	0.985	0.977	0.923	0.963	0.952
0.7	0.5	0.981	0.971	0.926	0.959	0.959
	0.6	0.978	0.968	0.927	0.953	0.967
	0.7	0.972	0.962	0.927	0.948	0.973
	0.8	0.959	0.950	0.915	0.935	0.977
	0.9	0.961	0.948	0.916	0.933	0.984
	0.1	0.998	0.995	0.966	0.987	0.993
	0.2	0.997	0.994	0.968	0.987	0.988
	0.3	0.999	0.996	0.972	0.988	0.987
	0.4	0.997	0.995	0.975	0.990	0.988
0.8	0.5	0.997	0.993	0.974	0.987	0.986
	0.6	0.992	0.988	0.969	0.980	0.984
	0.7	0.992	0.989	0.971	0.982	0.987
	0.8	0.989	0.986	0.964	0.977	0.991
	0.9	0.986	0.982	0.961	0.973	0.994
	0.1	0.999	0.998	0.985	0.994	0.997
	0.2	1.000	0.999	0.989	0.996	0.998
	0.3	1.000	1.000	0.992	0.998	0.997
	0.4	1.000	0.999	0.992	0.998	0.997
0.9	0.5	0.999	0.999	0.990	0.998	0.995
	0.6	0.999	0.999	0.989	0.997	0.995
	0.7	0.998	0.997	0.986	0.992	0.994
	0.8	0.997	0.996	0.987	0.992	0.995
	0.9	0.995	0.994	0.985	0.990	0.997

Table 2.A11: Power of tests against second order spatial moving average errors, R = 127, normal distribution.

θ_1	θ_2	Moran	LM-ERR	K-R	LM-EL	LM-ERR(2)
	0.1	0.152	0.127	0.109	0.117	0.125
	0.2	0.165	0.140	0.117	0.124	0.211
	0.3	0.183	0.159	0.139	0.155	0.336
	0.4	0.201	0.174	0.153	0.163	0.491
0.1	0.5	0.218	0.190	0.169	0.181	0.657
	0.6	0.248	0.220	0.193	0.207	0.779
	0.7	0.267	0.243	0.208	0.229	0.886
	0.8	0.288	0.262	0.226	0.242	0.951
	0.9	0.312	0.290	0.256	0.279	0.974
	0.1	0.358	0.315	0.226	0.291	0.281
	0.2	0.391	0.342	0.244	0.314	0.366
	0.3	0.398	0.354	0.262	0.322	0.466
	0.4	0.419	0.380	0.284	0.353	0.612
0.2	0.5	0.431	0.393	0.308	0.371	0.722
	0.6	0.442	0.400	0.311	0.371	0.816
	0.7	0.469	0.430	0.357	0.397	0.899
	0.8	0.457	0.426	0.362	0.402	0.945
	0.9	0.481	0.447	0.387	0.430	0.973
	0.1	0.631	0.581	0.410	0.542	0.513
	0.2	0.640	0.587	0.425	0.549	0.558
	0.3	0.641	0.599	0.445	0.559	0.642
	0.4	0.647	0.604	0.463	0.557	0.719
0.3	0.5	0.661	0.625	0.475	0.584	0.798
	0.6	0.662	0.624	0.491	0.581	0.869
	0.7	0.656	0.621	0.498	0.581	0.916
	0.8	0.664	0.634	0.509	0.591	0.960
	0.9	0.663	0.632	0.530	0.598	0.973
	0.1	0.848	0.821	0.629	0.770	0.754
	0.2	0.849	0.816	0.627	0.774	0.765
	0.3	0.851	0.824	0.646	0.779	0.805
	0.4	0.833	0.804	0.644	0.760	0.838
0.4	0.5	0.835	0.808	0.649	0.774	0.883
	0.6	0.829	0.800	0.650	0.766	0.920
	0.7	0.832	0.806	0.667	0.769	0.943
	0.8	0.833	0.806	0.676	0.765	0.970
	0.9	0.832	0.806	0.686	0.765	0.984
	0.1	0.948	0.933	0.792	0.905	0.894
	0.2	0.948	0.933	0.807	0.911	0.902
0.5	0.3	0.946	0.931	0.801	0.902	0.909
	0.4	0.940	0.924	0.798	0.903	0.923
	0.5	0.935	0.923	0.801	0.888	0.938

Table 2.A11: continued.

θ_1	θ_2	Moran	LM-ERR	K-R	LM-EL	LM-ERR(2)
	0.6	0.940	0.925	0.802	0.893	0.957
0.5	0.7	0.927	0.914	0.798	0.881	0.970
	0.8	0.920	0.902	0.805	0.879	0.979
	0.9	0.914	0.898	0.801	0.877	0.987
	0.1	0.989	0.983	0.909	0.973	0.969
	0.2	0.990	0.985	0.910	0.970	0.968
	0.3	0.986	0.978	0.899	0.967	0.967
	0.4	0.984	0.979	0.902	0.965	0.969
0.6	0.5	0.983	0.978	0.897	0.967	0.977
	0.6	0.979	0.973	0.899	0.959	0.981
	0.7	0.976	0.971	0.902	0.955	0.986
	0.8	0.975	0.967	0.899	0.951	0.990
	0.9	0.972	0.964	0.896	0.950	0.994
	0.1	0.998	0.996	0.966	0.993	0.992
	0.2	0.998	0.997	0.962	0.993	0.993
	0.3	0.998	0.996	0.965	0.990	0.992
	0.4	0.996	0.995	0.959	0.990	0.991
0.7	0.5	0.995	0.993	0.958	0.989	0.992
	0.6	0.995	0.992	0.954	0.987	0.993
	0.7	0.994	0.993	0.953	0.986	0.996
	0.8	0.992	0.990	0.948	0.984	0.996
	0.9	0.990	0.988	0.952	0.981	0.997
	0.1	0.999	0.999	0.987	0.998	0.998
	0.2	0.999	0.999	0.991	0.999	0.998
	0.3	0.999	0.999	0.987	0.998	0.998
	0.4	0.999	0.999	0.987	0.997	0.998
0.8	0.5	0.999	0.998	0.985	0.997	0.997
	0.6	0.999	0.999	0.985	0.996	0.999
	0.7	0.998	0.997	0.982	0.996	0.997
	0.8	0.999	0.998	0.982	0.995	0.998
	0.9	0.998	0.997	0.980	0.995	0.998
	0.1	1.000	1.000	0.998	1.000	1.000
	0.2	1.000	1.000	0.996	1.000	0.999
	0.3	1.000	1.000	0.996	1.000	1.000
	0.4	1.000	1.000	0.996	1.000	1.000
0.9	0.5	1.000	1.000	0.996	0.999	0.999
	0.6	1.000	1.000	0.994	0.999	0.999
	0.7	1.000	0.999	0.993	0.999	1.000
	0.8	1.000	1.000	0.993	0.998	0.999
	0.9	1.000	1.000	0.993	0.999	1.000

Table 2.A12: Power of tests against spatial ARMA process, R = 40, normal distribution.

ρ	θ_1	Moran	LM-ERR	K-R	LM-EL	SARMA	LM-LAG	LM-LE
	0.1	0.173	0.119	0.108	0.060	0.201	0.237	0.178
	0.2	0.275	0.203	0.165	0.090	0.273	0.299	0.189
	0.3	0.408	0.318	0.246	0.158	0.357	0.353	0.201
	0.4	0.544	0.461	0.348	0.238	0.491	0.416	0.202
0.1	0.5	0.666	0.583	0.443	0.342	0.583	0.471	0.207
	0.6	0.772	0.695	0.551	0.445	0.690	0.518	0.223
	0.7	0.852	0.797	0.644	0.564	0.778	0.576	0.235
	0.8	0.911	0.861	0.736	0.641	0.848	0.617	0.240
	0.9	0.942	0.912	0.808	0.729	0.897	0.656	0.236
	0.1	0.339	0.259	0.207	0.060	0.532	0.602	0.481
	0.2	0.475	0.385	0.293	0.095	0.603	0.644	0.480
	0.3	0.606	0.515	0.395	0.151	0.664	0.688	0.462
	0.4	0.712	0.627	0.495	0.222	0.751	0.739	0.465
0.2	0.5	0.808	0.741	0.601	0.318	0.801	0.771	0.465
	0.6	0.870	0.815	0.691	0.420	0.857	0.802	0.460
	0.7	0.920	0.883	0.768	0.497	0.911	0.828	0.442
	0.8	0.951	0.922	0.819	0.567	0.934	0.839	0.454
	0.9	0.972	0.952	0.877	0.640	0.954	0.850	0.423
	0.1	0.565	0.474	0.371	0.047	0.838	0.881	0.784
	0.2	0.691	0.602	0.464	0.076	0.856	0.905	0.785
	0.3	0.767	0.690	0.567	0.134	0.893	0.911	0.751
	0.4	0.844	0.778	0.655	0.196	0.913	0.931	0.747
0.3	0.5	0.899	0.858	0.739	0.273	0.941	0.931	0.718
	0.6	0.930	0.897	0.806	0.338	0.955	0.947	0.708
	0.7	0.961	0.940	0.862	0.420	0.969	0.952	0.679
	0.8	0.976	0.963	0.892	0.480	0.980	0.956	0.658
	0.9	0.985	0.975	0.923	0.540	0.986	0.958	0.651
	0.1	0.770	0.691	0.571	0.041	0.964	0.984	0.947
	0.2	0.839	0.780	0.658	0.063	0.972	0.985	0.929
	0.3	0.899	0.850	0.738	0.104	0.973	0.983	0.914
	0.4	0.928	0.897	0.804	0.150	0.984	0.985	0.891
0.4	0.5	0.959	0.935	0.847	0.200	0.988	0.986	0.877
	0.6	0.970	0.953	0.887	0.244	0.990	0.988	0.854
	0.7	0.986	0.975	0.923	0.304	0.994	0.989	0.843
	0.8	0.991	0.984	0.940	0.353	0.994	0.991	0.823
	0.9	0.994	0.990	0.959	0.408	0.997	0.989	0.799
	0.1	0.926	0.887	0.792	0.029	0.997	0.998	0.988
	0.2	0.946	0.914	0.824	0.044	0.997	0.999	0.983
0.5	0.3	0.964	0.939	0.870	0.065	0.997	0.999	0.976
	0.4	0.974	0.958	0.903	0.097	0.998	0.998	0.966
	0.5	0.982	0.970	0.923	0.117	0.998	0.998	0.954

Table 2.A12: continued.

ρ	θ_1	Moran	LM-ERR	K-R	LM-EL	SARMA	LM-LAG	LM-LE
	0.6	0.991	0.985	0.950	0.163	0.998	0.999	0.944
0.5	0.7	0.995	0.992	0.961	0.209	1.000	0.998	0.926
	0.8	0.996	0.993	0.971	0.218	0.998	0.999	0.911
	0.9	0.998	0.995	0.980	0.273	1.000	0.999	0.892
	0.1	0.981	0.965	0.918	0.014	1.000	1.000	0.998
	0.2	0.989	0.978	0.938	0.020	1.000	1.000	0.997
	0.3	0.990	0.982	0.950	0.032	1.000	1.000	0.995
	0.4	0.996	0.991	0.963	0.046	1.000	1.000	0.986
0.6	0.5	0.997	0.992	0.976	0.063	0.999	1.000	0.984
	0.6	0.998	0.996	0.980	0.087	1.000	1.000	0.975
	0.7	0.999	0.998	0.988	0.103	1.000	1.000	0.967
	0.8	1.000	0.998	0.989	0.131	1.000	1.000	0.953
	0.9	0.999	0.999	0.993	0.151	1.000	1.000	0.947
	0.1	0.998	0.996	0.982	0.005	1.000	1.000	1.000
	0.2	0.998	0.996	0.985	0.008	1.000	1.000	0.999
	0.3	0.998	0.998	0.987	0.013	1.000	1.000	0.997
	0.4	1.000	0.999	0.992	0.024	1.000	1.000	0.995
0.7	0.5	0.999	0.998	0.993	0.031	1.000	1.000	0.991
	0.6	1.000	0.999	0.997	0.041	1.000	1.000	0.987
	0.7	1.000	1.000	0.997	0.049	1.000	1.000	0.985
	0.8	1.000	1.000	0.998	0.062	1.000	1.000	0.975
	0.9	1.000	1.000	0.998	0.076	1.000	1.000	0.966
	0.1	1.000	0.999	0.998	0.003	1.000	1.000	1.000
	0.2	1.000	1.000	0.998	0.003	1.000	1.000	0.999
	0.3	1.000	1.000	0.998	0.005	1.000	1.000	0.997
	0.4	1.000	1.000	0.998	0.009	1.000	1.000	0.997
0.8	0.5	1.000	1.000	0.999	0.013	1.000	1.000	0.994
	0.6	1.000	1.000	0.999	0.014	1.000	1.000	0.989
	0.7	1.000	1.000	0.999	0.024	1.000	1.000	0.988
	0.8	1.000	1.000	0.999	0.025	1.000	1.000	0.980
	0.9	1.000	1.000	1.000	0.037	1.000	1.000	0.973
	0.1	1.000	1.000	1.000	0.004	1.000	1.000	0.999
	0.2	1.000	1.000	1.000	0.004	1.000	1.000	0.998
	0.3	1.000	1.000	1.000	0.005	1.000	1.000	0.996
	0.4	1.000	1.000	1.000	0.009	1.000	1.000	0.993
0.9	0.5	1.000	1.000	1.000	0.008	1.000	1.000	0.987
	0.6	1.000	1.000	1.000	0.016	1.000	1.000	0.982
	0.7	1.000	1.000	1.000	0.015	1.000	1.000	0.974
	0.8	1.000	1.000	1.000	0.020	1.000	1.000	0.971
	0.9	1.000	1.000	1.000	0.018	1.000	1.000	0.957

Table 2.A13: Power of tests against spatial ARMA process, R = 81 (queen), normal distribution.

ρ	θ_1	Moran	LM-ERR	K-R	LM-EL	SARMA	LM-LAG	LM-LE
	0.1	0.225	0.167	0.144	0.080	0.298	0.357	0.285
	0.2	0.366	0.286	0.239	0.146	0.395	0.429	0.292
	0.3	0.515	0.431	0.349	0.245	0.512	0.482	0.304
	0.4	0.677	0.592	0.488	0.393	0.634	0.556	0.304
0.1	0.5	0.793	0.724	0.617	0.542	0.741	0.589	0.302
	0.6	0.878	0.824	0.727	0.669	0.824	0.630	0.309
	0.7	0.928	0.893	0.817	0.768	0.887	0.690	0.316
	0.8	0.960	0.942	0.882	0.853	0.936	0.720	0.315
	0.9	0.984	0.972	0.932	0.914	0.968	0.767	0.320
	0.1	0.497	0.405	0.342	0.115	0.783	0.849	0.766
	0.2	0.636	0.551	0.473	0.204	0.827	0.874	0.759
	0.3	0.763	0.686	0.582	0.319	0.868	0.888	0.731
	0.4	0.845	0.786	0.686	0.457	0.904	0.904	0.731
0.2	0.5	0.915	0.875	0.789	0.595	0.946	0.918	0.722
	0.6	0.950	0.924	0.861	0.709	0.965	0.924	0.695
	0.7	0.976	0.956	0.908	0.802	0.974	0.938	0.682
	0.8	0.985	0.975	0.943	0.865	0.986	0.950	0.671
	0.9	0.993	0.988	0.965	0.922	0.992	0.956	0.680
	0.1	0.793	0.724	0.648	0.209	0.983	0.993	0.979
	0.2	0.869	0.815	0.742	0.311	0.988	0.991	0.973
	0.3	0.928	0.889	0.814	0.430	0.992	0.993	0.968
	0.4	0.953	0.928	0.874	0.563	0.994	0.994	0.955
0.3	0.5	0.974	0.960	0.920	0.675	0.995	0.995	0.949
	0.6	0.988	0.977	0.950	0.766	0.997	0.995	0.936
	0.7	0.992	0.984	0.965	0.839	0.997	0.995	0.923
	0.8	0.997	0.994	0.982	0.891	0.999	0.994	0.907
	0.9	0.998	0.997	0.989	0.934	0.999	0.995	0.901
	0.1	0.960	0.938	0.895	0.341	0.999	1.000	0.999
	0.2	0.978	0.959	0.926	0.465	1.000	1.000	0.998
	0.3	0.984	0.974	0.950	0.563	0.999	1.000	0.998
	0.4	0.993	0.985	0.967	0.671	1.000	1.000	0.997
0.4	0.5	0.997	0.994	0.978	0.754	1.000	1.000	0.994
	0.6	0.998	0.996	0.989	0.808	1.000	1.000	0.992
	0.7	0.999	0.998	0.991	0.870	1.000	1.000	0.988
	0.8	0.999	0.999	0.996	0.911	1.000	1.000	0.985
	0.9	1.000	1.000	0.998	0.938	1.000	1.000	0.979
	0.1	0.997	0.995	0.990	0.518	1.000	1.000	1.000
	0.2	0.998	0.997	0.993	0.610	1.000	1.000	1.000
0.5	0.3	0.999	0.997	0.991	0.682	1.000	1.000	1.000
	0.4	1.000	0.999	0.996	0.752	1.000	1.000	1.000
	0.5	1.000	0.999	0.997	0.809	1.000	1.000	1.000

Table 2.A13: continued.

ρ	θ_1	Moran	LM-ERR	K-R	LM-EL	SARMA	LM-LAG	LM-LE
	0.6	1.000	0.999	0.998	0.863	1.000	1.000	1.000
0.5	0.7	1.000	1.000	0.998	0.892	1.000	1.000	0.998
	0.8	1.000	1.000	1.000	0.925	1.000	1.000	0.998
	0.9	1.000	1.000	0.999	0.944	1.000	1.000	0.997
	0.1	1.000	1.000	1.000	0.688	1.000	1.000	1.000
	0.2	1.000	1.000	0.999	0.744	1.000	1.000	1.000
	0.3	1.000	1.000	1.000	0.786	1.000	1.000	1.000
	0.4	1.000	1.000	1.000	0.825	1.000	1.000	1.000
0.6	0.5	1.000	1.000	1.000	0.868	1.000	1.000	1.000
	0.6	1.000	1.000	1.000	0.886	1.000	1.000	1.000
	0.7	1.000	1.000	1.000	0.911	1.000	1.000	1.000
	0.8	1.000	1.000	1.000	0.927	1.000	1.000	1.000
	0.9	1.000	1.000	1.000	0.943	1.000	1.000	1.000
	0.1	1.000	1.000	1.000	0.809	1.000	1.000	1.000
	0.2	1.000	1.000	1.000	0.835	1.000	1.000	1.000
	0.3	1.000	1.000	1.000	0.861	1.000	1.000	1.000
	0.4	1.000	1.000	1.000	0.873	1.000	1.000	1.000
0.7	0.5	1.000	1.000	1.000	0.899	1.000	1.000	1.000
	0.6	1.000	1.000	1.000	0.906	1.000	1.000	1.000
	0.7	1.000	1.000	1.000	0.916	1.000	1.000	1.000
	0.8	1.000	1.000	1.000	0.933	1.000	1.000	1.000
	0.9	1.000	1.000	1.000	0.946	1.000	1.000	1.000
	0.1	1.000	1.000	1.000	0.866	1.000	1.000	1.000
	0.2	1.000	1.000	1.000	0.866	1.000	1.000	1.000
	0.3	1.000	1.000	1.000	0.878	1.000	1.000	1.000
	0.4	1.000	1.000	1.000	0.890	1.000	1.000	1.000
0.8	0.5	1.000	1.000	1.000	0.895	1.000	1.000	1.000
	0.6	1.000	1.000	1.000	0.891	1.000	1.000	1.000
	0.7	1.000	1.000	1.000	0.903	1.000	1.000	1.000
	0.8	1.000	1.000	1.000	0.897	1.000	1.000	1.000
	0.9	1.000	1.000	1.000	0.900	1.000	1.000	1.000
	0.1	1.000	1.000	1.000	0.351	1.000	1.000	1.000
	0.2	1.000	1.000	1.000	0.357	1.000	1.000	1.000
	0.3	1.000	1.000	1.000	0.375	1.000	1.000	1.000
	0.4	1.000	1.000	1.000	0.369	1.000	1.000	1.000
0.9	0.5	1.000	1.000	1.000	0.361	1.000	1.000	1.000
	0.6	1.000	1.000	1.000	0.359	1.000	1.000	1.000
	0.7	1.000	1.000	1.000	0.372	1.000	1.000	1.000
	0.8	1.000	1.000	1.000	0.365	1.000	1.000	1.000
	0.9	1.000	1.000	1.000	0.377	1.000	1.000	1.000

Table 2.A14: Power of tests against spatial ARMA process, R = 81 (rook), normal distribution.

ρ	θ_1	Moran	LM-ERR	K-R	LM-EL	SARMA	LM-LAG	LM-LE
	0.1	0.225	0.167	0.144	0.080	0.298	0.357	0.285
	0.2	0.366	0.286	0.239	0.146	0.395	0.429	0.292
	0.3	0.515	0.431	0.349	0.245	0.512	0.482	0.304
	0.4	0.677	0.592	0.488	0.393	0.634	0.556	0.304
0.1	0.5	0.793	0.724	0.617	0.542	0.741	0.589	0.302
	0.6	0.878	0.824	0.727	0.669	0.824	0.630	0.309
	0.7	0.928	0.893	0.817	0.768	0.887	0.690	0.316
	0.8	0.960	0.942	0.882	0.853	0.936	0.720	0.315
	0.9	0.984	0.972	0.932	0.914	0.968	0.767	0.320
	0.1	0.497	0.405	0.342	0.115	0.783	0.849	0.766
	0.2	0.636	0.551	0.473	0.204	0.827	0.874	0.759
	0.3	0.763	0.686	0.582	0.319	0.868	0.888	0.731
	0.4	0.845	0.786	0.686	0.457	0.904	0.904	0.731
0.2	0.5	0.915	0.875	0.789	0.595	0.946	0.918	0.722
	0.6	0.950	0.924	0.861	0.709	0.965	0.924	0.695
	0.7	0.976	0.956	0.908	0.802	0.974	0.938	0.682
	0.8	0.985	0.975	0.943	0.865	0.986	0.950	0.671
	0.9	0.993	0.988	0.965	0.922	0.992	0.956	0.680
	0.1	0.793	0.724	0.648	0.209	0.983	0.993	0.979
	0.2	0.869	0.815	0.742	0.311	0.988	0.991	0.973
	0.3	0.928	0.889	0.814	0.430	0.992	0.993	0.968
	0.4	0.953	0.928	0.874	0.563	0.994	0.994	0.955
0.3	0.5	0.974	0.960	0.920	0.675	0.995	0.995	0.949
	0.6	0.988	0.977	0.950	0.766	0.997	0.995	0.936
	0.7	0.992	0.984	0.965	0.839	0.997	0.995	0.923
	0.8	0.997	0.994	0.982	0.891	0.999	0.994	0.907
	0.9	0.998	0.997	0.989	0.934	0.999	0.995	0.901
	0.1	0.960	0.938	0.895	0.341	0.999	1.000	0.999
	0.2	0.978	0.959	0.926	0.465	1.000	1.000	0.998
	0.3	0.984	0.974	0.950	0.563	0.999	1.000	0.998
	0.4	0.993	0.985	0.967	0.671	1.000	1.000	0.997
0.4	0.5	0.997	0.994	0.978	0.754	1.000	1.000	0.994
	0.6	0.998	0.996	0.989	0.808	1.000	1.000	0.992
	0.7	0.999	0.998	0.991	0.870	1.000	1.000	0.988
	0.8	0.999	0.999	0.996	0.911	1.000	1.000	0.985
	0.9	1.000	1.000	0.998	0.938	1.000	1.000	0.979
	0.1	0.997	0.995	0.990	0.518	1.000	1.000	1.000
	0.2	0.998	0.997	0.993	0.610	1.000	1.000	1.000
0.5	0.3	0.999	0.997	0.991	0.682	1.000	1.000	1.000
	0.4	1.000	0.999	0.996	0.752	1.000	1.000	1.000
	0.5	1.000	0.999	0.997	0.809	1.000	1.000	1.000

Table 2.A14: continued.

ρ	θ_1	Moran	LM-ERR	K-R	LM-EL	SARMA	LM-LAG	LM-LE
	0.6	1.000	0.999	0.998	0.863	1.000	1.000	1.000
0.5	0.7	1.000	1.000	0.998	0.892	1.000	1.000	0.998
	0.8	1.000	1.000	1.000	0.925	1.000	1.000	0.998
	0.9	1.000	1.000	0.999	0.944	1.000	1.000	0.997
	0.1	1.000	1.000	1.000	0.688	1.000	1.000	1.000
	0.2	1.000	1.000	0.999	0.744	1.000	1.000	1.000
	0.3	1.000	1.000	1.000	0.786	1.000	1.000	1.000
	0.4	1.000	1.000	1.000	0.825	1.000	1.000	1.000
0.6	0.5	1.000	1.000	1.000	0.868	1.000	1.000	1.000
	0.6	1.000	1.000	1.000	0.886	1.000	1.000	1.000
	0.7	1.000	1.000	1.000	0.911	1.000	1.000	1.000
	0.8	1.000	1.000	1.000	0.927	1.000	1.000	1.000
	0.9	1.000	1.000	1.000	0.943	1.000	1.000	1.000
	0.1	1.000	1.000	1.000	0.809	1.000	1.000	1.000
	0.2	1.000	1.000	1.000	0.835	1.000	1.000	1.000
	0.3	1.000	1.000	1.000	0.861	1.000	1.000	1.000
	0.4	1.000	1.000	1.000	0.873	1.000	1.000	1.000
0.7	0.5	1.000	1.000	1.000	0.899	1.000	1.000	1.000
	0.6	1.000	1.000	1.000	0.906	1.000	1.000	1.000
	0.7	1.000	1.000	1.000	0.916	1.000	1.000	1.000
	0.8	1.000	1.000	1.000	0.933	1.000	1.000	1.000
	0.9	1.000	1.000	1.000	0.946	1.000	1.000	1.000
	0.1	1.000	1.000	1.000	0.866	1.000	1.000	1.000
	0.2	1.000	1.000	1.000	0.866	1.000	1.000	1.000
	0.3	1.000	1.000	1.000	0.878	1.000	1.000	1.000
	0.4	1.000	1.000	1.000	0.890	1.000	1.000	1.000
0.8	0.5	1.000	1.000	1.000	0.895	1.000	1.000	1.000
	0.6	1.000	1.000	1.000	0.891	1.000	1.000	1.000
	0.7	1.000	1.000	1.000	0.903	1.000	1.000	1.000
	0.8	1.000	1.000	1.000	0.897	1.000	1.000	1.000
	0.9	1.000	1.000	1.000	0.900	1.000	1.000	1.000
	0.1	1.000	1.000	1.000	0.351	1.000	1.000	1.000
	0.2	1.000	1.000	1.000	0.357	1.000	1.000	1.000
	0.3	1.000	1.000	1.000	0.375	1.000	1.000	1.000
	0.4	1.000	1.000	1.000	0.369	1.000	1.000	1.000
0.9	0.5	1.000	1.000	1.000	0.361	1.000	1.000	1.000
	0.6	1.000	1.000	1.000	0.359	1.000	1.000	1.000
	0.7	1.000	1.000	1.000	0.372	1.000	1.000	1.000
	0.8	1.000	1.000	1.000	0.365	1.000	1.000	1.000
	0.9	1.000	1.000	1.000	0.377	1.000	1.000	1.000

Table 2.A15: Power of tests against spatial ARMA process, R = 127, normal distribution.

ρ	θ_1	Moran	LM-ERR	K-R	LM-EL	SARMA	LM-LAG	LM-LE
	0.1	0.411	0.358	0.258	0.121	0.673	0.727	0.590
	0.2	0.696	0.642	0.470	0.311	0.806	0.805	0.590
	0.3	0.876	0.846	0.672	0.561	0.911	0.851	0.580
	0.4	0.964	0.955	0.840	0.787	0.965	0.883	0.569
0.1	0.5	0.992	0.988	0.928	0.910	0.992	0.917	0.573
	0.6	0.998	0.998	0.974	0.968	0.998	0.939	0.547
	0.7	1.000	1.000	0.996	0.991	1.000	0.948	0.528
	0.8	1.000	1.000	0.998	0.998	1.000	0.961	0.519
	0.9	1.000	1.000	1.000	1.000	1.000	0.975	0.515
	0.1	0.801	0.760	0.635	0.164	0.994	0.998	0.991
	0.2	0.931	0.913	0.809	0.379	0.998	0.998	0.984
	0.3	0.983	0.976	0.909	0.604	0.999	0.999	0.979
	0.4	0.992	0.989	0.960	0.780	0.999	0.999	0.968
0.2	0.5	0.998	0.998	0.988	0.902	1.000	0.999	0.965
	0.6	1.000	1.000	0.997	0.966	1.000	1.000	0.950
	0.7	1.000	1.000	0.999	0.989	1.000	0.999	0.944
	0.8	1.000	1.000	1.000	0.996	1.000	0.999	0.922
	0.9	1.000	1.000	1.000	0.999	1.000	0.999	0.917
	0.1	0.985	0.978	0.940	0.209	1.000	1.000	1.000
	0.2	0.995	0.992	0.972	0.421	1.000	1.000	1.000
	0.3	0.999	0.999	0.990	0.619	1.000	1.000	1.000
	0.4	1.000	1.000	0.997	0.789	1.000	1.000	1.000
0.3	0.5	1.000	1.000	0.999	0.888	1.000	1.000	1.000
	0.6	1.000	1.000	1.000	0.947	1.000	1.000	1.000
	0.7	1.000	1.000	1.000	0.976	1.000	1.000	0.999
	0.8	1.000	1.000	1.000	0.990	1.000	1.000	0.997
	0.9	1.000	1.000	1.000	0.994	1.000	1.000	0.995
	0.1	1.000	1.000	0.997	0.227	1.000	1.000	1.000
	0.2	1.000	1.000	0.999	0.407	1.000	1.000	1.000
	0.3	1.000	1.000	1.000	0.559	1.000	1.000	1.000
	0.4	1.000	1.000	1.000	0.697	1.000	1.000	1.000
0.4	0.5	1.000	1.000	1.000	0.823	1.000	1.000	1.000
	0.6	1.000	1.000	1.000	0.895	1.000	1.000	1.000
	0.7	1.000	1.000	1.000	0.941	1.000	1.000	1.000
	0.8	1.000	1.000	1.000	0.970	1.000	1.000	1.000
	0.9	1.000	1.000	1.000	0.979	1.000	1.000	1.000
	0.1	1.000	1.000	1.000	0.176	1.000	1.000	1.000
	0.2	1.000	1.000	1.000	0.277	1.000	1.000	1.000
0.5	0.3	1.000	1.000	1.000	0.400	1.000	1.000	1.000
	0.4	1.000	1.000	1.000	0.540	1.000	1.000	1.000
	0.5	1.000	1.000	1.000	0.651	1.000	1.000	1.000

Table 2.A15: continued.

ρ	θ_1	Moran	LM-ERR	K-R	LM-EL	SARMA	LM-LAG	LM-LE
	0.6	1.000	1.000	1.000	0.723	1.000	1.000	1.000
0.5	0.7	1.000	1.000	1.000	0.792	1.000	1.000	1.000
	0.8	1.000	1.000	1.000	0.853	1.000	1.000	1.000
	0.9	1.000	1.000	1.000	0.891	1.000	1.000	1.000
	0.1	1.000	1.000	1.000	0.056	1.000	1.000	1.000
	0.2	1.000	1.000	1.000	0.093	1.000	1.000	1.000
	0.3	1.000	1.000	1.000	0.164	1.000	1.000	1.000
	0.4	1.000	1.000	1.000	0.246	1.000	1.000	1.000
0.6	0.5	1.000	1.000	1.000	0.325	1.000	1.000	1.000
	0.6	1.000	1.000	1.000	0.393	1.000	1.000	1.000
	0.7	1.000	1.000	1.000	0.462	1.000	1.000	1.000
	0.8	1.000	1.000	1.000	0.553	1.000	1.000	1.000
	0.9	1.000	1.000	1.000	0.595	1.000	1.000	1.000
	0.1	1.000	1.000	1.000	0.003	1.000	1.000	1.000
	0.2	1.000	1.000	1.000	0.011	1.000	1.000	1.000
	0.3	1.000	1.000	1.000	0.016	1.000	1.000	1.000
	0.4	1.000	1.000	1.000	0.036	1.000	1.000	1.000
0.7	0.5	1.000	1.000	1.000	0.055	1.000	1.000	1.000
	0.6	1.000	1.000	1.000	0.090	1.000	1.000	1.000
	0.7	1.000	1.000	1.000	0.126	1.000	1.000	1.000
	0.8	1.000	1.000	1.000	0.163	1.000	1.000	1.000
	0.9	1.000	1.000	1.000	0.207	1.000	1.000	1.000
	0.1	1.000	1.000	1.000	0.000	1.000	1.000	1.000
	0.2	1.000	1.000	1.000	0.000	1.000	1.000	1.000
	0.3	1.000	1.000	1.000	0.001	1.000	1.000	1.000
	0.4	1.000	1.000	1.000	0.001	1.000	1.000	1.000
0.8	0.5	1.000	1.000	1.000	0.003	1.000	1.000	1.000
	0.6	1.000	1.000	1.000	0.003	1.000	1.000	1.000
	0.7	1.000	1.000	1.000	0.011	1.000	1.000	1.000
	0.8	1.000	1.000	1.000	0.013	1.000	1.000	1.000
	0.9	1.000	1.000	1.000	0.023	1.000	1.000	1.000
	0.1	1.000	1.000	1.000	0.000	1.000	1.000	1.000
	0.2	1.000	1.000	1.000	0.000	1.000	1.000	1.000
	0.3	1.000	1.000	1.000	0.000	1.000	1.000	1.000
	0.4	1.000	1.000	1.000	0.000	1.000	1.000	1.000
0.9	0.5	1.000	1.000	1.000	0.000	1.000	1.000	1.000
	0.6	1.000	1.000	1.000	0.000	1.000	1.000	1.000
	0.7	1.000	1.000	1.000	0.000	1.000	1.000	1.000
	0.8	1.000	1.000	1.000	0.001	1.000	1.000	1.000
	0.9	1.000	1.000	1.000	0.001	1.000	1.000	1.000

3 Spatial Correlation:
A Suggested Alternative to the Autoregressive Model

Harry H. Kelejian[1] and Dennis P. Robinson[2]
[1] University of Maryland, College Park, MD, U.S.A.
[2] U.S. Army Corps of Engineers, Alexandria, VA, U.S.A.

3.1 Introduction

Issues relating to spatially autocorrelated disturbance terms are often considered in regional econometric models.[1] Although various models have been suggested to describe such spatial correlation, one of the most widely used models is a spatial autoregressive (AR) model which was originally suggested by Whittle (1954) and then extensively studied by Cliff and Ord (1973).[2] In the model the regression disturbance vector is viewed as the sum of two parts. One of these parts involves the product of a spatial weighting matrix and a scalar parameter, say ρ; the other is a random vector whose elements are typically assumed to be independent and identically distributed (i.i.d.) with zero mean and finite variance. We will henceforth refer to this random vector as the innovation vector, so as to distinguish it from the disturbance vector.

---

[1] There is an extensive discussion of this issue in the regional science and geography literature. The classic reference is Cliff and Ord (1973, 1981); other excellent studies of the issues involved are given by Arora and Brown (1977), Anselin (1980, 1988, 1990), Bennett and Hordijk (1986), Anselin and Griffith (1988), Odland (1988), and Haining (1990). In contrast, spatial autocorrelation is rarely considered in the econometrics literature. For example, among other texts there is no formal discussion of the issues and models involved in Schmidt (1976), Maddala (1977), Dhrymes (1978), Intrilligator (1978), Chow (1983), Fomby et al. (1984), Judge et al. (1985), Kmenta (1986), and Kelejian and Oates (1989). There is a discussion of spatial autocorrelation in Johnston's (1984) text but that discussion is limited.

[2] See Anselin (1990) and the references cited therein.

The purpose of this chapter is twofold. First, we critically evaluate the spatial AR model. In doing this, we focus both on certain interpretations of ρ given by Cliff and Ord, and by others in the literature, as well as on basic specification issues and consequent problems. Second, we suggest an alternative model of spatial autocorrelation which does not have these 'consequent' problems.

As a preview of our results, in the spatial AR model $\rho = 0$ implies the absence of spatially autocorrelated regression disturbances; non-zero values of ρ imply such correlation. These results, and perhaps, the similarity in 'appearance' of the spatial AR model to a standard time series AR(1) model involving a scalar disturbance term has lead some researchers to interpret the parameter ρ as if it were an autocorrelation coefficient;[3] for a variety of other reasons, other researchers place bounds on the practical and, in some cases, the logical values of the parameter ρ. These bounds are typically $|\rho| < 1$, and in some cases $\rho < 1$.[4]

Our results suggest that the 'correlation coefficient' interpretation of ρ is 'reasonable' only for special cases of the weighting matrix and parameter space. For example, we show that except for a set of singular points, the spatial AR model can generally be defined for all $-\infty < \rho < \infty$. The number of singular points is at most equal to the sample size; the set of singular points depends upon both the weighting matrix and the sample size.

For a given sample size we show that for many of the weighting matrices considered in practice, a certain number of correlation coefficients relating the disturbance terms to each other approach unity in absolute value as ρ approaches a singular point. For most weighting matrices, less than 100 percent of the correlation coefficients approach unity in this manner; the exact percentage of such correlation coefficients varies across the singular points. We also show that for

---

[3] For the case in which the spatial weighting matrix is 'row-normalized' (i.e., the values of each row sum to unity), Cliff and Ord (1973, 1981) interpret ρ as an autocorrelation coefficient. Other analysts refer to ρ as an autocorrelation parameter but do not give an exact definition [see, for example, Bartels (1979), Johnston (1984), and Odland (1988)]. Still other researchers have referred to ρ as an autoregressive parameter as in a standard AR(1) model [for example, see Bennett and Hordijk (1986), and Getis (1990)].

[4] There are at least two arguments given for these restrictions. One is a claim that $|\rho| < 1$ ensures stationarity [see, e.g., Griffith (1988)]. Our discussion below will demonstrate that stationarity is not ensured by this restriction. A second argument is related to issues of computational convenience. For example, let λ_{min} and λ_{max} be the minimum and maximum eigenvalues of the weighting matrix. Then, the Jacobian in the likelihood function will be positive, without having to take absolute values, if:

$$-1/|\lambda_{min}| < \rho < 1/\lambda_{max}$$

[see, e.g., Anselin (1982)]. For row-normalized matrices $\lambda_{max} = 1$; typically, however, $|\lambda_{min}| < 1$. For still other arguments, see Hepple (1979), and Upton and Fingleton (1985).

many of these weighting matrices there are multiple positive as well as negative singular points. Unless special cases of the weighting matrix are considered, the largest negative singular point and the smallest positive singular point need not be plus and minus unity.[5] Indeed, for many cases of the weighting matrix the absolute value of the largest negative singular point is larger than the smallest positive singular point. All of this suggests that interpretations of ρ as a correlation coefficient may not be meaningful for most weighting matrices considered in practice.

For certain weighting matrices, however, the smallest positive and largest negative singular points are plus and minus unity. For these weighting matrices, all of the correlations relating the disturbances to each other approach unity as ρ approaches either plus or minus unity. Therefore, in these cases, if the possible values of ρ are restricted to the set $|\rho| < 1$, ρ can be interpreted as being 'similar' to a correlation coefficient in that $\rho = 0$ implies the absence of correlation, and $|\rho| = 1$ implies perfect correlation. However, we show that even in these cases the generality of the spatial AR model is reduced by the parameter restriction $|\rho| < 1$.

Finally, we note that estimation of an econometric model along with associated inferences requires a specification of the parameter space. We give results which imply that the spatial AR model, as it is typically specified, is inconsistent at the singular points of ρ. Because of this, these points should be excluded from the parameter space. We argue below that this could lead to large sample theory problems concerning the estimation of such models.[6]

The spatial AR model is specified in section two and certain issues concerning the singular points of ρ are discussed in section three. Corresponding parameter space issues are considered in section four. An alternative model of spatial autocorrelation is given in section five.

---

[5] The literature is both sparse and unclear concerning the largest negative and smallest positive singular points of ρ. An exception is given by Haining (1990, p. 82) who demonstrates that these values converge to $\pm$ 0.25 as the sample size increases if the weighting matrix has a rook design (see the appendix to this chapter for definitions of various weighting matrices). Upton and Fingleton (1985, pp. 281-282) also consider this issue. Interestingly, for certain case studies of weighting matrices, they calculate the largest negative and smallest positive singular points of ρ. Although the pairs of values they calculate are not $\pm$ 1, they nevertheless suggest the restriction $|\rho| < 1$.

[6] Among other things, note that certain complexities could arise if isolated points are excluded from the parameter space because that space would not be compact [see, e.g., Amemiya (1985, pp. 105-154)].

3.2 The Spatial AR Model of Autocorrelation

Consider the model:

$$\varepsilon = \rho M \varepsilon + U \tag{1}$$

where ε is a $(T \times 1)$ vector of disturbance terms, ρ is a scalar parameter, M is a nonstochastic $(T \times T)$ weighting matrix, and U is a $(T \times 1)$ vector of stochastic elements, i.e., innovations. The assumptions concerning U are:

$$E(U) = 0 \quad ; \quad E(UU') = \Omega_U \tag{2}$$

where Ω_U^{-1} exists.[7]

The expression in (1) can be rewritten as:

$$(I - \rho M)\varepsilon = U \tag{3}$$

If $(I\text{-}\rho M)$ is nonsingular:

$$\varepsilon = (I - \rho M)^{-1} U \tag{4}$$

so that:

$$E(\varepsilon) = 0 \quad ; \quad E(\varepsilon\varepsilon') \equiv \Omega_\varepsilon = (I - \rho M)^{-1} \Omega_U (I - \rho M')^{-1} \tag{5}$$

In this case the model determines ε in terms of U. Clearly if $\rho = 0$ and Ω_U is diagonal, the elements of ε will not be autocorrelated.[8]

3.3 The Singularity of $(I\text{-}\rho M)$

3.3.1 Theoretical Issues

For many specifications of T and M in practice, the matrix $(I\text{-}\rho M)$ will be singular for certain real values of ρ. Let $S(M,T)$ denote the set of values of ρ for which $(I$-

---

[7] In the literature Ω_U is typically taken to be $\Omega_U = \sigma_U^2 I$ where σ_U^2 scalar [see, e.g., Cliff and Ord (1981), Johnston (1984), and Anselin and Rey (1991) among others].

[8] In reference to footnote 4 we note that if $\Omega_U = \sigma_U^2 I$, M is a row-normalized double queen weighting matrix, and $\rho = 0.5$, the diagonal elements of Ω_ε are not all the same. Therefore $|\rho| < 1$ does not ensure stationarity.

$\rho \boldsymbol{M}$) is singular. Then, for $\rho \in S(\boldsymbol{M},T)$ the result in (4) does not hold, and so Ω_ε is no longer given by the expression in (5).

The result in (3) still holds, however, for $\rho \in S(\boldsymbol{M},T)$ if ε is generated as in (1). Because the variance-covariance matrix of the left-hand-side of (3) is equal to the variance-covariance matrix of the right-hand-side we have:

$$(\boldsymbol{I}-\rho\boldsymbol{M})\Omega_\varepsilon(\boldsymbol{I}-\rho\boldsymbol{M}') = \Omega_U \tag{6}$$

For all $\rho \in S(\boldsymbol{M},T)$, the left-hand-side of (6) is singular. Therefore for these values of ρ, Ω_U must also be singular. It follows that if Ω_U is assumed to be nonsingular, or (more passively) if Ω_U is specified in such a way that its inverse exists (e.g., $\Omega_U = \sigma^2\boldsymbol{I}$) the model must be inconsistent for all values of $\rho \in S(\boldsymbol{M},T)$.

Consider again the model in (1)-(5) for values of ρ which are not in $S(\boldsymbol{M},T)$ and a limiting result relating the correlations of the elements of ε to ρ. Let ρ_o be an element of $S(\boldsymbol{M},T)$ so that $(\boldsymbol{I}-\rho_o\boldsymbol{M})$ is singular; assume that $(\boldsymbol{I}-\rho\boldsymbol{M})$ is nonsingular for all ρ such that:

$$0 < |\rho-\rho_0| < \delta \tag{7}$$

where $\delta > 0$. Let $C_{ij}(\rho|\boldsymbol{M},T)$ be the correlation coefficient between the i-th and j-th elements of ε corresponding to a value of ρ in the interval (7), when the weighting matrix is $\boldsymbol{M}$ and the sample size is T. This correlation coefficient is based on Ω_ε in (5). Let $\max(\rho|\boldsymbol{M},T)$ be the largest of these correlations:

$$\max(\rho|\boldsymbol{M},T) = \max_{i,j} C_{i,j}(\rho|\boldsymbol{M},T) \tag{8}$$

Then, our calculations described in Table 3.1 below suggest that for all cases considered:

$$\lim_{\rho\to\rho_0} \max(\rho|\boldsymbol{M},T) = 1 \tag{9}$$

That is, if ρ is 'near' a singular point, at least one correlation relating disturbance terms is 'near' unity. The results in the table also suggest that the proportion of such limiting unitary correlations varies across singular points, and depends upon the weighting matrix and sample size. This implies that the non-linearities involved near the critical points are related to the type and size of the spatial weighting matrix, and are not the same at all of the points. For example, all of the results in Table 3.1 are based on ρ-values within 10^{-6} of the true critical points for each matrix considered. However, the proportion of computed correlation coefficients greater than 0.999 varies from 0.006 to 1.0.

Table 3.1: Percent of correlations greater than 0.999 using ρ values within 0.000001 of the two largest negative and two smallest positive singular points for each of the spatial weighting matrices considered.

	(25 × 25) weighting matrix		(100 × 100) weighting matrix	
	ρ	% Cor > 0.999	ρ	% Cor > 0.999
	ROW-NORMALIZED			
Linear	-1.008627960580	92.0	-1.000502711214	95.9
	-0.999999000000	100.0	-0.999999000000	100.0
	0.999999000000	100.0	0.999999000000	100.0
	1.008627960580	92.0	1.000502711214	95.9
Rook	-1.151145568595	9.3	-1.029438931733	3.6
	-0.999999000000	100.0	-0.999999000000	100.0
	0.999999000000	100.0	0.999999000000	100.0
	1.151145568595	9.3	1.029438931733	3.6
Bishop	-1.414212562373	6.7	-1.064176772476	2.6
	-0.999999000000	48.0	-0.999999000000	49.5
	0.999999000000	48.0	0.999999000000	49.5
	1.414212562373	6.7	1.064176772476	2.6
Dble Linear	-1.716968390513	92.0	-1.711701867529	29.1
	-1.707434386376	100.0	-1.711701765886	29.0
	0.999999000000	100.0	0.999999000000	100.0
	1.022720072147	92.0	1.001271804736	99.9
Dble Rook	-1.806946972762	22.7	-1.722176613344	9.3
	-1.645750311064	40.0	-1.704747123513	100.0
	0.999999000000	100.0	0.999999000000	100.0
	1.463742292133	9.3	1.081093004204	3.9
Queen	-2.129954639676	40.0	-1.971537331458	100.0
	-2.058441749303	100.0	-1.970428085334	3.9
	0.999999000000	100.0	0.999999000000	100.0
	1.249755696751	9.3	1.045221885637	5.5
Dble Queen	-4.726493576283	9.3	-5.194731245847	4.2
	-4.383571539057	100.0	-4.937299737704	5.5
	0.999999000000	100.0	0.999999000000	100.0
	2.535516711068	9.3	1.182197366184	3.9

Table 3.1: continued.

	(25 × 25) weighting matrix		(100 × 100) weighting matrix	
	ρ	% Cor > 0.999	ρ	% Cor > 0.999
Queen w Dist	-2.242982772511	20.0	-2.120428736242	6.0
	-2.156739643305	100.0	-2.100060182060	100.0
	0.999999000000	100.0	0.999999000000	100.0
	1.229504347151	9.3	1.042314090880	5.5
Dble Queen w Dist	-5.212620921672	9.3	-4.551527832443	5.7
	-3.879482321619	100.0	-4.408439486611	100.0
	0.999999000000	100.0	0.999999000000	100.0
	1.902801965452	9.3	1.134205692133	3.6
NOT ROW-NORMALIZED				
Linear	-0.514962915474	92.0	-0.500968076016	96.1
	-0.503671338432	100.0	-0.500240975902	100.0
	0.503671338432	100.0	0.500240975902	100.0
	0.514962915474	92.0	0.500968076016	96.1
Rook	-0.366024403784	9.3	-0.277661623923	3.3
	-0.288674134594	100.0	-0.260553279056	100.0
	0.288674134594	100.0	0.260553279056	100.0
	0.366024403784	9.3	0.277661623923	3.3
Bishop	-0.577349269189	6.7	-0.309720467890	2.6
	-0.333332333333	48.0	-0.271553129338	49.5
	0.333332333333	48.0	0.271553129338	49.5
	0.577349269189	6.7	0.309720467890	2.6
Dble Linear	-0.455340244256	92.0	-0.445155733021	20.7
	-0.454015229069	100.0	-0.445138396192	25.6
	0.254432769260	100.0	0.250298341406	100.0
	0.268483078509	92.0	0.251199690130	99.7
Dble Rook	-0.323532723563	9.3	-0.296102030596	0.6
	-0.309015994375	40.0	-0.249999000000	2.3
	0.170331660901	100.0	0.137413408371	100.0
	0.281414358904	9.3	0.158988537909	3.9

Table 3.1: continued.

	(25 × 25) weighting matrix		(100 × 100) weighting matrix	
	ρ	% Cor > 0.999	ρ	% Cor > 0.999
Queen	-0.405826419558	9.3	-0.288583772760	6.3
	-0.333332333333	48.0	-0.271553129338	49.5
	0.154699538379	100.0	0.132969254328	100.0
	0.224008237739	9.3	0.146407587938	3.3
Dble Queen	-0.368310679306	9.3	-0.252321269717	3.4
	-0.354670866821	100.0	-0.250672971459	100.0
	0.063154229917	100.0	0.042687914592	100.0
	0.192518572910	9.3	0.056922339928	3.6
Queen w Dist	-0.511038558576	9.3	-0.396900030212	6.3
	-0.471403520791	48.0	-0.384034532629	49.5
	0.179036502877	100.0	0.155232761878	100.0
	0.252728754360	9.3	0.169935993055	3.3
Dble Queen w Dist	-0.622074162894	40.0	-0.470180898253	3.9
	-0.501909585263	100.0	-0.417254238824	100.0
	0.104505397585	100.0	0.076233499069	100.0
	0.234322555864	9.3	0.095606744850	3.6

3.3.2 Independent Corroborative Evidence

The result in (9) explains certain asymmetric Monte Carlo results reported by Anselin and Rey (1991). For example, Anselin and Rey's model 2 is based upon a row-normalized queen specification of the weighting matrix $\boldsymbol{M}$ (see Appendix 1). The sample sizes they considered range from $T = 25$ to $T = 225$. Our results in Table 3.2 imply that the largest negative value of ρ for which $(\boldsymbol{I}\text{-}\rho\boldsymbol{M})$ is singular for the row normalized queen specification is -2.058 when $T = 25$; this value slowly increases as the sample size increases and is -1.934 when $T = 225$. On the other hand, the smallest positive value of ρ for which $(\boldsymbol{I}\text{-}\rho\boldsymbol{M})$ is singular is unity for the row normalized queen specification of $\boldsymbol{M}$ for all values of the sample size. The implication of all this is that in Anselin and Rey's model 2, values of ρ near +1 imply 'very' large correlations of the elements of ε (see Table 3.1); on the other hand, values of ρ near -1 do not imply such large correlations.[9] Ceteris paribus, if the correlations of the disturbance terms are high, the power of a test for correlation should also be high, and vice versa. This is precisely the picture conveyed in Anselin and Rey's Figure 3. Anselin and Rey's model 3 is based upon a row-normalized double queen specification of the weighting matrix $\boldsymbol{M}$ (see Appendix 1). A glance at our results in Tables 3.1 and 3.2 relating to this specification of $\boldsymbol{M}$ should explain,[10] in a similar fashion, the asymmetric results described in Anselin and Rey's Figure 4.

3.4 The Parameter Space

Consider again the spatial AR model in (1) and (2). Since Ω_U is typically assumed to be nonsingular, the parameter space should not contain those points for which $(\boldsymbol{I}-\rho\boldsymbol{M})$ is singular. For a given weighting matrix $\boldsymbol{M}$, these points can be determined as the roots of the polynomial equation:[11]

$$|\boldsymbol{I}-\rho\boldsymbol{M}| = 0 \tag{10}$$

In general, there will be $T_1 \leq T$ real distinct roots to (10) since $(\boldsymbol{I}-\rho\boldsymbol{M})$ is a $(T \times T)$ matrix. Clearly, these values will depend upon T as well as $\boldsymbol{M}$. Illustrative calculations concerning these roots are given in Tables 3.2 and 3.3.

Certain points should be noted about the results in Tables 3.2 and 3.3. First, the singular points of ρ, i.e., roots of (10), very definitely vary with the sample size T and the weighting matrix $\boldsymbol{M}$. Second, for all nine cases considered involving a row-normalized weighting matrix, the smallest positive root is +1.[12]

---

[9] For example, if $\Omega_U = \boldsymbol{I}$ in the queen specification when T = 25 and ρ = -1, the largest correlation (in absolute value) is 0.3470; when T = 100 and ρ = -1, the absolute value of this correlation is 0.3464.

[10] In the double queen specification when T = 25 and ρ = -1, the absolute value of the largest correlation is 0.1324; when T = 100 the absolute value of this largest correlation is 0.1177. These results correspond to the case in which $\Omega_U = \boldsymbol{I}$.

[11] The result in (10) implies $|\boldsymbol{M}-\rho^{-1}\boldsymbol{I}| = 0$ and so the roots of (10) can also be determined as the reciprocals of the characteristic roots of the weighting matrix $\boldsymbol{M}$. Because we are concerned with parameter space issues, we do not consider possible complex roots of (10).

[12] That unity is a root of (10) if $\boldsymbol{M}$ is row-normalized is a known result in the literature and should be clear. For example, in this case $(\boldsymbol{I}-\boldsymbol{M})$ is a matrix each of whose rows sums to zero; $(\boldsymbol{I}-\boldsymbol{M})$ is therefore singular since $(\boldsymbol{I}-\boldsymbol{M})e_T = 0$ where e_T is a $(T \times 1)$ vector of unit elements.

Table 3.2: Negative and positive singular points of ρ closest to zero for each spatial weighting matrix considered.

ROW-NORMALIZED

	Linear		Rook		Bishop	
T	-ρ	+ρ	-ρ	+ρ	-ρ	+ρ
4	-1.0000	1.0000	-1.0000	1.0000	-1.0000	1.0000
9	-1.0000	1.0000	-1.0000	1.0000	-1.0000	1.0000
16	-1.0000	1.0000	-1.0000	1.0000	-1.0000	1.0000
25	-1.0000	1.0000	-1.0000	1.0000	-1.0000	1.0000
36	-1.0000	1.0000	-1.0000	1.0000	-1.0000	1.0000
49	-1.0000	1.0000	-1.0000	1.0000	-1.0000	1.0000
64	-1.0000	1.0000	-1.0000	1.0000	-1.0000	1.0000
81	-1.0000	1.0000	-1.0000	1.0000	-1.0000	1.0000
100	-1.0000	1.0000	-1.0000	1.0000	-1.0000	1.0000
225	-1.0000	1.0000	-1.0000	1.0000	-1.0000	1.0000

	Double Linear		Double Rook		Queen	
T	-ρ	+ρ	-ρ	+ρ	-ρ	+ρ
4	-1.5000	1.0000	-1.0000	1.0000	-3.0000	1.0000
9	-1.6460	1.0000	-2.0000	1.0000	-2.2087	1.0000
16	-1.6903	1.0000	-1.5000	1.0000	-2.1682	1.0000
25	-1.7074	1.0000	-1.6458	1.0000	-2.0584	1.0000
36	-1.7112	1.0000	-1.6277	1.0000	-2.0471	1.0000
49	-1.7115	1.0000	-1.6277	1.0000	-2.0027	1.0000
64	-1.7117	1.0000	-1.6969	1.0000	-1.9975	1.0000
81	-1.7117	1.0000	-1.6460	1.0000	-1.9734	1.0000
100	-1.7117	1.0000	-1.7047	1.0000	-1.9704	1.0000
225	-1.7117	1.0000	-1.7098	1.0000	-1.9343	1.0000

	Double Queen		Queen w. Distance		Double Queen w. Dis.	
T	-ρ	+ρ	-ρ	+ρ	-ρ	+ρ
4	-3.0000	1.0000	-2.0938	1.0000	-2.0938	1.0000
9	-8.0000	1.0000	-2.2373	1.0000	-3.9213	1.0000
16	-5.0000	1.0000	-2.1875	1.0000	-3.9856	1.0000
25	-4.3836	1.0000	-2.1567	1.0000	-3.8795	1.0000
36	-4.7564	1.0000	-2.1396	1.0000	-4.2074	1.0000
49	-4.8820	1.0000	-2.1246	1.0000	-4.3015	1.0000
64	-4.7530	1.0000	-2.1146	1.0000	-4.2833	1.0000
81	-4.8460	1.0000	-2.1064	1.0000	-4.3557	1.0000
100	-4.9373	1.0000	-2.1001	1.0000	-4.4084	1.0000
225	-5.0738	1.0000	-2.0817	1.0000	-4.4813	1.0000

Table 3.2: continued.

NOT ROW-NORMALIZED

	Linear		Rook		Bishop	
T	-ρ	+ρ	-ρ	+ρ	-ρ	+ρ
4	-0.6180	0.6180	-0.5000	0.5000	-1.0000	1.0000
9	-0.5257	0.5257	-0.3536	0.3536	-0.5000	0.5000
16	-0.5087	0.5087	-0.3090	0.3090	-0.3820	0.3820
25	-0.5037	0.5037	-0.2887	0.2887	-0.3333	0.3333
36	-0.5018	0.5018	-0.2775	0.2775	-0.3080	0.3080
49	-0.5010	0.5010	-0.2706	0.2706	-0.2929	0.2929
64	-0.5006	0.5006	-0.2660	0.2660	-0.2831	0.2831
81	-0.5004	0.5004	-0.2629	0.2629	-0.2764	0.2764
100	-0.5002	0.5002	-0.2606	0.2606	-0.2716	0.2716
225	-0.5000	0.5000	-0.2549	0.2549	-0.2599	0.2599

	Double Linear		Double Rook		Queen	
T	-ρ	+ρ	-ρ	+ρ	-ρ	+ρ
4	-0.6404	0.3904	-0.5000	0.5000	-1.0000	0.3333
9	-0.5000	0.2802	-0.5000	0.2500	-0.5000	0.2071
16	-0.4654	0.2603	-0.3202	0.1952	-0.3820	0.1708
25	-0.4540	0.2544	-0.3090	0.1703	-0.3333	0.1547
36	-0.4495	0.2522	-0.2775	0.1571	-0.3080	0.1460
49	-0.4472	0.2512	-0.2663	0.1491	-0.2929	0.1407
64	-0.4461	0.2507	-0.2597	0.1438	-0.2831	0.1372
81	-0.4455	0.2505	-0.2500	0.1401	-0.2764	0.1347
100	-0.4451	0.2503	-0.2500	0.1374	-0.2716	0.1330
225	-0.4446	0.2501	-0.2349	0.1308	-0.2599	0.1287

	Double Queen		Queen w. Distance		Double Queen w. Dist.	
T	-ρ	+ρ	-ρ	+ρ	-ρ	+ρ
4	-1.0000	0.3333	-0.7735	0.3694	-0.7735	0.3694
9	-1.0000	0.1250	-0.7071	0.2357	-0.6849	0.1786
16	-0.4540	0.0802	-0.5402	0.1966	-0.5710	0.1264
25	-0.3547	0.0632	-0.4714	0.1790	-0.5019	0.1045
36	-0.3142	0.0546	-0.4355	0.1695	-0.4726	0.0930
49	-0.2836	0.0496	-0.4142	0.1637	-0.4535	0.0861
64	-0.2700	0.0464	-0.4004	0.1598	-0.4366	0.0816
81	-0.2579	0.0442	-0.3909	0.1572	-0.4250	0.0785
100	-0.2507	0.0427	-0.3840	0.1552	-0.4173	0.0762
225	-0.2265	0.0389	-0.3675	0.1505	-0.3956	0.0708

Table 3.3: All real roots (ρ) for which (**I**-ρ***M***) is singular.

(25 × 25) ***M***-matrix row-normalized		
Rook	Queen	Double Queen
4.567	4.698	23.950
2.795	3.179	15.106
1.809	3.130	15.038
1.414	1.601	10.387
1.151	1.250	2.536
1.000	1.000	1.000
-1.000	-2.058	-4.384
-1.151	-2.130	-4.726
-1.414	-2.202	-6.448
-1.809	-3.180	-6.500
-2.795	-3.333	-7.500
-4.567	-3.534	-8.114
	-3.961	-8.128
	-5.000	-8.308
	-6.410	-8.606
	-6.647	-9.834
	-7.067	-11.136
	-15.331	

(25 × 25) ***M***-matrix not row-normalized		
Rook	Queen	Double Queen
1.366	1.000	1.572
1.000	0.577	1.000
0.577	0.333	0.985
0.500	0.224	0.768
0.366	0.155	0.193
0.289	-0.333	0.063
-0.289	-0.406	-0.355
-0.366	-0.500	-0.368
-0.500	-0.577	-0.432
-0.577	-0.584	-0.434
-1.000	-0.897	
-1.366	-1.000	
	-2.155	

(49 × 49) ***M***-matrix row-normalized		
Rook	Queen	Double Queen
8.497	6.999	173.979
5.118	4.059	39.856
3.989	3.983	16.771
3.250	3.360	13.629
2.268	2.545	13.321
2.000	1.736	11.499
1.847	1.544	8.374
1.460	1.540	5.285
1.299	1.225	5.233
1.155	1.104	2.643
1.066	1.000	1.487
1.000	-2.003	1.000
-1.000	-2.035	-4.882
-1.066	-2.056	-5.119
-1.155	-2.522	-5.455
-1.299	-2.553	-5.480
-1.460	-2.648	-5.630
-1.847	-2.747	-5.738
-2.000	-2.893	-5.766
-2.268	-3.417	-6.849
-3.250	-3.493	-7.153
-3.989	-4.163	-7.671
-5.118	-4.340	-7.713
-8.497	-4.476	-7.722
	-5.057	-8.022
	-5.458	-9.715
	-6.040	-9.790
	-7.385	-10.395
	-8.342	-11.287
	-9.307	-12.678
	-10.671	-16.562
	-11.096	-17.450
	-14.362	-20.936
	-19.687	-22.098
	-19.873	-75.594
	-27.004	-115.962
		-118.743

Of these nine cases the largest negative root is -1 in only three cases. These three cases are the simplest in that they involve the least interaction (see Appendix 1). If the parameter space for ρ is restricted to the set $|\rho|< 1$ for these three cases, interpretations of ρ in the literature as an indicator of the strength of the autocorrelation in a manner which is similar to the autoregressive coefficient in an AR(1) disturbance process is 'reasonable'. On the other hand, restricting the parameter space to $|\rho|< 1$ may be unnecessarily restrictive because the model in (1) is well defined for $|\rho|> 1$ as long as $(\boldsymbol{I}-\rho\boldsymbol{M})$ is nonsingular. For example, let $\rho-$ and $\rho+$ be, respectively, the largest negative and the smallest positive roots of (10); let $\rho^* > \rho+$ or $\rho^* < \rho-$. Then, in general, there does not exist a value of ρ in the interval $\rho-$, $\rho+$ such that $\Omega_\varepsilon(\rho^*) = \Omega_\varepsilon(\rho)$.[13]

For the remaining six cases involving a row-normalized weighting matrix, and for all nine cases in which the weighting matrix is not row-normalized, the interpretation of ρ as 'similar' to a correlation coefficient is not a good one (see Table 3.2).

Third, as noted, the singular points of ρ vary with the sample size for each of the weighting matrices considered. Furthermore, these singular points are determined by a T-th degree polynomial equation, or as the reciprocals of the characteristic roots of the weighting matrix, and so for large values of T can, generally, only be determined numerically. Since the parameter space should not include possible values of ρ for which the model is inconsistent, this raises a problem. Specifically, if large sample theory is considered, the specification of the parameter space is 'difficult' because the singular points of ρ for all sample sizes $T\to\infty$ cannot be calculated. Perhaps problems which are associated with this sample space issue can be resolved because there would only be, at most, a countable infinity of such points. We leave this for future research.

3.5 A Suggested Variation of the Spatial AR Model

3.5.1 The Suggested Model

In this section we suggest a variation of the spatial AR model. The model we suggest is such that there are two stochastic shocks generated within each region. One of them is unique to the region in that it does not lead to spill-over effects in other regions. The other is not region specific in that it does lead to such spill-over

---

13 As an illustration, take $\Omega_U = \boldsymbol{I}$ so that $\Omega_\varepsilon(\rho) = (\boldsymbol{I}-\rho\boldsymbol{M})^{-1}(\boldsymbol{I}-\rho\boldsymbol{M}')^{-1}$, and denote its i,j-th element as $\Omega_{ij}(\rho)$. Then, for the case of a (16 × 16) row-normalized rook weighting matrix, and $\rho^* = 13$, it turns out that:

$$\min_{\rho-<\rho<\rho+} \sum_{i\leq j} (\Omega_{ij}(\rho) - \Omega_{ij}(13))^2 \doteq 11$$

effects. The regression disturbance term corresponding to each region is viewed as the sum of two parts. One is the region specific shock. The other is a linear combination of the spill-over shocks generated in other regions. The model is general enough so that for a given region, this linear combination could entail all of these spill-over shocks, or just the spill-over shocks of related regions, e.g., neighboring regions, industrially similar regions, etcetera. In the model this linear combination is determined via a weighting matrix. It will become clear that this weighting matrix also determines the pattern of spatial correlation. Finally, the model is easily generalized to include 'macro' disturbances which are distributed over the regions. We give one such generalization.[14]

Let V be the $(T \times 1)$ vector of regression disturbances. Then, we suggest the following spatial correlation model for V. Assume:

$$V = \boldsymbol{W}U + \psi \tag{11}$$

where $\boldsymbol{W}$ is a nonstochastic $(T \times T)$ weighting matrix, and U and ψ are $(T \times 1)$ stochastic vectors. Assume $E(U) = E(\psi) = 0$, and second moment matrices:

$$E(UU') = \sigma_U^2 \boldsymbol{I} \quad ; \quad E(\psi\psi') = \sigma_\psi^2 \boldsymbol{I} \quad ; \quad E(U\psi') = 0 \tag{12}$$

where $\sigma_\psi^2 > 0$.

These assumptions imply:

$$E(V) = 0 \quad ; \quad E(VV') \equiv \Omega_V = \sigma_U^2 \boldsymbol{WW'} + \sigma_\psi^2 \boldsymbol{I} \tag{13}$$

so that the model determines the first two moments of V.[15]

Since $\sigma_\psi^2 \boldsymbol{I}$ is positive definite, and $\sigma_U^2 \boldsymbol{WW'}$ is at least positive semidefinite, Ω_V is positive definite and therefore invertible. Thus, problems relating to singular variance-covariance matrices do not arise in this model. In passing it should be clear that U represents the vector of spill-over shocks, and ψ represents the vector of region specific shocks.

If $\sigma_U^2 = 0$, the elements of V are uncorrelated; if $\sigma_U^2 \neq 0$, at least some elements of V will be correlated unless $\boldsymbol{WW'}$ is diagonal, a case which we suspect would not

---

[14] Recently Case (1991) suggested a spatial correlation model which has error components and is therefore somewhat similar to the model suggested in this section. However a major difference between our model and that of Case's is that she maintains the spatial AR specification that leads to the singular matrix issues.

[15] If U and ψ are also assumed to be jointly normally distributed, then the distribution of V is completely determined, i.e., $V \sim N(0, \Omega_V)$.

arise for 'most' reasonable specifications of $\boldsymbol{W}$. Thus, in a sense the parameter σ_U^2 is 'somewhat similar' to ρ in the spatial AR model (see equation (1) above).[16]

3.5.2 Some Limiting Correlations

Let e_i' be a $(1 \times T)$ row vector whose i-th element is unity and whose remaining T-1 elements are zero. Let $C_{s,t}$ be the correlation coefficient between the s and t-th elements of V, say V_s and V_t. Then it is not difficult to show that:

$$C_{s,t}^2 = \frac{\sigma_U^4 (e_s' \boldsymbol{W}\boldsymbol{W}' e_t')^2}{(\sigma_U^2 e_t' \boldsymbol{W}\boldsymbol{W}' e_t + \sigma_\Psi^2)(\sigma_U^2 e_s' \boldsymbol{W}\boldsymbol{W}' e_s + \sigma_\Psi^2)}$$

$$= \frac{(e_s' \boldsymbol{W}\boldsymbol{W}' e_t)^2}{\left[e_t' \boldsymbol{W}\boldsymbol{W}' e_t + \frac{\sigma_\Psi^2}{\sigma_U^2}\right]\left[e_s' \boldsymbol{W}\boldsymbol{W}' e_s + \frac{\sigma_\Psi^2}{\sigma_U^2}\right]} \tag{14}$$

since $V_s = e_s' V$ and $V_t = e_t' V$. It follows from (14) that $C_{s,t}^2$ decreases as $\sigma_\Psi^2 / \sigma_U^2$ increases. This has a clear interpretation in that, for any given weighting matrix $\boldsymbol{W}$, large values of $\sigma_\Psi^2 / \sigma_U^2$ imply that the region specific shocks are 'large' relative to the spill-overshocks. It follows that the largest possible value of $C_{s,t}^2$ implied by the model corresponds to the limiting case $\rho \equiv (\sigma_\Psi^2 / \sigma_U^2) \rightarrow 0$ and is given as:

$$\lim_{r \to 0} C_{s,t}^2 = \frac{(e_s' \boldsymbol{W}\boldsymbol{W}' e_t)^2}{(e_t' \boldsymbol{W}\boldsymbol{W}' e_t)(e_s' \boldsymbol{W}\boldsymbol{W}' e_s)} \tag{15}$$

Let $\bar{C}_{s,t}^2$ be the limiting value of $C_{s,t}^2$ in (15). Then $\bar{C}_{s,t}^2$ can be viewed as the square of a correlation coefficient which corresponds to the variance-covariance matrix, $\boldsymbol{W}\boldsymbol{W}'$. Since rank($\boldsymbol{W}\boldsymbol{W}'$) = rank($\boldsymbol{W}$), it follows that $\bar{C}_{s,t}^2 < 1$ if rank($\boldsymbol{W}$) = T. Alternatively, if $\bar{C}_{s,t}^2 = 1$ then rank($\boldsymbol{W}$) < T.

In order to get a feel for the relationships involved, let $\boldsymbol{W}_{ij}$ be the i,j-th element of $\boldsymbol{W}$. Then, as an illustration, $\bar{C}_{s,t}^2 = 1$ if:

[16] Let L_t be the number of nonzero elements in the t-th row of $\boldsymbol{W}$, $t = 1, \ldots, T$. Then, as a point of interest we note that the Kelejian-Robinson (1992) test for spatial correlation is compatible with (11)-(13) if $L_t < L$ for all t even as $T \rightarrow \infty$, where L is a finite constant. Another test for spatial autocorrelation which is compatible with (11)-(13) is described in Kelejian and Robinson (1993).

$$e_s' W = e_t' W \tag{16}$$

or more generally if:

$$\left[\sum_{j=1}^{T} W_{tj} W_{sj}\right]^2 = \left[\sum_{j=1}^{T} W_{tj}^2\right]\left[\sum_{j=1}^{T} W_{sj}^2\right] \tag{17}$$

Therefore, one case in which unity is the limiting value of all of the squared correlation coefficients relating to the elements of V is the one in which $W_{ij} = 1$, for all $i, j = 1, ..., T$. In this case all of the regions would be viewed as identical concerning the (infinitely important) spill-over effects.

3.5.3 A Generalization

The model in (11) can be generalized to incorporate overall or 'macro' shocks. For example (11) can be expanded to:

$$V = WU + H\phi + \psi \tag{18}$$

where W, U, and ψ are as previously defined, ϕ is a $(q \times 1)$ stochastic vector of macro shocks, and H is a $(T \times q)$ nonstochastic matrix which allocates the elements of ϕ to each of the regions. In this framework it would be reasonable to assume that $q < T$ since the same macro shock may effect more than one region.

One specification concerning ϕ is that its mean is zero, $E(\phi) = 0$, its elements are uncorrelated with those of U and ψ, $E(\phi U') = E(\phi\psi') = 0$, and its variance-covariance matrix is $E(\phi\phi') = \sigma_\phi^2 I$, $\sigma_\phi^2 \geq 0$. Under these conditions the first and second moments of V are:

$$E(V) = 0 \quad ; \quad E(VV') = \sigma_U^2 WW' + \sigma_\psi^2 HH' + \sigma_\psi^2 I \tag{19}$$

Since $\sigma_\psi^2 I$ is positive definite if $\sigma_\psi^2 > 0$, while $\sigma_U^2 WW'$ and $\sigma_\phi^2 HH'$ are at least positive semidefinite, $E(VV')$ must be positive definite. Thus, again, problems concerning singular variance-covariance matrices do not arise.

Finally, we note that the elements of W and H could be known or parametrically specified. If parametrically specified along reasonable lines, the parameters could be estimated along with those of the regression model.

3.6 Suggestions for Further Work

Under the spatial AR specifications, it would be interesting to explicitly account for parameter space issues in large sample analyses of various estimators which might typically be considered in regional models. One of these estimators is the maximum likelihood estimator. Another is the feasible generalized least squares estimator of the regression coefficients which is based on an estimated value of ρ. It would also be interesting to determine the empirical usefulness of the suggested model in (11)-(13). Finally, it would be interesting to develop a test which discriminates between the spatial AR model in (1) and (2), and the suggested model in (11)-(13).

Acknowledgements

We would like to thank, without implicating, Ingmar Prucha and an anonymous referee for constructive comments. We would also like to thank George Antle of the Army Corps of Engineers for support and encouragement. Finally, the views expressed in this study are solely those of the authors, and are not necessarily those of the Army Corps of Engineers.

References

Amemiya, T., *Advanced Econometrics*, Cambridge: Cambridge University Press, 1985.

Anselin, L., *Estimation Methods for Spatial Autoregressive Structures*, Ithaca, N.Y.: Cornell University, Regional Science Dissertation and Monograph Series no. 8, 1980.

Anselin, L., A Note on Small Sample Properties of Estimators in a First-Order Spatial Autocorrelative Model, *Environment and Planning A*, 14, 1023-1030, 1982.

Anselin, L., Model Validation in Spatial Econometrics: A Review and Evaluation of Alternative Approaches, *International Regional Science Review*, 11, 279-316, 1988.

Anselin, L., Spatial Dependence and Spatial Structural Instability in Applied Regression Analysis, *Journal of Regional Science*, 30, 185-207, 1990.

Anselin, L. and D. Griffith, Do Spatial Effects Really Matter in Regression Analysis, *Papers of the Regional Science Association*, 65, 11-34, 1988.

Anselin, L. and S. Rey, Properties of Tests for Spatial Dependence in Linear Regression Models, *Geographical Analysis*, 23, 110-131, 1991.

Arora, S. and M. Brown, Alternative Approaches to Spatial Autocorrelation: An Improvement over Current Practice, *International Regional Science Review*, 2, 67-78, 1977.

Bartels, C., Operational Statistical Methods for Analyzing Spatial Data, in: C. Bartels and R. Ketellapper (eds.), *Exploratory and Explanatory Statistical Analysis of Spatial Data*, Boston: Martinus Nijhoff, 1977.

Bennett, R. and L. Hordijk, Regional Econometric and Dynamic Models, in: P. Nijkamp (ed.), *Handbook of Regional and Urban Economics*, vol. 1, Amsterdam: North Holland, 1986.

Case, A., Spatial Patterns in Household Demand, *Econometrica*, 59, 953-966, 1991.

Chow, G., *Econometrics*, New York: McGraw-Hill, 1983.

Cliff, A. and J. Ord, *Spatial Autocorrelation*, London: Pion, 1973.

Cliff, A. and J. Ord, *Spatial Processes, Models and Applications*, London: Pion, 1981.

Dhrymes, P., *Introductory Econometrics*, New York: Springer, 1978.

Fomby, T., R. Hill and S. Johnson, *Advanced Econometric Methods*, New York: Springer, 1984.

Getis, A., Screening for Spatial Dependence in Regression Analysis, *Papers of the Regional Science Association*, 69, 69-81, 1990.

Griffith, D., *Advanced Spatial Statistics*, Dordrecht: Kluwer, 1988.

Haining, R., *Spatial Data Analysis in the Social and Environmental Sciences*, New York: Cambridge University Press, 1990.

Hepple, L., Bayesian Analysis of the Linear Model with Spatial Dependence, in: C. Bartels and R. Ketellapper (eds.), *Exploratory and Explanatory Statistical Analysis of Spatial Data*, Boston: Martinus Nijhoff, 1979.

Intrilligator, M., *Econometric Models, Techniques, and Applications*, Englewood Cliffs: Prentice-Hall, 1978.

Johnston, J., *Econometric Methods*, New York: McGraw-Hill, 1984.

Judge, G. et al., *The Theory and Practice of Econometrics*, New York: John Wiley & Sons, 1985 (second ed.).

Kelejian, H. and W. Oates, *Introduction to Econometrics: Principles and Applications*, New York: Harper and Row, 1989.

Kelejian, H. and D. Robinson, Spatial Autocorrelation: A New Computationally Simple Test with an Application to Per Capita County Police Expenditures, *Regional Science and Urban Economics*, 22, 317-331, 1992.

Kelejian, H. and D. Robinson, A Suggested Method of Estimation for Spatial Interdependent Models with Autocorrelated Errors, and an Application to a County Expenditure Model, *Papers in Regional Science*, 72, 297-312, 1993.

Kmenta, J., *Elements of Econometrics*, New York: Macmillan, 1986.

Maddala, G.S., *Econometrics*, New York: MacGraw-Hill, 1977.

Odland, J., *Spatial Autocorrelation*, London: Sage, 1988.

Upton, G. and B. Fingleton, *Spatial Data Analysis by Example*, New York: John Wiley & Sons, 1985.

Schmidt, P., *Econometrics*, New York: Marcel Dekker, 1976.

Whittle, P., On Stationary Processes in the Plane, *Biometrica*, 41, 434-449, 1954.

Appendix 1: Spatial Weighting Matrices

For the reader's convenience, we define all of the weighting matrices referenced in the text. Although these matrices are standard ones, their definitions seem to be 'scattered' in the literature.

The weighting matrices used in the text are all based on locational sites which have a geographic configuration in the form of a 'checker board' of squares. Such a (7 × 7) configuration is described in Figure 3.A1. Larger and smaller configurations, some of which are referenced in the text, are obvious variations of that described in Figure 3.A1.

In the simplest case of binary connectiveness, the elements of the weighting matrix, $\boldsymbol{M}$, are $m_{ij} = 1$ if site i is spatially related to site j, and $m_{ij} = 0$ otherwise. This is the case in which $\boldsymbol{M}$ is not 'row-normalized'. The corresponding case in which $\boldsymbol{M}$ is row-normalized is the one in which $m_{ij} = 1/\Sigma_j m_{ij}$ if site i is spatially related to site j and $m_{ij} = 0$, otherwise.

The $\boldsymbol{M}$ matrix can be defined in terms of distance measures. Let d_{ij} be a distance measure between sites i and j. Then, one such scheme for defining $\boldsymbol{M}$ is to take $m_{ij} = 1/d_{ij}$ if site i is spatially related to site j and $m_{ij} = 0$, otherwise. Again, the $\boldsymbol{M}$ matrix can be row-normalized by taking its i,j element as $m_{ij}/\Sigma_j m_{ij}$, where m_{ij} is as defined above.

In Table 3.A1 we define the nine weighting matrices referenced in the text. In each case we illustrate the definition in reference to Figure 3.A1.

1	2	3	4	5	6	7
8	9	10	11	12	13	14
15	16	17	18	19	20	21
22	23	24	25	26	27	28
29	30	31	32	33	34	35
36	37	38	39	40	41	42
43	44	45	46	47	48	49

Figure 3.A1: A spatial configuration using a square (7 × 7) geographic grid.

Table 3.A1: Weighting matrix definitions for geographic grids composed of unit squares.

Linear: $m_{ij} = 1$ for cells that share a common edge to the immediate right or left of the reference location i, $m_{ij} = 0$ otherwise. For reference location $i = 25$, $m_{ij} = 1$ for $j = 24$ and 26, and $m_{ij} = 0$, otherwise.

Rook: $m_{ij} = 1$ for cells that share a common side with the reference location i, $m_{ij} = 0$ otherwise. For reference location $i = 25$, $m_{ij} = 1$ for $j = 18$, 24, 26, and 32, and $m_{ij} = 0$, otherwise.

Bishop: $m_{ij} = 1$ for cells that share a common vertex with the reference location i, $m_{ij} = 0$, otherwise. For reference location $i = 25$, $m_{ij} = 1$ for $j = 17$, 19, 31, and 33, and $m_{ij} = 0$, otherwise.

Double Linear: $m_{ij} = 1$ for two cells to the immediate right or left of the reference location i, $m_{ij} = 0$, otherwise. For reference location $i = 25$, $m_{ij} = 1$ for $j = 23$, 24, 26, and 27, and $m_{ij} = 0$, otherwise.

Double Rook: $m_{ij} = 1$ for two cells to the right, left, north, and south of the reference location i, $m_{ij} = 0$, otherwise. For reference location $i = 25$, $m_{ij} = 1$ for $j = 11$, 18, 23, 24, 26, 27, 32, and 39, and $m_{ij} = 0$, otherwise.

Queen: $m_{ij} = 1$ for cells that share a common side or vertex with the reference location i, $m_{ij} = 0$, otherwise. For reference location $i = 25$, $m_{ij} = 1$ for $j = 17$, 18, 19, 24, 26, 31, 32, and 33, and $m_{ij} = 0$, otherwise.

Double Queen: $m_{ij} = 1$ for cells that fall within three units of the reference location i, $m_{ij} = 0$ otherwise. In the text we used geographic grids with unit squares. For example, for reference location $i = 25$, $m_{ij} = 1$ for $j = 4$, 9, 10, 11, 12, 13, 16, 17, 18, 19, 20, 22, 23, 24, 26, 27, 28, 30, 31, 32, 33, 34, 37, 38, 39, 40, 41, and 46, and $m_{ij} = 0$, otherwise.

Queen with Distance: $m_{ij} = 1/d_{ij}$ for cells that share a common side or vertex with the reference location i, $m_{ij} = 0$ otherwise. d_{ij} is the distance between sites i and j. For reference location $i = 25$, $m_{ij} = 1/d_{ij}$ for $j = 17$, 18, 19, 24, 26, 31, 32, and 33, and $m_{ij} = 0$, otherwise. In the text we used geographic grids with unit squares. In Figure 3.A1, for example, $d_{ij} = 1$ if $i = 25$ and $j = 26$, and $d_{ij} = \sqrt{2}$ if $i = 25$ and $j = 19$, etcetera.

Double Queen with Distance: $m_{ij} = 1/d_{ij}$ for cells that fall within three units of the reference location i, $m_{ij} = 0$ otherwise. d_{ij} is the distance between sites i and j. For reference location $i = 25$, $m_{ij} = 1/d_{ij}$ for $j = 4$, 9, 10, 11, 12, 13, 16, 17, 18, 19, 20, 22, 23, 24, 26, 27, 28, 30, 31, 32, 33, 34, 37, 38, 39, 40, 41, and 46, and $m_{ij} = 0$, otherwise. In the text we used geographic grids with unit squares. In Figure 3.A1, for example, $d_{ij} = 1$ if $i = 25$ and $j = 26$, and $d_{ij} = \sqrt{2}$ if $i = 25$ and $j = 19$, etcetera.

---

4 Spatial Autoregressive Error Components in Travel Flow Models:

An Application to Aggregate Mode Choice

Denis Bolduc[1], Richard Laferrière[2] and Gino Santarossa[1]
[1] Université Laval, Sainte-Foy, Québec, Canada
[2] Université de Montréal, Montréal, Canada

4.1 Introduction

In this chapter we use empirical examples to demonstrate the usefulness of the generalized error component framework suggested in Bolduc et al. (1992) for dealing with the problem of correlation among the errors of a regression based on travel flow data. This methodology augments standard error component decompositions with first-order spatial autoregressive processes, i.e., SAR(1), with the purpose of allowing for the different sources of misspecification generally associated with this type of model. The error component approach splits the error term into a sum of one component related to the zones in origin, one component associated with the zones in destination and a remainder. The interdependencies among the errors are modeled with the help of SAR(1) processes. This decompositional approach extends the previous works by Brandsma and Ketellapper (1979) and Bolduc et al. (1989) which also relied on spatial autoregressive processes to model the error correlation.

As indicated in our earlier paper on the subject, the Brandsma-Ketellapper (1979) solution ignores the network related factors contained in the error term while Bolduc et al. (1989) extends their proposed formulation to incorporate these missing effects into the model. They capture the interdependencies between the flow errors using a first-order spatial autoregressive error process based on a general weight structure that can sometimes be rather complicated to interpret. The error component approach differs from these methodologies in offering a simpler and more natural framework to handle the spatial correlation in the errors.

In the next section, we quickly summarize the main concepts employed in the study field of flow data and describe the travel flow model formulation using SAR(1) error components, i.e., EC-SAR(1). We use a notation which can accomodate both balanced and unbalanced data. Section three describes the maximum likelihood algorithm used to address the practical implementation of this technique in the context of a large sample size. The last section presents the results of the application of the technique to the aggregate binary logit share model considered in Bolduc et al. (1989). To better valuate the size of the gain that can be obtained when the right model specification and the right error correlation structure is employed in estimation, an illustration based on synthetic data is also discussed.

4.2 The First-Order Spatially Autoregressive Error Components Formulation

A regression model based on travel flow data may be written as:

$$y_{ij} = \beta_0 + \boldsymbol{N}_{ij}\beta_1 + \boldsymbol{S}_i\beta_2 + \boldsymbol{S}_j\beta_3 + \varepsilon_{ij} \tag{1}$$

where y_{ij} denotes a dependent variable which represents a flow between a given pair of zones i and j. The indices i (= 1, ..., N) and j (= 1, ..., T) refer to zones belonging to a set of N zones at the origin and T zones at the destination, respectively. Passengers' travel flow regression models typically contain three kinds of variables: network variables, ($\boldsymbol{N}_{ij}$), socio-economic variables associated with the origin ($\boldsymbol{S}_i$) and socio-economic variables associated with the destination ($\boldsymbol{S}_j$). Network variables usually refer to levels of service (in-vehicle travel time, out-of-vehicle travel time, fare, frequency, etcetera) of the relevant modes, and socio-economic variables (population, income, employment, etcetera) characterize the generation and attraction capacity of the two zones involved in the flow. β_0 is a constant term and β_1, β_2 and β_3 are vectors of fixed coefficients associated with the variables described above. Note that a given observation of the $\boldsymbol{N}_{ij}$, $\boldsymbol{S}_i$ and $\boldsymbol{S}_j$ variables combined usually refers only to the corresponding *ij* market, that is network variables such as travel time or generalized cost for the competing markets *is* (same origin but alternative destination) or *rj* (same destination but alternative origin) do not appear in (1). The same is true for socio-economic variables of zones other than i and j. It might be conceivable that socio-economic and network variables adjacent to market *ij* should also be incorporated in the relationship explaining y_{ij}, but to do this in a proper way could well lead to an intractable model. The omission of those variables and the frequent lack of data describing the geographic structure of the region gives rise to some spatial autocorrelation in the regression errors. Sources of spatial autocorrelation among errors are model misspecification and omission of explanatory variables capturing effects related to the physical and economic structure (distance between zones, size of zones, length of frontier between adjacent zones, etcetera) of the region. For these reasons, the error term ε_{ij} in equation (1) will reflect this situation. In the following, we use an error structure that takes into account the several sources of misspecification discussed above. The formulation writes the ε_{ij} error term in equation (1) as:

$$\varepsilon_{ij} = \alpha_i + \lambda_j + u_{ij} \tag{2}$$

This decomposition of the unexplained flow between zones i and j is consistent with the specification of the fixed part of the regression model.[1] The error term α_i is interpreted as capturing unmeasured effects associated with origin i (it contains omitted variables from $\boldsymbol{S}_i$), λ_j picks up those missing effects related to destination j (it contains omitted variables from $\boldsymbol{S}_j$) and finally u_{ij} incorporates all the network variables not included into the model specification. To allow for the presence of spatial correlation among the errors associated with the origin (α_i), the destination (λ_j) and the flow (u_{ij}), we postulate a normalized first-order generalized spatial autoregressive, i.e., SAR(1), generating process for each component. We write it as:

$$\alpha_i = \rho_1 \sum_{r=1}^{N} w_{1,\,ir} \alpha_r + \xi_i \tag{3}$$

$$\lambda_j = \rho_2 \sum_{s=1}^{T} w_{2,\,js} \lambda_s + \zeta_j \tag{4}$$

$$u_t = \rho_3 \sum_{l=1}^{L} w_{3,\,tl} u_l + \upsilon_t \tag{5}$$

where for notational simplicity, t designates a given flow ($i \rightarrow j$) and l an alternative flow ($r \rightarrow s$). The formulations in equations (3)-(5) translate the interdependencies discussed above into an operational framework. We postulate that $\xi_i \sim N(0,\sigma_\xi^2)$, $\zeta_j \sim N(0,\sigma_\zeta^2)$ and $\upsilon_t \sim N(0,\sigma_\upsilon^2)$, $\forall i, j$, and t. Also, these random terms are assumed to be uncorrelated with each other. The parameter ρ_1 is a spatial autocorrelation coefficient which measures the degree of linear dependency between α_i and a weighted sum of each other 'neighbor' α_r in origin. The parameters ρ_2 and ρ_3 are defined analogously.

To describe the weights, we start with the autoregressive process associated with the origin. The weights $w_{1,\,ir}$ are *estimable* parametric functions specified by the analyst to describe the influence of each origin specific error upon the others. In this chapter, we write this function as:

---

[1] Several issues typically raised when using error component structures, such as random versus fixed effects modeling, are discussed in detail in Bolduc et al. (1992).

$$w_{1,\,ir} = \frac{w^{*}_{1,\,ir}}{\sum_{r=1}^{N} w^{*}_{1,\,ir}} \tag{6}$$

with $w_{1,\,ii} = 0$, and

$$w^{*}_{1,\,ir} = (d_{ir})^{-\theta_1} \quad ; \quad w^{*}_{1,\,ii} = 0 \; , \quad \forall i \tag{7}$$

where $\theta_1 > 0$ and d_{ir} is the distance between zones i and r. With this weight specification, when the distance between zones i and r is larger, the errors α_i and α_r are less related. The process is said to be generalized because the θ_1 parameter is estimated along with the other unknown model parameters. The weights $w_{2,\,ir}$ involved in defining the inter-relationships in destination are defined similarly. The interpretation of the SAR(1) process pertaining to the network specific effects is slightly different from the one given above. The function used in the empirical application considered later in the chapter takes into consideration *cross distance* effects between two origins and two destinations. For a given pair of travel flows say $(i \rightarrow j)$ and $(r \rightarrow s)$, the formulation implies that for a given $\theta_3 > 0$, everything else being equal, the higher those distances are, the smaller is $w_{3,\,tl}$:

$$w_{3,\,tl} = \frac{w^{*}_{3,\,tl}}{\sum_{l=1}^{L} w^{*}_{3,\,tl}} \tag{8}$$

with $w_{3,\,tt} = 0$, and

$$w^{*}_{3,\,tl} = (d_{is} + d_{rj})^{-\theta_3} \quad ; \quad w^{*}_{3,\,tt} = 0 \; , \; \forall t \tag{9}$$

In order to illustrate the use of the above spatial autoregressive process, consider Figure 4.1 which contains four different configurations of two given flows ($f_1 : i \rightarrow j$ and $f_2 : r \rightarrow s$). More configurations are possible, but those retained here are the most important. Because of the weight function defined in equations (8) and (9), f_1 and f_2 are more origin-correlated in cases *A*, *B* and *D* than in case *C*. Flows f_1 and f_2 are also more intensively destination-correlated in cases *A*, *B* and *C* than in case *D*. This specific weight function rightly implies that the network-correlation of f_1 and f_2 is greater in case *A* than in case *B*. Furthermore, according to those two equations, the network-correlation of f_1 and f_2 in cases *B*, *C* and *D* is very similar since cross distances (*i* to *s*, *r* to *j*) are equal. However, one would expect network-correlation to be decreasing as one moves from case *B* to case *D*. This suggests that in order to capture this phenomenon, a spatial localization measure of flows should ideally be incorporated into the weight function. Since no geocoding information is available in our particular empirical example, cases *B*, *C* and *D* would be perceived as equivalent by the model. When the information is available, the generalized error component approach can potentially allow for a very complete treatment of network-correlation.

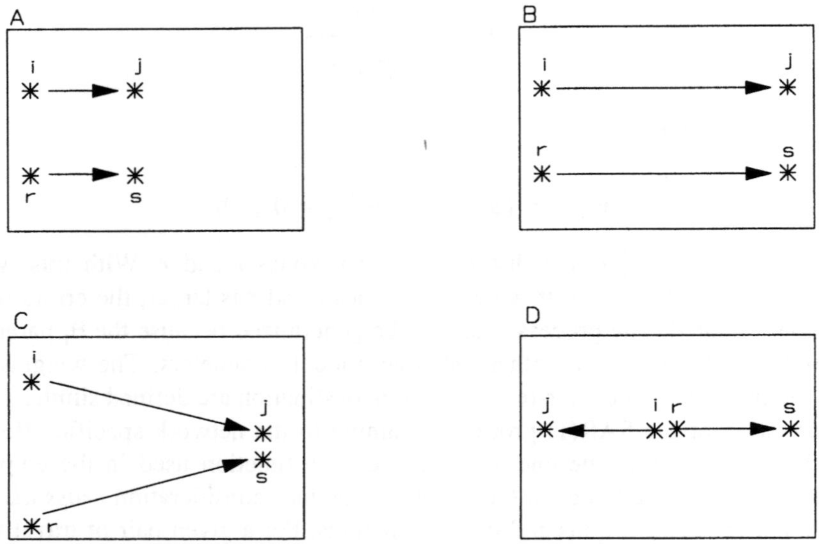

Figure 4.1: Different configurations for pairs of travel flows.

4.3 Estimation Issues

In order to discuss estimation related issues, it is more appropriate to write the EC-SAR(1) formulation in vector form. Incorporating all network specific and socio-economic variables into a single matrix denoted by $\boldsymbol{X}$, the regression model with EC-SAR(1) errors is written as:

$$y = \boldsymbol{X}\beta + \varepsilon \tag{10}$$

$$\varepsilon = \boldsymbol{C}\alpha + \boldsymbol{D}\lambda + u \tag{11}$$

with

$$\alpha = \rho_1 \boldsymbol{W}_1 \alpha + \xi \quad , \quad \xi \sim N(0, \sigma_\xi^2 \boldsymbol{I}_N) \tag{12}$$

$$\lambda = \rho_2 \boldsymbol{W}_2 \lambda + \zeta \quad , \quad \zeta \sim N(0, \sigma_\zeta^2 \boldsymbol{I}_T) \tag{13}$$

$$u = \rho_3 \boldsymbol{W}_3 u + \upsilon \quad , \quad \upsilon \sim N(0, \sigma_\upsilon^2 \boldsymbol{I}_L) \tag{14}$$

where L denotes the total number of observations in the sample. The vectors y, ε, u and υ are of dimension $(L \times 1)$. The vectors α and ξ are N-dimensional whereas λ and ζ are T-dimensional. The matrices $\boldsymbol{C}$ and $\boldsymbol{D}$ are defined in Bolduc et al. (1992) for both cases of balanced and unbalanced data settings. Because of the assumptions previously made, the vectors ξ, ζ and υ are assumed to be distributed independently of each other. $\boldsymbol{X}$ is an $(L \times K)$ matrix of fixed regressors and β is a $(K \times 1)$ vector of parameters. The weights defined in equations (6)-(9) are incorporated into matrices $\boldsymbol{W}_1$, $\boldsymbol{W}_2$ and $\boldsymbol{W}_3$ of dimension $(N \times N)$, $(T \times T)$ and $(L \times L)$, respectively. These matrices are row-normalized so that the corresponding SAR(1) processes are stable over the unit (-1, +1) interval of the spatial correlation coefficient associated with the process. Our distributional assumptions imply that $\varepsilon \sim N(0, \Phi)$, where Φ is the error variance-covariance matrix of size $(L \times L)$. Call $\eta = (\beta', \sigma_\xi^2, \sigma_\zeta^2, \sigma_\upsilon^2, \rho_1, \rho_2, \rho_3, \theta_1', \theta_2', \theta_3')'$ the vector of parameters considered. To estimate the η vector, we use the maximum likelihood framework where the log-likelihood function can be written as:

$$L(\eta \mid y) = -\frac{L}{2}\ln(2\pi) + \ln|\Phi| - \frac{1}{2}(y - \boldsymbol{X}\beta)'\Phi^{-1}(y - \boldsymbol{X}\beta) \qquad (15)$$

and where $|\Phi|$ denotes the determinant of Φ. This model formulation is a special case of the very familiar general linear model with non-scalar error covariance matrix considered in Magnus (1978) and Heijmans and Magnus (1986a,b). Conditions for the consistency and asymptotic normality of the maximum likelihood (ML) estimator $\hat{\eta}$ of η may be found there. For similar conditions with special reference to spatial models, see Mardia and Marshall (1984). For the model considered here, these conditions essentially amount to the following requirements: the existence of the log-likelihood for all values taken by $\hat{\eta}$; continuous differentiability of the log-likelihood function and positive definiteness of the error covariance matrix. Typically, if the weight structure is a well defined continuous function based on a fixed number of identifiable parameters (i.e., fixed as $L \to \infty$), if they are normalized as proposed in this chapter and if the autocorrelation coefficients associated with each process are restricted to lie on the unit interval, they should be satisfied.

Although the log-likelihood function in (15) looks quite standard, one must realize that even with a small balanced setting involving 20 origins and 20 destinations, the covariance matrix Φ that needs to be inverted at every iteration in the estimation process would be of size (400×400). While still computationally tractable, for larger problems, the maximum likelihood solution may become numerically unreachable in a reasonable amount of computing time. For these reasons, Bolduc et al. (1992) suggested a more efficient approach where the inverse and the determinant of Φ is based on a much smaller matrix of size $\left((N+T)\times(N+T)\right)$. Their proposed approach is used in the two applications described in the next section.

4.4 Empirical Example

This section presents estimation results from applying a few EC-SAR(1) specifications to the model formulation previously estimated in Bolduc et al. (1989). This model is a two-mode share formulation for automobile and bus trips with logit shares respectively denoted by:

$$p_A = \frac{e^{V_A}}{e^{V_A} + e^{V_B}} \quad \text{and} \quad p_B = \frac{e^{V_B}}{e^{V_A} + e^{V_B}} \tag{16}$$

where V_A and V_B are mode specific linear in parameters functions containing characteristics of the modes and socioeconomic variables describing the travelers associated with the origin and the destination. The transformation $y = \ln(p_B/p_A)$ and the addition of a residual error term lead to the following linear model:

$$y = X\beta + u \tag{17}$$

where $X\beta = V_B - V_A$. If specification (17) is true, the OLS estimation method has some optimal properties when u is a vector of i.i.d. components, since the previous transformation is such that $y \in (-\infty, +\infty)$.

The data used in the testing of this model were collected in the city of Winnipeg, Canada during the summer of 1976. The sample represents twenty percent of the total population. Only observations pertaining to peak a.m. home origin based trips to work were retained in the present study. All the information was aggregated at the level of 147 zones. Most of the level of service variables were calculated using different procedures of assignment that aim to minimize the travel time. For estimation, we kept only those observations for which the total of the flows by car and bus is larger than 60. The final estimation sample contains 211 observations. The estimated model is written as:

$$\ln\left(\frac{p_B}{p_A}\right) = \beta_0 + \beta_1\ \mathrm{TZIMP1} + \beta_2\ \mathrm{AUTIM1} + \beta_3\ \mathrm{TAUT} + \beta_4\ \mathrm{OINC} + \beta_5\ \mathrm{OCAR} + \beta_6\ \mathrm{MEN1} \tag{18}$$

This model specification uses a combination of levels of service (LOS) variables (TZIMP1, AUTIM1 and TAUT) and socioeconomic variables (OINC, OCAR and MEN1). Their definitions, as well as their respective sample means, follow:

- TZIMP1: Travel time (in minutes) by bus, which is the sum of the in-vehicle travel time, the waiting, the access and the egress time (76.91 minutes);
- AUTIM1: Travel time by car, expressed in minutes (20.90 minutes),

- TAUT: Travel cost by car defined as: (0.058 × distance) + (parking cost at the destination, 0.53 dollars);
- OINC: Average income of households at the zone of origin of the trip (16,355 dollars);
- OCAR: Average number of cars per household at the origin (0.96 cars); and
- MEN1: Proportion of males traveling on the O-D segment (0.60).

The travel cost by bus, which is the same for everyone, is implicitly included in the constant term with a coefficient assumed equal to β_3. Since (18) is defined as the ratio of bus to auto shares, the following relationships are anticipated. The effect of the auto (bus) LOS variables is anticipated to be positive (negative), respectively. The bus mode share is expected to be inversely related to the number of cars and the level of income at the origin. The variable MEN1 represents the inclination of males to use their car for trips to work. The reader must keep in mind that this data was collected in 1976.

OLS estimates are presented in column 1 of Table 4.1. All the coefficient estimates have the expected sign. AUTIM1 and OINC are, however, not significantly different from zero at a reasonable level of significance. Based on the discussion in section two and the results obtained in Bolduc et al. (1989), it is fairly reasonable to assume that the data are spatially correlated. To model it, we use the EC-SAR(1) error structure defined in equations (3)-(9). The non-normalized weighting structure used to model the correlation in origin and in destination is based on an inverse function of distance $w^* = d^{-\theta}$ between the centroids of pairs of zones in origin and in destination, respectively where $\theta = \theta_1$ for the origin effect and $\theta = \theta_2$ for the destination effect. The non-normalized weight structure related to the network effect is based on equations (8)-(9) where only cross distance effects are accounted for. Sub-models, such as the standard GLS 3-way error component model and the general GLS EC-SAR(1) of the general specification proposed in this paper, are presented in columns 2-3 of Table 4.1. These estimations are based on a maximum likelihood Davidon-Fletcher-Powell iterative procedure which uses analytical gradients. Computations were made on a SUN SparcStation 2 using the FORTRAN language.[2] We now summarize the major findings. In this particular application, the standard 3-way error component formulation (column 2) does not seem to be robust for capturing the spatial correlation in origin and in destination.

The variances σ_ξ^2 and σ_ζ^2 are not found to be significantly different from zero at a resonable level of significance. Based on the log-likelihood value, the improvement is very marginal over the OLS results. In column 3 of Table 4.1, the results regarding the full EC-SAR(1) specification are presented.

[2] We also have available upon request to the first author, the GAUSS386-PC code required to do the estimation. In our experiments, we decided upon using the SUN platform instead of the PC because of the memory requirements implied by the present formulation.

Table 4.1: OLS and GLS estimation results based on actual data.[a]

Variables Parameters	OLS	GLS 3-way E.C.	GLS EC-SAR(1)
Constant	2.592**	2.447**	1.607**
	(9.218)	(7.888)	(2.590)
TZIMP1	-0.017**	-0.016**	-0.011**
	(-4.863)	(-4.231)	(-2.700)
AUTIM1	0.003	0.008	0.016
	(0.255)	(0.675)	(1.263)
TAUT	1.416**	1.258**	0.478
	(4.941)	(3.771)	(1.272)
OINC	0.000005	0.00001	0.00001
	(1.551)	(0.485)	(0.475)
OCAR	-1.903**	-1.942**	-1.654**
	(-4.080)	(-3.854)	(-3.102)
MEN1	-1.769**	-1.725**	-1.842**
	(-4.994)	(-4.792)	(-5.247)
σ^2_ξ		0.039	0.031*
		(1.120)	(1.643)
σ^2_ζ		0.035	0.031*
		(0.651)	(1.6443)
σ^2_υ		0.496**	0.457**
		(8.407)	(9.177)
ρ_1			0.898**
			(11.444)
ρ_2			0.898**
			(11.444)
ρ_3			0.898**
			(1.444)
θ_1			2.320**
			(3.056)
θ_2			2.320**
			(3.056)
θ_3			2.320**
			(3.056)
Log-likelihood	-238.716	-236.946	-231.94

[a] Please note that *t*-statistics are given in parentheses, and ** indicates 95% significance and * 90% significance.

The lack of variability in origin effects and destination effects, as suggested by the value of the estimated variances displayed in column 2, made it necessary to constrain the value of the origin and destination related variances to be equal. For a similar reason, we imposed the following constraints: $\rho_1 = \rho_2 = \rho_3$ and $\theta_1 = \theta_2 = \theta_3$. With a fully saturated version, where every parameter is freely estimated, a corner solution was always obtained. As one will see in the application based on simulated data, the fully saturated model is estimable. The estimation problems encountered here are in our opinion due to the data in hand. The estimation presented in column 3 of Table 4.1 clearly show the presence of some significant correlation in both the origin and the destination of the flows and in the network component as well. Analogously to Bolduc et al. (1989), the major change compared to the OLS estimates are for variables AUTIM1 and TAUT. The estimated values for AUTIM1 and TZIMP1 seem to indicate that the travel time variable should be postulated to be generic. A similar conclusion was also reached in Bolduc et al. (1989). The auto fare effect TAUT decreased by a large amount and is now not significantly different from zero. This may be caused by the fact that some of the variability in this variable is captured by the postulated error structure. In this particular estimation, the size of the improvement in the log-likelihood function of the model based on four free nuisance parameters ($\sigma_\xi^2 = \sigma_\zeta^2$, σ_υ^2, $\rho_1 = \rho_2 = \rho_3$, $\theta_1 = \theta_2 = \theta_3$) clearly justifies the use of the EC-SAR(1) as a valuable tool for capturing the spatial autocorrelation in this data.

4.4.1 An Illustration Based on Synthetic Data

To provide an even more convincing empirical example, we use synthetic data to estimate the following statistical model:

$$y_{ij} = \beta_0 + x_{1,ij}\beta_1 + x_{2,ij}\beta_2 + \alpha_i + \lambda_j + u_{ij} \tag{19}$$

with $\beta_0 = 5$, $\beta_1 = 3.5$, $\beta_2 = 4$, and i and $j = (1, ..., 25)$. The values taken by the two x variables are randomly generated from a uniform distribution. The three SAR(1) processes associated with the origin, the destination and the network dimensions are based on the following simple weight structure:

$$w_{rs}^{*} = (d_{rs})^{-\theta} \tag{20}$$

where d_{rs} represents a distance between unit r and unit s. These variables are random uniform variables taking values between 0 and 100. The distance values taken in the three processes are different. Each process defined in equations (6)-(9) with row-normalized weighting matrix is based on the following true values $\rho_1 = \rho_2 = \rho_3 = 0.7$, $\sigma_\xi^2 = \sigma_\zeta^2 = \sigma_\upsilon^2 = 2$, $\theta_1 = \theta_2 = \theta_3 = 1.5$. Only one replication was made because the idea here is to demonstrate empirically that this model formulation is fully estimable and to show that huge gains can be obtained when the right error structure is used in estimation. The three ρ coefficients are constrained to lie on the unit interval. This condition combined with the normalized weight structure used essentially ensure that the estimators will be consistent and asymptotically normal in large samples. The estimation results are presented in Table 4.2.

Table 4.2: OLS and GLS estimation results based on synthetic data, with *t*-statistics in parentheses.

Parameters	true values	OLS	GLS 3-way E.C.	GLS EC-SAR(1)
β_0	5	4.730	4.767	8.687
		(11.86)	(7.41)	(3.82)
β_1	3.5	3.272	3.773	3.467
		(6.47)	(12.91)	(16.90)
β_2	4	4.203	3.608	4.071
		(8.34)	(12.59)	(20.76)
σ_ξ^2	2		5.856	1.883
			(3.41)	(3.28)
σ_ζ^2	2		3.192	1.509
			(3.34)	(2.98)
σ_υ^2	2		3.847	2.040
			(16.97)	(15.81)
ρ_1	0.7			0.964
				(30.06)
ρ_2	0.7			0.938
				(17.19)
ρ_3	0.7			0.692
				(16.09)
θ_1	1.5			0.674
				(4.58)
θ_2	1.5			1.089
				(4.41)
θ_3	1.5			1.561
				(9.70)
Log-likelihood		-1680.70	-1390.88	-1247.13

Even with such a small case problem, β_1 and β_2 are already very close to their true values. A similar comment holds for the nuisance parameters associated with the third error component which refers to 625 observations. The estimates obtained can therefore be viewed as reliable. The estimation results for the nuisance parameters belonging to the first two components are pointing in the right direction. However, the sample size used in each process (25 observations) is too small to produce more reliable estimates. With large scale problems, one could be confident to get consistent estimates for all the parameters. When N and/or T are considered to be too small, to use a mixed dummy variable EC-SAR(1) model strategy where the random variables α_i and λ_j associated with the problematic dimension(s) are fixed parameters is probably a very good idea.

4.5 Conclusions

To capture the spatial correlation among the errors of a regression model on flow data, we used an error generating process based on a components decomposition and spatial autoregressive processes of order 1. The error term is broken down into a sum of one component related to the origin zone, one component related to the destination zone and a remainder. Each component is assumed to result from a first-order spatial autoregressive generating process. An estimation approach based on maximum likelihood which addresses the practical implementation of such a model with a large sample size was employed. The two empirical applications gave encouraging results. Clearly, more testing with real data applications are needed to further demonstrate the applicability of the approach.

References

Bolduc, D., M.G. Dagenais and M.J. Gaudry, Spatially Autocorrelated Errors in Origin-Destination Models: A New Specification Applied to Urban Travel Demand in Winnipeg, *Transportation Research B*, 23, 361-372, 1989.

Bolduc, D., R. Laferrière and G. Santarossa, Spatial Autoregressive Error Components in Travel Flow Models, *Regional Science and Urban Economics*, 22, 371-385, 1992.

Brandsma, A.S. and R.H. Ketellapper, A Biparametric Approach to Spatial Autocorrelation, *Environment and Planning A*, 11, 51-58, 1979.

Heijmans, R.D.H. and J.R. Magnus, Asymptotic Normality of Maximum Likelihood Estimators Obtained from Normally Distributed but Dependent Observations, *Econometric Theory*, 2, 374-412, 1986a.

Heijmans, R.D.H. and J.R. Magnus, Consistent Maximum Likelihood Estimation with Dependent Observations: The General (Non-Normal) Case and the Normal Case, *Journal of Econometrics*, 32, 253-285, 1986b.

Magnus, J., Maximum Likelihood Estimation of the GLS Model with Unknown Parameters in the Disturbance Covariance matrix, *Journal of Econometrics*, 7, 281-312, 1978 (Corrigenda, *Journal of Econometrics*, 10, 261, 1981).
Mardia, K.V. and R.J. Marshall, Maximum Likelihood Estimation of Models for Residual Covariance in Spatial Regression, *Biometrika*, 71, 135-146, 1984.

PART I-B

Spatial Effects in Linear Regression Models

Spatial Data and Model Transformations

5 The Impacts of Misspecified Spatial Interaction in Linear Regression Models

Raymond J.G.M. Florax[1] and Serge Rey[2]
[1] Wageningen Agricultural University, Wageningen, The Netherlands
[2] San Diego State University, San Diego, CA, U.S.A.

5.1 Introduction

In most applied empirical studies in the realm of amongst others regional economics, urban planning and environmental economics, the estimation of an *a priori* specified model is based on observations for a finite set of spatial units. This causes a number of persistent problems, which to varying extents have been dealt with in the literature. Essentially, the more fundamental problems boil down to the following issues.

First, the modifiable areal unit problem, which concerns the aggregation of observations over space. Unless the observations of a phenomenon at a disaggregated level are homogeneously distributed over space, inferences based on an analysis at a higher level of spatial aggregation are meaningless, and generalizations to a spatial process are invalid [Anselin (1988, pp. 26-28)].

Second, border or edge effects, pertaining to the problem that inferences are based on a finite set of observations, whereas the spatial process extends to spatial units not represented in the data set. As shown by e.g., Griffith (1985) no acceptable solution for this problem exists to date.

Third, the specification of the spatial interaction structure, which is typically represented by a spatial weight matrix. Generally, the use of a spatial weights matrix leads to a model specification that comprises a variate consisting of a (mostly linear) transformation of the observations on contiguous spatial units or spatial units at a fixed distance, either modeled explicitly as an additional explanatory variable or implicitly in the error term.

Fourth, testing for spatial effects, by means of measures of spatial association or correlation. Upon Moran's (1950) initial suggestion of a special form of a cross-product statistic, many others have developed similar statistics, e.g. the Geary statistic and, more recently, the class of *G*-statistics [Cliff and Ord (1981) and for a more up to date review Getis (1991)]. In the context of regression analysis large sample tests for spatial dependence have been developed following the general rules of misspecification testing in a maximum likelihood framework (i.e., Wald, Likelihood Ratio and Lagrange Multiplier tests [Burridge (1980), Anselin (1988, pp. 65-73), Bera and Yoon (1992), Anselin (1994a)]. Kelejian and Robinson (1992) defined a large sample test that avoids the assumption of normality for the error distribution. Anselin and Rey (1991) as well as Anselin and Florax (1995) concisely present an overview of the power of various tests for a number of different configurations.

Finally, for the estimation of spatial models adjusted estimators are needed because of the multidirectional nature of spatial dependence [Cliff and Ord (1981), Anselin (1988)]. Among the important results in this context are the recognition that the spatial counterparts of the estimated generalized least squares (EGLS) estimators developed in the time-series literature are not consistent, and the derivation of maximum likelihood (ML) estimators for the spatial lag model and the spatial autoregressive error model.

This paper focuses on the investigation of the last three issues mentioned above, in particular on the effects of misspecification of the spatial interaction structure on the power of various test statistics for spatial dependence and on the finite sample properties of various estimators for spatial models.

The remainder of this chapter appears as follows. Section two discusses the misspecification of the spatial interaction structure from a theoretical point of view. In section three the experimental design is laid out, and several tests for spatial autocorrelation and spatial estimators are discussed. Section four is concerned with the presentation and the analysis of the experimental simulation results. The experiments are concerned with the impact of misspecification of the weight matrix on the size and power of various tests as well as the small sample properties of a number of (spatial) estimations. Finally, section five is aimed at making some general inferences that may be of use for applied research.

5.2 Aggregation and the Identification of Spatial Interaction

The issue of the possible impacts of misspecification of the spatial weights matrix has not yet received ample treatment in the literature. Noteworthy exceptions are Stetzer (1982), Anselin (1985) and Anselin and Rey (1991).

Stetzer (1982) is concerned with a comparison of the effect of alternative weight specifications on the accuracy of parameter estimation and the forecasting of future data values, in a Monte Carlo investigation of space-time forecasting models. As a general result it turns out that the correct functional specification of the weights is important, especially when considering small sample sizes and autocorrelated data. However, an adequate match of the effective area is even more important than matching the functional form.[1] Two major limitations of Stetzer's pioneering article are: the issue of explicit testing for spatial dependence is not addressed and, given the absence of spatial simultaneity in the spatial forecasting model, the only estimator used is ordinary least squares (OLS).

Anselin's (1985) paper centers on the impact on the small sample distribution of maximum likelihood and OLS estimators caused by functional misspecification of the spatial interaction structure in conjunction with incorrect distributional assumptions regarding the error term. The paper concludes that the favorable asymptotic characteristics of maximum likelihood carry over to the finite sample case, even when the underlying distributional assumptions are not correct. Moreover, there is slight evidence that an inverse distance interaction structure is more robust to functional misspecification than a binary contiguity matrix. However, as in the case of Stetzer (1982) diagnostic checking for spatial dependence is not addressed, and it is therefore assumed that the 'correct' estimator is known beforehand, with OLS added as a sort of bench-mark case.[2]

The use of diagnostic tests for spatial dependence is, however, crucial in assessing the impacts of functional misspecification of the spatial interaction structure on the finite sample distribution of estimators. This may be inferred from the actual practice in estimating spatial models in applied research. Assume the standard spatial model with autoregressive disturbances, which for r (= 1, 2, ..., R) spatial cross-sections reads as:

$$y = \boldsymbol{X}\beta + \varepsilon \tag{1}$$

with:

$$\varepsilon = \lambda \boldsymbol{W}\varepsilon + \mu \tag{2}$$

where y is the $(R \times 1)$ stochastic variate, $\boldsymbol{X}$ is a $(R \times k)$ matrix of regressors, β is a $(k \times 1)$ parameter vector, and $\mu \sim N(0, \sigma^2 \boldsymbol{I})$. In a different fashion this model can be written as:

---

[1] The effective area, taken as shallower or larger than the true area, is determined on the basis of the cut-off point for the inverse distance and the negative exponential decay function. On a square regular grid Stetzer (1982) makes a similar distinction by determining contiguity on the basis of first-order contiguity and first- and second-order contiguity using the queen criterion. (Readers not familiar with these concepts may find a concise overview in Appendix 1 of chapter three.)

[2] Anselin (1986) considers the issue of functional misspecification of the weights matrix in non-nested tests based on augmented regressions. He concludes that one can have some doubt as to the appropriateness of tests on non-nested hypothesis in this context, as the power of the tests seems to depend on the similarity between the alternative hypotheses as well as on the nature of the true underlying spatial weights structure [Anselin (1986, p. 282)].

$$y = X\beta + (I - \lambda W)^{-1}\,\mu \tag{3}$$

It is common practice to start the estimation procedure by OLS estimation of the initial model given in (1) and to subsequently test for a spatial autoregressive component in the vector u of the estimated residuals. It is obvious that in proceeding thus, a first potential for drawing inappropriate conclusions arises, as the specified weight matrix may not be the true weight matrix, or in other words, the weight matrix may be misspecified. Apart from the differences in power intrinsic to the various test statistics available [see Anselin and Rey (1991) and Anselin and Florax (1995), for an assessment], the misspecification of the weight matrix can influence the power of the test statistics to a differing extent. Anselin and Rey's (1991) paper on the power of Moran's I and LM statistics in the context of a regular lattice structure[3] shows that the power functions are remarkably different (under the alternative as well as under the null hypothesis), and misspecification of the weights matrix substantially lowers the power of the tests.

Further, the actual treatment of the model given in (1)-(3) above is a case of a pre-test estimator (i.e., OLS in combination with spatial AR estimators such as spatial EGLS or ML). Consequently, misspecification of the weights matrix may also have an impact on the small sample distribution of the estimator used. This is an issue that has not been investigated, as in Anselin (1985) it is assumed that the correct estimator is known beforehand. In this chapter the pre-test aspect is incorporated through the investigation of the effect of misspecification of the weights matrix on the mean square error of the coefficient estimates resulting from two different pre-test estimators.[4]

In order to avoid the estimation of spatial (S)AR(MA) models, in which an *a priori* and exogenously specified weights matrix is needed for reasons of identification, models based on space-time series data can be used. The richness of information available in the latter type of data makes it possible to estimate the spatial dependencies as unparameterized covariance structures or, alternatively by means of stochastic elements incorporated in the weight matrix [e.g., Bolduc et al. (1995), Brett et al. (1994)]. However, for most economic applications the dimension of time-series that are available is too small for efficient estimation of the covariances or the spatial parameters.

The fundamental problem that is circumvented through the use of a weighting structure is the issue of identification. This may be seen as follows. If it is assumed that substantial spatial dependence in the endogenous variable is present, the standard procedure is to estimate a model with a spatially lagged endogenous variable,

---

[3] The weights matrices are based on rook, queen and distance based contiguity.

[4] One is a weighted average of the spatial Cochrane-Orcutt estimator (CO) and OLS, and the other a weighted average of the maximum likelihood estimator for spatial autoregressive error processes (MLERR) and OLS; see the next section for more details.

with specific weights assigned to every region and an average spatial interaction parameter ρ to be estimated. The weighting structure is thus *a priori* and exogenously determined. However, the true spatial interaction structure among the elements of the dependent variable may differ from the imposed structure with an average autocorrelation parameter. The spatial lag model is in fact a rather crude generalization of the following system of equations, where the subscripts refer to the spatial cross-sections:

$$\begin{aligned} y_1 &= \rho_1 w_1' y + x_1'\beta + \varepsilon_1 \\ y_2 &= \rho_2 w_2' y + x_2'\beta + \varepsilon_2 \\ &\vdots \qquad\qquad\qquad\qquad \vdots \\ y_R &= \rho_R w_R' y + x_R'\beta + \varepsilon_R \end{aligned} \tag{4}$$

where all w_r' are unit vectors although with element $w_r = 0$. This system of equations per region can be summarized in the following equation which pertains to all spatial cross-sections r:

$$y_r = \rho_r w_r' y + x_r'\beta + \varepsilon_r \tag{5}$$

with $w_r = 0 \; \forall \; w_r'$, or in matrix notation:

$$y = (\boldsymbol{W}y)'\boldsymbol{Z} + \boldsymbol{X}\beta + \varepsilon \tag{6}$$

where $\boldsymbol{W}$ is a matrix of units with zero diagonal elements, and $\boldsymbol{Z} = \text{diag}(\rho)$ so that the elements $z_{rr'}$ are as follows:

$$\begin{cases} z_{rr'} = \rho_r & \textit{if} \quad r \neq r' \\ z_{rr'} = 0 & \textit{if} \quad r = r' \end{cases} \tag{7}$$

The specification in (7) obviously results, even when the term $\boldsymbol{X}$ is excluded, in the incidental parameter problem.

With respect to the distribution of the error term two types of misspecification may occur, viz. functional and distributional misspecification. The issue of distributional misspecification is mostly investigated in experimental settings by incorporating the exponential and/or lognormal distribution [Anselin (1985), Anselin and Rey (1991), Anselin and Florax (1995)]. As documented in Anselin and Rey (1991) Moran's I and the LM test for spatial autoregressive errors seriously underreject the null hypothesis of no spatial dependence for distributions that differ from the standard normal, and misspecification of the weights matrix substantially lowers the power of the tests.[5]

The latter already refers to a specific type of functional misspecification. The problem of functional misspecification is in fact identical to the modifiable areal unit problem, which includes two differing types of problems: the problem of aggregation and, presupposing a given configuration of areal units, the problem of the functional form of the error process (including the contiguity or distance decay pattern between the units). The aggregation problem is concerned with both the level of spatial disaggregation as well as with the way in which the areas are organized in space. With respect to the aggregation problem Stetzer (1982) comes to the conclusion, on the basis of simulation experiments, that it may constitute a serious source of misspecification error. This problem is not addressed in the remaining part of this chapter. The problem of the functional form of the error process starts from the peculiarities of most tests for spatial dependence in the error process. First, tests like Moran's *I* and the Kelejian-Robinson test [see Kelejian and Robinson (1992)] are general tests for the null hypothesis of no spatial correlation among the residuals, but the alternative hypothesis is merely spatial correlation among the estimated residuals without a proper specification of the underlying generating process.[6] Second, tests based on the Lagrange Multiplier principle in most instances also apply to more than one alternative situation; the LM test for a spatial error process is the same for an autoregressive as for a moving average process [see Anselin (1994a)]. Consequently, even if tests for spatial dependence lead to the rejection of the null hypothesis of no spatial correlation, it remains unclear what the exact alternative specification should be. Moreover, in order to test for spatial dependence the contiguity or distance decay pattern has to be specified *a priori*. In the simulation experiments to follow misspecification of the contiguity pattern is explicitly addressed.

5.3 Experimental Design

5.3.1 Sample Size

As is well known, the sampling properties of tests for spatial dependence as well as maximum likelihood estimators are only known in asymptotic settings. Conse-

[5] See chapter two for more elaborate results.

[6] A similar process has been observed in the time-series context, where the Durbin-Watson test has been shown to have power against, various types of misspecification [Thursby and Schmidt (1977)].

quently, a problem facing the spatial analyst working in non-asymptotic contexts is that the sampling properties of these estimators and tests are unknown. In order to provide some insight as to these properties simulations for four small sample sizes (R = 16, 25, 36, 49) are carried out, where the observations are taken from a regular lattice, similar to the design employed by Anselin and Rey (1991).

5.3.2 Spatial Interaction Structures

Three different binary contiguity weight matrices are considered. The first two ($\boldsymbol{W}_1$, $\boldsymbol{W}_2$) are based on the familiar rook and queen contiguity criteria. For the third matrix ($\boldsymbol{W}_3$), all observations separated by less than a Euclidean distance of 2 units are taken as contiguous. These different definitions of contiguity give rise to the following number of potential joins for an internal location in the lattice: 4 joins for $\boldsymbol{W}_1$, 8 joins for $\boldsymbol{W}_2$ and 12 joins for $\boldsymbol{W}_3$.

While the degree of connectivity obviously increases across the three different matrices, an important issue is the change in the magnitude of the spatial connectedness that inevitably occurs when changing the sample size for a particular weight matrix. This is largely neglected in the spatial simulation literature up to now, and constitutes another noteworthy peculiarity that makes the spatial case so distinct from the time-series case. Obviously, the importance of the choice of the spatial weight matrix – and thus of a particular magnitude of spatial connectedness – implies that one should be careful in generalizing the results of the simulation experiments.

Table 5.1 displays four summary measures for the weighting structures for the different sample sizes. The first measure, the percentage of non-zero cells in the off diagonal portion of the matrix, tends to decrease with increasing sample size for each of the weighting structures. This suggest that increasing the sample size decreases the degree of connectivity for the same definition of contiguity. The second measure, the largest eigenvalue of the matrix, tends to increase with increasing sample size and such a pattern is often interpreted as an indication of increasing connectivity [Cliff and Ord (1981), Griffith (1988)]. However, as argued by Boots and Royle (1991) the principal eigenvalue is in part dependent on the size of the weight matrix and generally increases with the size of the matrix, or in this case the sample size considered. Boots and Royal suggest the maximum value for a principal eigenvalue on a planar graph can be approximated by $0.5\times(3+\sqrt{8n-15})$ where n is the sample size. This planar equivalent is interpreted as the principal eigenvalue for a planar graph with maximum connectivity and provides some insight as to the role of increasing sample size on the value of the principal eigenvalue. The values of this planar equivalent do in fact increase with larger sample sizes for each of the weighting structures considered here, suggesting that the increase in the principal eigenvalues with increasing sample size for each weight matrix may not be pointing to increasing connectivity.

Table 5.1: Summary measures for spatial weight matrices.

n	W_1	W_2	W_3
	Percentage non-zero cells (off-diagonal)		
16	0.192	0.333	0.458
25	0.133	0.240	0.340
36	0.094	0.173	0.248
49	0.071	0.133	0.192
	Largest eigenvalue (non standardized W)		
16	3.154	5.626	7.376
25	3.464	6.464	8.835
36	3.597	6.837	9.555
49	3.696	7.110	10.104
	Planar equivalent		
16	6.815	6.815	6.815
25	8.301	8.301	8.301
36	9.761	9.761	9.761
49	11.208	11.208	11.208
	Largest eigenvalue/planar equivalent		
16	0.463	0.825	1.082
25	0.417	0.779	1.064
36	0.368	0.700	0.979
49	0.330	0.634	0.902

The fourth summary measure is an attempt to control for the effect of the sample size on the principal eigenvalue as a measure of connectivity, and is obtained as the ratio of the principal eigenvalue to the planar equivalent. This ratio decreases in value with increasing sample size for each weight matrix, suggesting that the larger samples have spatial structures with lesser degrees of connectivity relative to smaller samples, for the same definition of contiguity.[7] Interestingly, ignoring the effect of increasing sample size on the value of the largest eigenvalue, would have resulted in the opposite conclusion if the conventional interpretation were used.

[7] For the two smallest sample sizes, the ratio of the principal eigenvalue to the approximation of the maximum value exceed unity. At first glance this may appear to represent an apparent inconsistency; however, it is important to keep in mind that the planar equivalent is an approximation of the maximum value and not an analytically derived form. Nevertheless, the fact that this ratio provides an indication of decreasing connectivity with increasing sample size, in agreement with the percentage of non-zero cells measure, suggests that is is useful in this context.

5.3.3 Spatial Models and Parameter Space

In the simulations two well-known underlying true models have been used. The base case is the model without any spatial effects, for which OLS is the best linear unbiased estimator (BLUE); formally:

$$y = \boldsymbol{X}\beta + \varepsilon \tag{8}$$

where y is the $(R \times 1)$ vector of observations on the dependent variable, $\boldsymbol{X}$ the $(R \times 2)$ matrix of exogenous variables including a constant term and the $(R \times 1)$ uniformly distributed variable on the interval (0, 1), β is the (2×1) vector of parameters ($\beta_1 = \beta_2 = 0.5$), ε is the $(R \times 1)$ normal, independently distributed error term with the expectation of 0 and variance $\sigma^2 = 4.0$. The second model includes spatial error dependence which is the most commonly encountered spatial econometric specification in practice [Anselin and Hudak (1992, p. 517)]:

$$y = \boldsymbol{X}\beta + (I - \lambda \boldsymbol{W})^{-1}\varepsilon \tag{9}$$

with ε as before. It is well known that OLS, although unbiased for β is inefficient, while the maximum likelihood estimator (i.e., the so-called MLERR) is consistent [Anselin (1988)].

Ten non-zero values for the spatial parameter λ are examined, ranging from -0.9 through +0.9 in increments of 0.2. Additionally, the case when λ is set to zero, and thus model (8) is obtained, is considered. The experiments consist of nine different combinations of weight matrices, for four different sample sizes and 11 values for λ resulting in 396 different cases. For each case we carry out 1000 replications keeping the matrix of explanatory variables constant.

5.3.4 Test Statistics and Estimators

The application of tests statistics and estimators is viewed in a sequential fashion, recognizing that the ultimate choice of estimator is often based on the outcome of a hypothesis test concerning a particular form of spatial dependence. Three familiar tests of spatial dependence in linear regression models are investigated, Moran's I, a Lagrange multiplier test for spatial error dependence (LMERR) and a Lagrange multiplier for an omitted spatially lagged dependent variable (LMLAG). All three tests are based on the OLS estimation of model (8). Moran's I can be expressed as:

$$I = \frac{R}{S_0} \frac{e'\boldsymbol{W}e}{e'e} \tag{10}$$

where R is the number of regions in the spatial system, S_0 the sum of the elements of the weight matrix $\boldsymbol{W}$, and e the vector of OLS residuals. Cliff and Ord (1981, pp. 200-203) prove that Moran's I is asymptotically normally distributed for OLS residuals whenever the population errors are random independent drawings from a normal population. They also present the first and second moments of the statistics used for hypotheses testing.

The LMERR statistics is essentially a scaled squared Moran coefficient, and reads in formal terms as:

$$\text{LMERR} = \frac{R^2}{\text{tr}(\boldsymbol{W}'\boldsymbol{W} + \boldsymbol{W}^2)} \left(\frac{e'\boldsymbol{W}e}{e'e} \right)^2 \tag{11}$$

with all symbols as before and tr for the trace operator.
The LMLAG statistic reads as:

$$\text{LMLAG} = \frac{R}{\dfrac{(\boldsymbol{WX}\beta)'\boldsymbol{M}(\boldsymbol{WX}\beta)}{(e'e)} + \text{tr}(\boldsymbol{W}'\boldsymbol{W} + \boldsymbol{W}^2)} \left(\frac{e'\boldsymbol{W}y}{e'e} \right)^2 \tag{12}$$

with β for the vector with OLS coefficient estimates, $\boldsymbol{M} = \boldsymbol{I} - \boldsymbol{X}(\boldsymbol{X}'\boldsymbol{X})^{-1}\boldsymbol{X}'$ is the projection matrix, and all other symbols as before.

The estimators used here are the iterative EGLS estimator based on the spatial Cochrane and Orcutt (CO) approach as outlined by Hordijk (1974), and the maximum likelihood estimator (MLERR) as described in Anselin (1988, pp. 60-65). Additionally, OLS is applied for all spatial parameter values. As is well known, the CO estimator is inconsistent for β, due to the reliance on an OLS estimation of the spatial parameter λ which is not consistent. In contrast, MLERR yields consistent estimates for both β and λ. The estimates for β based on OLS, although unbiased, will not have the usual form for their estimated variances, but instead the variances will be a complex function of λ. However, in finite samples, no exact results are available for MLERR, and Monte Carlo experiments in Anselin (1981) suggest the OLS may dominate MLERR in terms of bias and mean squared error. The extent to which these results are robust to misspecification of the weight matrix remains largely unknown and is examined below.

Two pre-test estimators are also implemented. The first is based on the Moran's I test and uses the EGLS estimator only if there is evidence of spatial dependence, otherwise the model is estimated using the initial OLS results. The second pre-test estimator is based on the LMERR test. If LMERR results in a significant test statistic the model is estimated using MLERR, otherwise the initial OLS estimates are used.

Several points need to be made with respect to these two pre-test estimators. First, as shown by Anselin and Rey (1991) Moran's I has power against both forms of spatial dependence (error and lagged dependent variable) yet it can not be used to discriminate between these two forms. In contrast the power of LMERR was found to be greater than that of LMLAG when the underlying model was the spatial error specification, while the respective powers were reversed for the spatial lag model. Florax and Folmer (1992) investigate the use of the LM testing strategy in a series of Monte Carlo simulations and find that it outperforms the conventional procedure based on Moran's *I*. In this study the focus is solely on a comparison of LMERR with Moran's *I* in the context of pre-test estimation. The sensitivity of the joint LM testing strategy to misspecification of the spatial weights matrix would require the consideration of the spatial lag model as a possible alternative, which is beyond the scope of the current study.

5.3.5 Forms of Misspecification

With respect to the weight matrix two forms of misspecification can be identified: overspecification and underspecification. Overspecification occurs when a more extensive weight matrix is used to carry out the tests, when the true weight matrix in the data generating process is less extensive (i.e., $\boldsymbol{W}_3$ vs. $\boldsymbol{W}_1$, $\boldsymbol{W}_3$ vs. $\boldsymbol{W}_2$, $\boldsymbol{W}_2$ vs. $\boldsymbol{W}_1$). In this case, observations are erroneously included in the neighborhood set, thus giving rise to an error of commission. Underspecification arises in cases where a less extensive matrix is used in place of the more extensive process matrix (i.e., $\boldsymbol{W}_1$ vs. $\boldsymbol{W}_3$, $\boldsymbol{W}_1$ vs. $\boldsymbol{W}_2$, $\boldsymbol{W}_2$ vs. $\boldsymbol{W}_3$). This could be considered an error of omission since relevant observations are excluded from the neighborhood set.

5.4 Empirical Results

The experiments generated a mass of results from which a select number of topics are considered in detail here.[8] First the size of the three tests under the null hypothesis of no spatial dependence is examined. Next, attention turns to the relative power of the three tests for spatial dependence. The effect of misspecification of the weight matrix on the power of the tests is then examined. This is followed by an examination of the impacts of misspecification of the weight matrix on parameter estimation.

5.4.1 Size of Tests Under the Null

The first issue examined is the extent to which the size of the tests are reflected in the empirical rejection frequencies observed in the experiments.

[8] The detailed simulation results are available from the authors upon request.

Table 5.2: Size of tests for spatial dependence.[a]

	W_1			W_2			W_3		
	MI	LMERR	LMLAG	MI	LMERR	LMLAG	MI	LMERR	LMLAG
$R = 16$	0.057	0.038	0.047	0.053	0.017	0.030	0.047	0.005	0.023
	0.955	-1.985	-0.448	0.423	-8.073	-3.708	-0.448	-20.175	5.696
$R = 25$	0.054	0.042	0.058	0.051	0.028	0.032	0.050	0.020	0.033
	0.560	-1.261	1.082	0.144	-4.217	-3.234	0.000	-6.776	-3.009
$R = 36$	0.054	0.044	0.053	0.051	0.040	0.034	0.053	0.029	0.040
	0.560	-0.925	0.423	0.144	-1.614	-2.792	0.423	-3.957	-1.614
$R = 49$	0.045	0.043	0.045	0.038	0.032	0.040	0.050	0.029	0.039
	-0.763	-1.091	-0.763	-1.985	-3.234	-1.614	0.000	-3.957	-1.797

[a] The first row contains rejection frequencies, the second z values for the null hypothesis that the frequency equals 0.05.

Table 5.2 reports both the rejection frequencies together with the z-scores for a hypothesis test in which the null is that the true rejection frequency for each test is equal to 0.05. The most striking result pertains to the discrepancies between the sizes of the tests across the different weight matrices. For the error model the two Lagrange Multiplier tests have empirical rejection frequencies that are significantly smaller than 0.05 for a number of sample sizes for the W_2 and W_3 matrices. LMERR significantly under-rejects for $R = 16$, 25, and 49 for W_2 and for all sample sizes for W_3, while LMLAG under-rejects for $R = 25$, 36 and 49 for W_2 and $R = 16$ and $R = 25$ for W_3. Moran's I has an empirical rejection frequency that is significantly different from the nominal frequency for only $R = 49$ and W_2. In contrast to the size of the tests under W_2 and W_3, there is only one case, LMERR $R = 16$, in which a test has an empirical rejection frequency significantly different from 0.05.

5.4.2 Power of Tests

Figures 5.1, 5.2, and 5.3 portray the comparative power functions for each of the tests and the three different weighting structures when $R = 49$. The weight matrix used to carry out the tests is the same one used in the data generating process. The general patterns are similar to the results in Anselin and Rey (1991), and here the focus is only on those results most relevant to the issue of weight matrix misspecification. Of particular importance is the asymmetry in the power functions of all three tests for the W_2 and W_3 matrices in Figures 5.2 and 5.3 in comparison to the power functions for W_1 displayed in Figure 5.1. For negative values of λ all three tests have relatively lower power for W_2 and W_3 than the case when λ takes on a positive value of the same absolute magnitude. Also, for all values of λ the power of the three tests is highest for W_1, followed by W_2 and then W_3. This asymmetry has been noted by several others [Anselin (1990), Anselin and Rey (1991)] and may be due to the inherent differences between the concepts of negative and positive spatial dependence which are magnified on the more extensive weight matrices (W_2 and W_3).

5.4.3 Misspecification Effects on the Power of Tests for Spatial Dependence

Next, attention turns to the impact that misspecification of the weight matrix can have on the power of the three tests for spatial error dependence. Considering overspecification first, Figures 5.4 and 5.5 display the difference between the power of the tests when the weight matrix is over and correctly specified.[9] The most striking finding is that the relative effects on the power of the tests are greater for negative error dependence than for positive error dependence. This can be attributed in part to the asymmetric nature of the power functions for the more extensive weight matrices with respect to negative dependence alluded to in the previous section. At the same time, the overspecification also results in the inclusion of observations which are not members of the neighborhood set. Moreover, the size of the reduction in power due to overspecification generally increases with the strength of dependence, although for the larger sample ($R = 49$), the effects appear from the moderate to the extreme values of the spatial parameter.[10]

While the general pattern of the reduction in power is similar across the three tests, the Lagrange Multiplier tests appear to be slightly more sensitive to overspecification of the weight matrix, particularly for strong levels of dependence. Both the LMERR and LMLAG display larger reductions in power relative to Moran's I, for both positive and negative dependence, and again this effect appears to become stronger as both the sample size and strength of spatial dependence increase.

Figures 5.6 and 5.7 portray the change in the power of three tests when the weight matrix is specified as W_1 and when the matrix in the data generating process is W_3. In general, when λ takes on a negative value, the effects of underspecification do not appear to be as strong as those associated with overspecification. For example, in Figure 5.7 when $\lambda = -0.7$ and $R = 49$, underspecification reduces the power of the three tests an average of -0.17.

---

[9] As noted in the text, overspecification also arises when the underlying matrix is W_2 yet the tests are based on W_3, or in cases where W_1 is the correct matrix but the tests are implemented using W_2. Both of these cases are less extreme forms of underspecification than those discussed in the text, yet qualitatively, the effects are similar. Due to space limitations, these results are not discussed in the text.

[10] We display only the smallest and largest sample size results due to space considerations. The intermediate sample sizes (R = 25, 36) reflect transitions between these two cases.

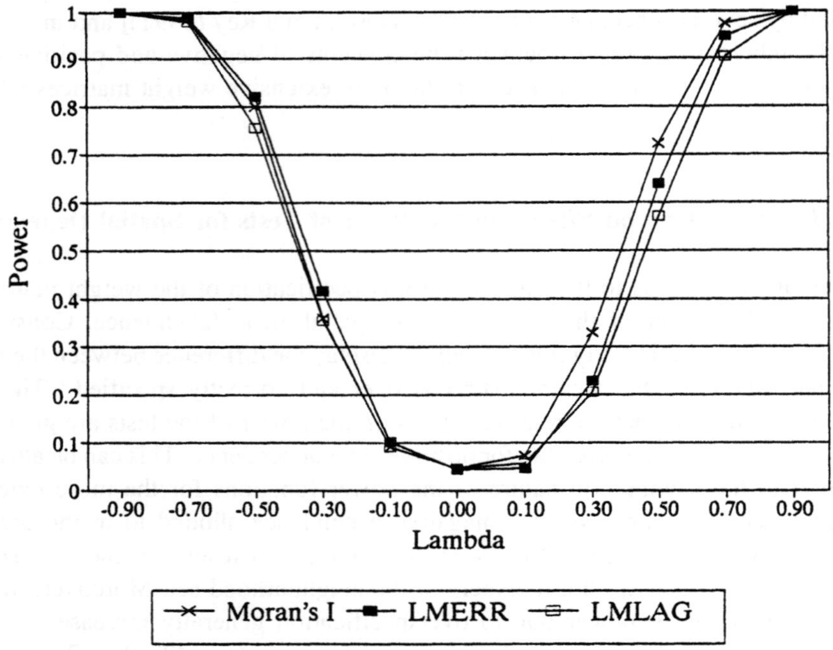

Figure 5.1: The power function for the error model, R = 49, $\boldsymbol{W}_1$.

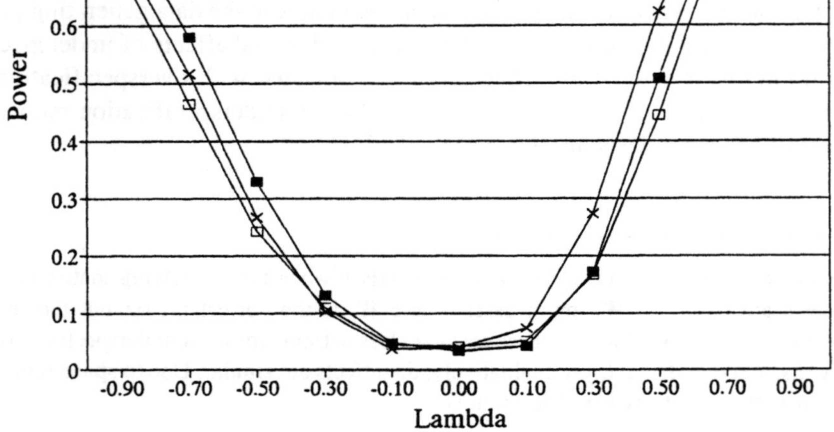

Figure 5.2: The power function for the error model, R = 49, $\boldsymbol{W}_2$.

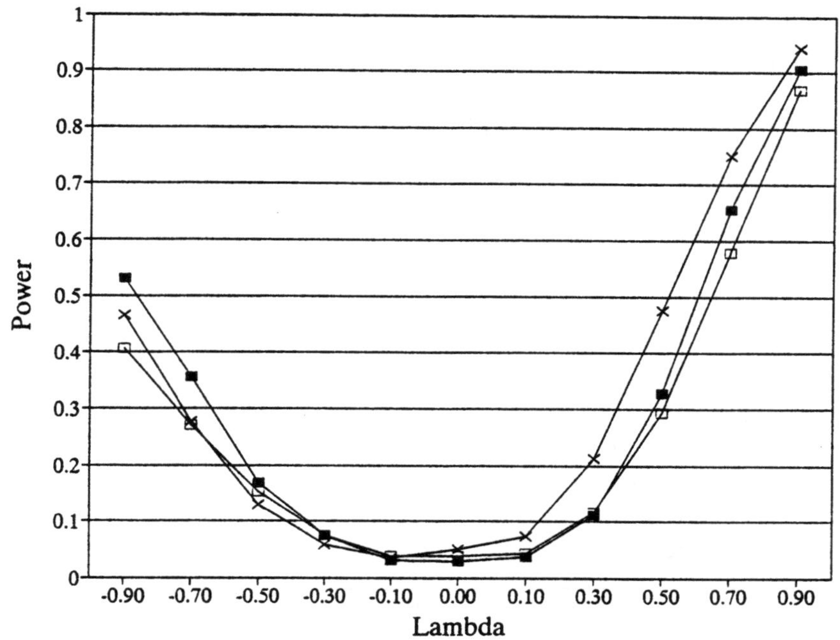

Figure 5.3: The power function for the error model, R = 49, $\boldsymbol{W}_3$.

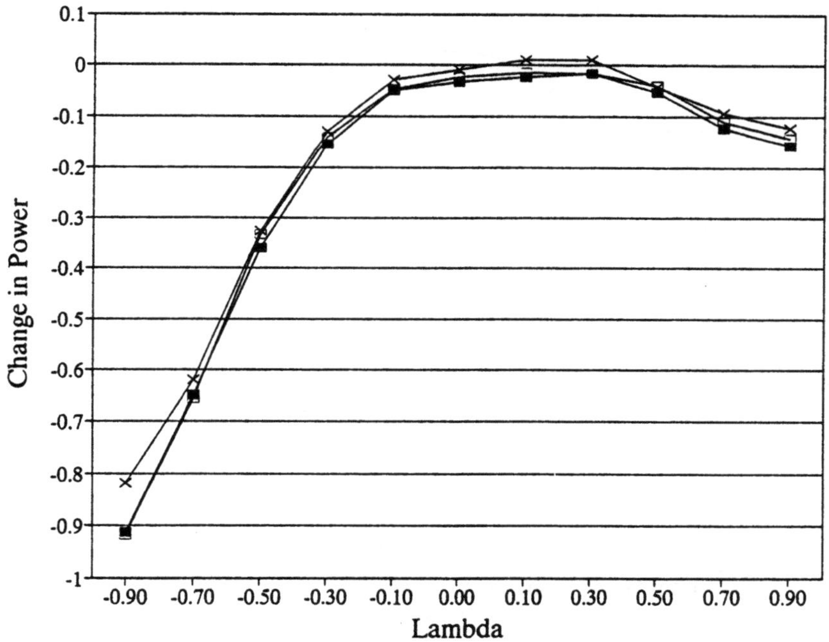

Figure 5.4: The misspecification effect of $\boldsymbol{W}_3$ vs. $\boldsymbol{W}_1$ for the error model, R = 16.

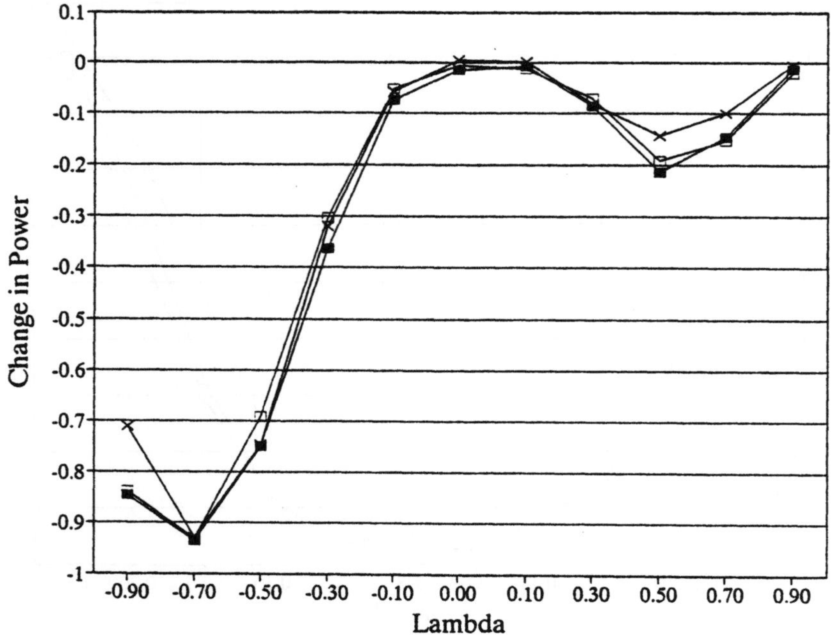

Figure 5.5: The misspecification effect of W_3 vs. W_1 for the error model, R = 49.

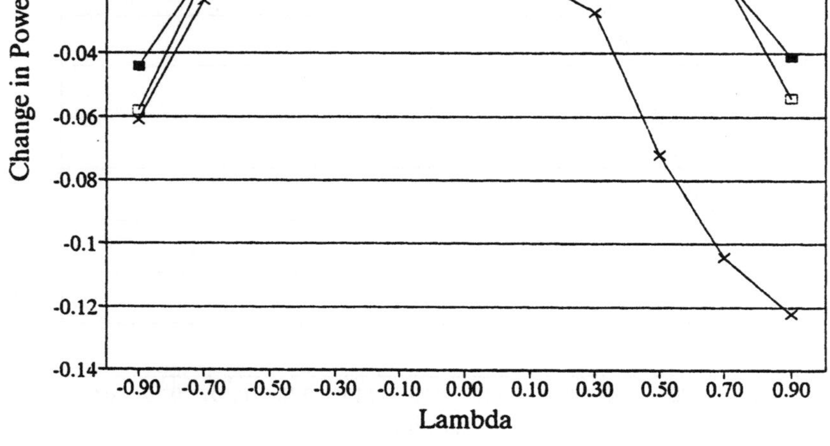

Figure 5.6: The misspecification effect of W_1 vs. W_3 for the error model, R = 16.

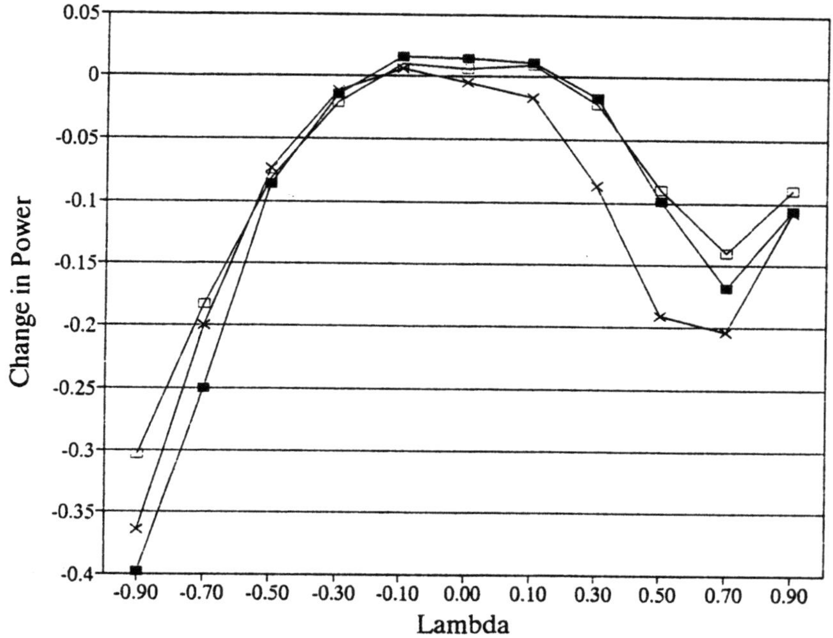

Figure 5.7: The misspecification effect of $\boldsymbol{W}_1$ vs. $\boldsymbol{W}_3$ for the error model, R = 49.

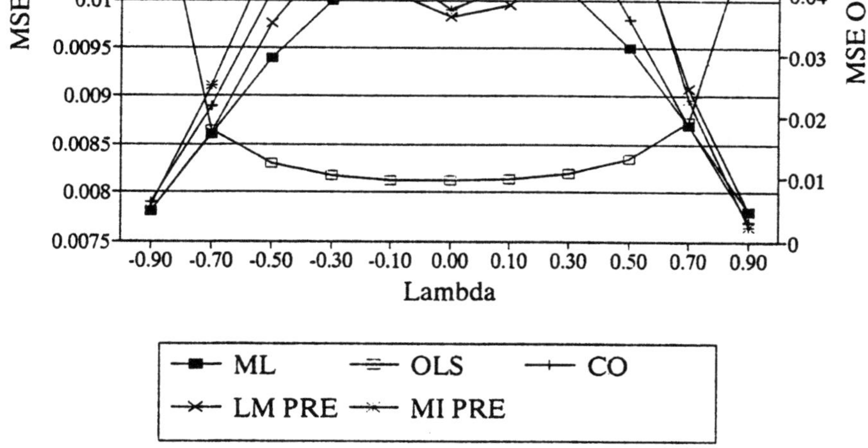

Figure 5.8: The MSE of β for the error model, R = 49, $\boldsymbol{W}_1$.

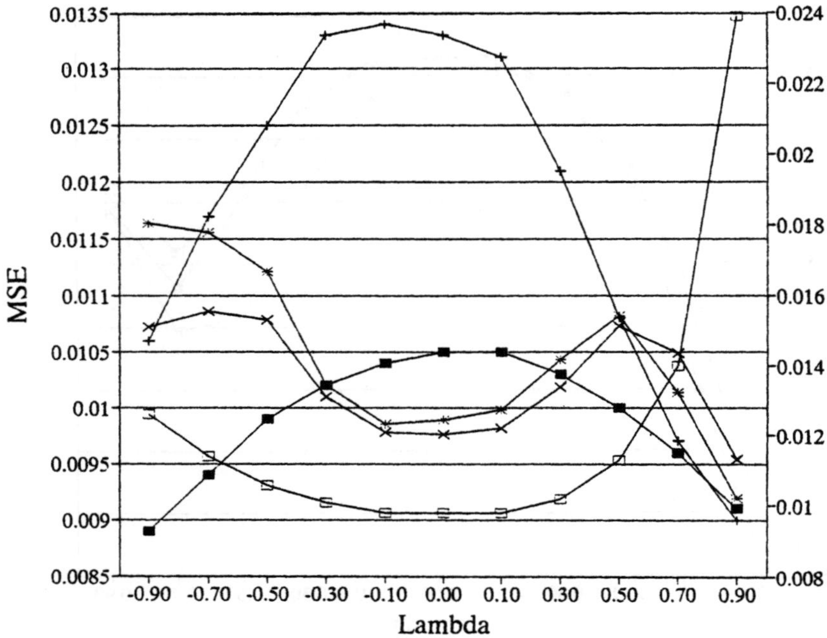

Figure 5.9: The MSE of β for the error model, R = 49, $\boldsymbol{W}_3$.

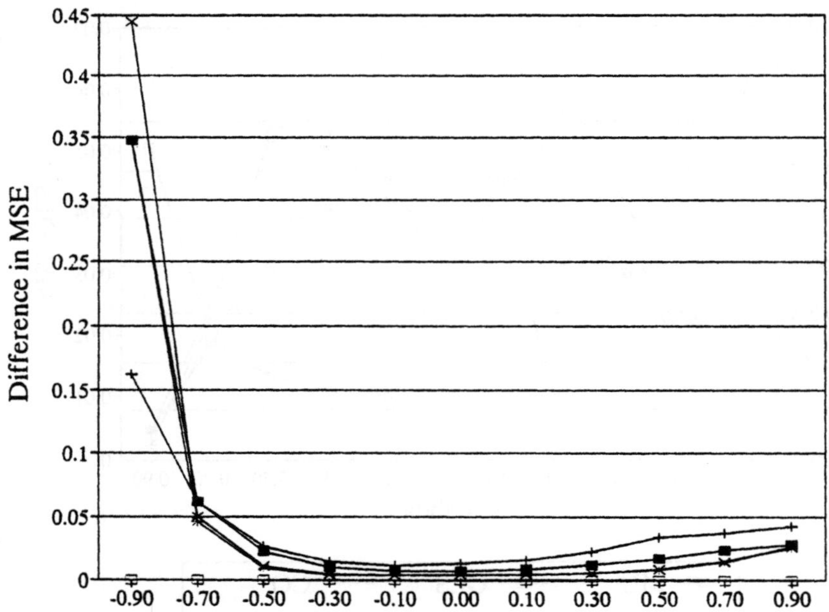

Figure 5.10: The misspecification effect of $\boldsymbol{W}_3$ vs. $\boldsymbol{W}_1$ in terms of the MSE of β for the error model, R = 16.

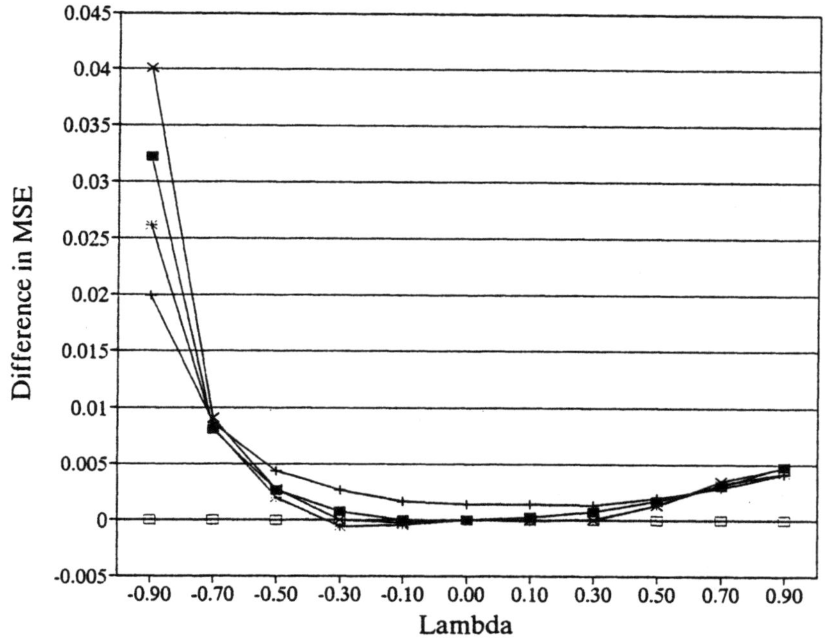

Figure 5.11: The misspecification effect of W_3 vs. W_1 in terms of the MSE of β for the error model, $R = 49$.

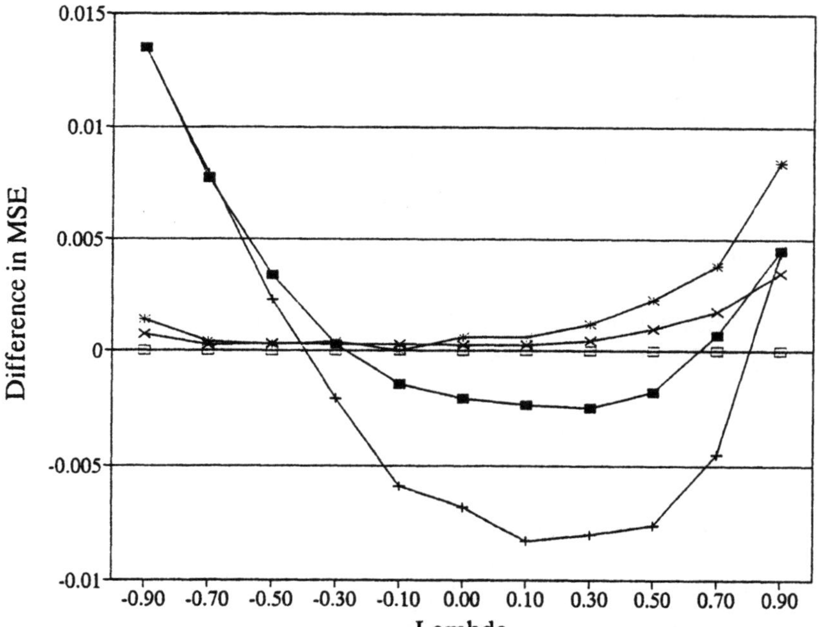

Figure 5.12: The misspecification effect of W_1 vs. W_3 in terms of the MSE of β for the error model, $R = 16$.

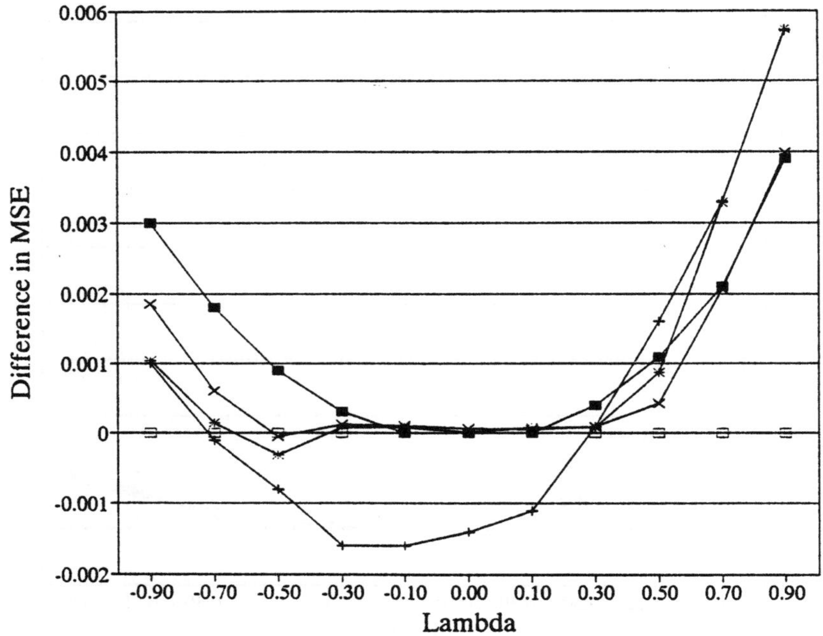

Figure 5.13: The misspecification effect of W_1 vs. W_3 in terms of the MSE of β for the error model, R = 49.

In the case of overspecification (Figure 5.5) for the same value of λ and sample size, the average reduction in power is -0.932. However, for positive values of λ, the effects of underspecification appear to be slightly larger relative to those associated with overspecification. For example, when λ = 0.70 and R = 49 using W_1 to carry out the tests when W_3 is the true weight matrix reduces the power by an average of 0.170 relative to when W_3 is used to carry out the test. In contrast, the power is reduced by an average of 0.133 when the tests are based on W_3 instead of W_1 in the case of overspecification.

The pattern of the relative effects of underspecification on the power of three tests is qualitatively different from that in the case of overspecification. Moran's I is clearly more sensitive to underspecification for positive values of λ, while the two Lagrange Multiplier tests display approximately the same levels of sensitivity. For negative λ, LMERR appears to be most sensitive to underspecification, followed by Moran's I and LMLAG.

Interestingly, for certain values of λ and in small samples, there is evidence of increases in power arising from an underspecification of the weight matrix. More specifically in Figure 5.6, for small (absolute) values of λ (-0.10 through 0.10) there power of the LMERR and LMLAG tests increase when W_1 is used to carry out the tests relative to when the true matrix W_3 is used. Although the increases are not enough to result in a bias, or over-rejection, for the two tests when λ = 0, the result is particularly intriguing. The increase in power can be attributed to the generally lower power displayed by the tests for the W_2 and W_3 matrices as discussed in section 5.4.2. However, the increase in power made possible by using W_1 rather than either of these two weight matrices may have implications for model specification and the relative precision with which the coefficients of the models will be estimated. We return to this issue in section 5.4.5.

5.4.4 Sensitivity of Parameter Estimation to Specification of Weight Matrix

Figures 5.8 and 5.9 display the average mean squared error (MSE) for the slope estimates from the five estimators for $R = 49$ when the weighting structures are correctly specified as W_1 and W_3 respectively. In the case of W_1 (Figure 5.8), ML and CO have concave MSE functions in λ, with MLERR dominating for all values of λ. For $\lambda \geq |0.50|$, the MSE of OLS increases with stronger spatial error dependence. These results suggest that the asymptotic properties may provide useful approximations to small sample settings similar to the ones examined here.

Turning to the pre-test estimators, LMPRE is found to dominate MIPRE for all values of λ. Together with the results of Florax and Folmer (1992) there is growing evidence that the LM based tests represent improvements to the conventional approaches to model specification in spatial econometrics. Moreover, for $\lambda \geq |0.10|$, LMPRE has the lowest MSE of all five estimators. As λ approaches its maximum absolute values, the MSE of the estimates become similar, with the exception of OLS.

When the underlying weights matrix is W_3 (Figure 5.9), the pattern of the MSE functions is qualitatively similar to the W_1 case, with a few exceptions. The MSE of the pre-test estimators do not decline with increasingly strong negative dependence, in contrast to the W_1 case. Additionally, MIPRE now does better than LMPRE for $\lambda \geq 0.70$, although LMPRE continues to dominate for the remainder of the parameter space. The MSE of OLS also declines relative to the case of W_1, and displays a markedly asymmetric form with lower MSE values for negative dependence.

5.4.5 Impact of Misspecification of Weight Matrix on Estimation

Figures 5.10 and 5.11 show the effects of overspecification on the MSE of the estimators for $R = 16$ and 49 respectively. The nature of the impact of overspecification appears to be similar across the estimators, with the largest effects occuring for $\lambda = -0.9$, with LMPRE displaying the largest change in MSE, followed by ML,

MIPRE and CO.[11] Over the positive range of the parameter pace, CO appears to be slightly more sensitive to overspecification than are the remaining estimators.

Figure 5.11 suggests that the effects of overspecification lessen as the sample size increases. Moreover for positive values of λ the effects are more similar across the estimators than compared with the small sample case (Figure 5.10). Nevertheless, the general pattern is quite similar to that for the smaller sample case, with the largest effects occuring for strong negative dependence. In general terms, overspecification of the weight matrix leads to small increases of the MSE of the estimators, although for strong negative dependence the effects can be more pronounced.

The impact of underspecification of the spatial interaction structure is of a qualitatively different nature however, as can be seen by comparing Figures 5.12 and 5.13 with 5.10 and 5.11. Underspecification when R = 16 results in a decrease in the MSE of CO for $-0.30 \leq \lambda \leq 0.70$ and ML for $-0.10 \leq \lambda \leq 0.50$. For other values of the parameter space, CO and ML exhibit increases in their MSE values due to underspecification. The two pre-test estimators display slight increases in MSE for all values of λ, with the largest increases occuring for positive dependence. As the sample size increases (Figure 5.13), however, these effects weaken, although there is still some evidence of decreases in the MSE of CO for. Of the two pre-test estimators, MIPRE appears to be slightly more sensitive to underspecification, particularly for positive spatial dependence and smaller samples.

5.5 General Inferences

Uncertainty with respect to the proper specification of the spatial weights matrix has long been recognized as a fundamental problem in applied spatial econometric modeling. In this study some insight is provided as to the potential costs of this uncertainty for parameter estimation and testing for spatial dependence in linear regression models. The experimental results suggest that the cause for concern over the unknown nature of the spatial weights matrix in practice is warranted since misspecification of this matrix can impact the performance of alternative estimators and diagnostic tests. The main results along these dimensions are summarized here, prior to the suggestion some directions for future research.

The performances of the three tests for spatial dependence in linear regression models were all found to be affected by misspecification of the spatial weights matrix. In the case were the weights matrix was overspecified, the power of the

[11] Because a weight matrix is not utilized when estimating a model using OLS, this estimator is unaffected by over or underspecification of the spatial structure, and consequently there is no change in the MSE function of OLS.

tests was reduced. On the other hand, underspecification was found to result in slight increases in the power of the tests in certain cases, however the extent to which this was do to the intrinsic properites of a negative spatial dependence regime requires further examination. Finally, the two pre-test estimators displayed similar patterns with respect to the underlying correct weight matrix as well when the spatial structure was misspecified. This would seem to indicate that the LM joint testing strategy does not result in model specifications that are more sensitive to misspecification of the weight matrix than those obtained using a traditional strategy based on Moran's *I*. In other words, the impacts of misspecification do not compromise the otherwise attractive properties of the LM based testing strategy relative to Moran's *I*.

With respect to the impacts of misspecification of the weight matrix on the performance of alternative estimators for spatial process models we find that, similar to the case of diagnostic tests for spatial dependence, all of the alternative approaches investigated were sensitive to incorrect specification of the weight matrix. In general terms, the effects of overspecification of the weight matrix on the MSE of the estimators tended to be larger than those associated with underspecification. However, the asymptotic properties of the alternative estimators were found to provide good guidance on the relative small sample properties of the estimators when the weight matrix was correctly specified. Even though, misspecification does affect the sampling properties of the estimators, the effects did not modify the relative performance of the alternative approaches.

Although the results do suggest that use of the wrong weight matrix can result in serious misspecification, the results do not reveal any empirical patterns which would serve as specific guidelines to use in a model discrimination framework. We feel, however, that given the importance of this form of misspecification, the development of tests for weight matrix misspecification represents an area were additional research is required. Earlier work by Anselin (1986) represents an initial effort in this regard. The potential of misspecification tests based on recently developed local indicators of spatial association (LISA) statistics [Getis and Ord (1992), Anselin (1994b)] also appears to be a promising direction for further research.

References

Anselin, L., Small Sample Properties of Estimators for the Linear Model with a Spatial Autoregressive Structure in the Disturbance, *Modeling and Simulation*, 12, 899-904, 1981.

Anselin, L., *Estimation and Inference in Misspecified Spatial Autoregressive Models, Some Small Sample Results*, Columbus: Ohio State University Research Foundation, 1985.

Anselin, L., Non-Nested Tests on the Weight Structure in Spatial Autoregressive Models: Some Monte Carlo Results, *Journal of Regional Science*, 26, 267-284, 1986.

Anselin, L., *Spatial Econometrics: Methods and Models*, Boston: Kluwer, 1988.

Anselin, L., Spatial Dependence and Spatial Structural Instability in Applied Region Analysis, *Journal of Regional Science*, 30, 185-207, 1990.

Anselin, L., *Testing for Spatial Dependence in Linear Regression Models: A Review*, Morgantown: West Virginia University, Regional Research Institute Research Paper 94-16, 1994a.

Anselin, L., *Local Indicators of Spatial Association - LISA*, Morgantown: West Virginia University, Regional Research Institute, Research Paper 93-31, 1994b.

Anselin, L. and R.J.G.M. Florax, Small Sample Properties of Tests for Spatial Dependence in Regression Models: Some further results, 1995 (this issue).

Anselin, L. and S. Hudak, Spatial Econometrics in Practice, A Review of Software Options, *Regional Science and Urban Economics*, 22, 509-536, 1992.

Anselin, L. and S. Rey, Properties of Tests for Spatial Dependence in Linear Regression Models, *Geographical Analysis*, 23, 112-131, 1991.

Bera, A.K. and M.J. Yoon, *Simple Diagnostic Tests for Spatial Dependence*, Champaign: University of Illinois, Department of Economics, 1992 (mimeo).

Bolduc, D., R. Laferrière and G. Santarossa, Spatial Autoregressive Error Components in Travel Flow Models: An Application to Aggregate Mode Choice, 1995 (this issue).

Boots, B.N and G.F. Royle, A Conjecture on the Maximum Value of the Principal Eigenvalue of a Planar Graph, *Geographical Analysis*, 23, 276-82, 1991.

Brett, C., J. Pinkse and M. E. Slade, *Spatial Price Competition, A Semiparametric Approach*, Paper Presented at the 41st North American Meetings of the Regional Science Association International, Niagara Falls, ON, 1994.

Burridge, P., On the Cliff-Ord Test for Spatial Autocorrelation, *Journal of the Royal Statistical Society B*, 42, 107-108, 1980.

Cliff, A. and J.K. Ord, *Spatial Processes: Models and Applications*, London: Pion, 1981.

Florax, R. and H. Folmer, Specification and Estimation of Spatial Linear Regression Models: Monte Carlo Evaluation of Pre-Test Estimators, *Regional Science and Urban Economics*, 22, 405-432, 1992.

Getis, A., Spatial Interaction and Spatial Autocorrelation: A Cross-Product Approach, *Environment and Planning A*, 23, 1269-1277, 1991.

Getis, A. and K. Ord, The Analysis of Spatial Association by Use of Distance Statistics, *Geographical Analysis*, 24, 189-206, 1992.

Griffith, D., An Evaluation of Correction Techniques for Boundary Effects and Missing Value Techniques, *Geographical Analysis*, 17, 81-88, 1985.

Griffith, D.A., *Advanced Spatial Statistics*, Dordrecht: Kluwer, 1988.

Hordijk, L., Spatial Correlation in the Disturbances of a Linear Interregional Model, *Regional and Urban Economics*, 4, 117-140, 1974.

Kelejian, H.H. and D.P. Robinson, Spatial Autocorrelation: A New Computationally Simple Test with an Application to per Capita County Police Expenditures, *Regional Science and Urban Economics*, 22, 317-331, 1992.

Moran, P.A.P., Notes on Continuous Stochastical Phenomena, *Biometrika*, 37, 17-23, 1950.

Stetzer, F., Specifying Weights in Spatial Forecasting Models: The Results of Some Experiments, *Environment and Planning A*, 14, 571-584, 1982.

Thursby, J.G. and P. Schmidt, Some Properties of tests for Specification Error in a Linear Regression Model, *Journal of the American Statistical Association*, 72, 635-641, 1977.

6 Computation of Box-Cox Transform Parameters:

A New Method and its Application to Spatial Econometrics

Reinaud A.J.J. van Gastel and Jean H.P. Paelinck
Erasmus University, Rotterdam, The Netherlands

6.1 Introduction

The reason for undertaking the present investigation is that, in spatial econometrics, as was argued in Paelinck and Klaassen (1979, pp. 6-9), econometric relations in space result more often than not in highly non-linear specifications. This is very probably even more the case if services or enterprise functions in space are considered [Paelinck (1987)].

One could have followed various paths to adress this difficulty, e.g., using so-called 'flexible forms' [Diewert and Wales (1987)]. It was judged however, and the cases treated below confirm this conjecture, that Box-Cox transforms [Box and Cox (1964) and earlier Tukey (1957)] are really very supple and allow for the treatment of extreme cases.

The real innovation has been not to use the initial specification:

$$y_i^{\sigma} = \sum_{k=1}^{K} a_k x_{ik}^{\rho_k} + b \tag{1}$$

where $i = 1, ..., N$, but to take elasticities instead (see section two) as a starting point for the new method to be developed. The first results were reported in Bailly et al. (1992). Some of those results, though representing an advancement of existing methods [see, for instance, Cook and Sanford (1982)], were judged to be not fully satisfactory, thus new research was undertaken, the first findings of which are being reported on here.

6.2 The Elasticity Method: Further Elaboration

The partial elasticities from (1) can be expressed as:

$$E_{ik} = \frac{a\rho_k}{\sigma} \cdot \frac{x_{ik}^{\rho_k}}{y_i^{\sigma}} \tag{2}$$

As y_i is endogenous, one could replace $y_i^{-\sigma}$ by an instrumental variable from a double logarithmic adjustment, $x_{ik}^{-a*\sigma}$, leading to:[1]

$$E_{ik} = a_k \rho_k \sigma^{-1} e^{(\rho_k - a_k^* \sigma) \ln x_{ik}} \tag{3}$$

and by linear expansion to:

$$E_{ik} = a_k \rho_k \sigma^{-1} + (a_k \rho_k \sigma^{-1})(\rho_k - a_k^* \sigma) \ln x_{ik} \tag{4}$$

As $\ln x_{ik}$ has been normalized (to zero), $a\rho_k\sigma^{-1}$ is the average elasticity when (4) is adjusted by OLS. Because the E_{ik}'s are bound to be rather erratic, a better estimate could be a_k*. From here on we will treat the one exogenous variable case (index k omitted), which gives:

$$a^* = a \rho \sigma^{-1} \tag{5}$$

or

$$\rho \sigma^{-1} = a^* a^{-1} \tag{6}$$

as a first result. This also suggests fitting (4) in first differences, with only one parameter α*, giving rise to:

$$\alpha^* = a^*(\rho - a^* \rho) \tag{7}$$

still leaving one parameter, a, to be estimated. It can be obtained from (1), given values for ρ and σ. Obvious starting values for an iterative process can be obtained for $a = a$*, leading to $\rho \equiv \sigma$. In fact an iteration was suggested in which two constraints were built in:

1) it should stop for a* inside the $a \pm t_a \sigma_a$ 95% bilateral interval; or
2) R^2 should not lie outside the 5% right-hand side F-interval for R^{*2}, belonging to the double-logarithmic adjustment; or both.

---

[1] a_k^* is the computed constant partial elasticity.

Table 6.1: Results from the estimation of (1).

	a^*	R^{*2}	α^*	ρ	σ	a	b	R^2	ρ_t	σ_t
(a)	0.856	0.989	0.156	1.264	1.264	0.857	0.111	0.979	1	1
(b)	0.483	0.991	0.205	0.819	0.819	0.497	0.462	0.976	0.6	1
(c)	0.320	0.919	-0.398	-1.827	-1.827	0.297	0.807	0.944	-1	-1
(d)	2.016	0.999	-0.271	0.132	0.132	1.938	-0.928	0.997	→0	→0

This procedure led to the results in Table 6.1, but the procedure stopped after only one iteration.[2] These results were rather encouraging, especially when interpreted qualitatively: (a) points to linear regression, (c) to inverse regression and (d) to double logarithmic regression; only (b) fails to produce a ρ/σ ratio different from unity.

The method can be extended to the multiple exogenous variable case. In order to do so (4) can be summed over k to give:

$$E_i \,\underline{\Delta}\, \sum_k \mathrm{E}_{ik} \tag{8}$$

$$= \sum_k a_k \rho_k \sigma^{-1} + \sum_k (a_k \rho_k \sigma^{-1})(\rho_k - a_k^* \sigma) \ln x_{ik} \tag{9}$$

$$\underline{\Delta}\, \theta + \sum_k \alpha_k^* \ln x_{ix} \tag{10}$$

where E_i is the local scale elasticity (Wicksell-Johnson theorem). In first backward differences:

$$\Delta' E_i = \sum_k \alpha_k^* \Delta' \ln x_{ik} \tag{11}$$

with again a 'best' estimate for $a_k\rho_k\sigma^{-1}$ being a_k* from the double logarithmic regression. So there are k relations for the α_k's and one for a_k* (cf. equations (5) and (6)), which together with the original function should allow computations of all the parameters, as explained above.[3] This case was not tested, but the method was applied to observed series.

---

[2] ρ_t and σ_t are the test values (see the data in section three).

[3] As a starting point for the iteration in E_i, an average E_i-series from σ and ρ_k, $\forall k$, put successively to -1 and 1, was taken.

In order to avoid the iteration problem mentioned above, the following path is now suggested,[4] exposed for the one exogenous variable case. From:

$$\Delta' E_i = a^*(\rho \Delta' \ln x_i - \sigma \Delta' \ln y_i) + \varepsilon_i \tag{12}$$

follows:

$$\Delta' E_i + a^* \rho \Delta' \ln y_i = a^* \rho \Delta' \ln x_i + \varepsilon_i \tag{13}$$

or:

$$z_i + a^* \sigma \tilde{y}_i = a^* \rho \tilde{x}_i + \varepsilon_i \tag{14}$$

or again:

$$(z_i \; \tilde{y}_i) \begin{pmatrix} 1 \\ a^* \sigma \end{pmatrix} = (\tilde{x}_i) \, a^* \rho + \varepsilon \tag{15}$$

In vector-matrix terms this becomes:

$$\boldsymbol{S} \mu = \boldsymbol{T} \nu + \varepsilon \tag{16}$$

The least squares estimator for ν is given by:

$$\hat{\nu} = (\boldsymbol{T}'\boldsymbol{T})^{-1} \boldsymbol{T}'\boldsymbol{S} \mu \tag{17}$$

for μ known. Substitution of ν in (16) gives:

$$\boldsymbol{S} \mu = \boldsymbol{T}(\boldsymbol{T}'\boldsymbol{T})^{-1} \boldsymbol{T}'\boldsymbol{S} \mu + \eta \tag{18}$$

and minimizing $\eta'\eta$ with a norm $\mu'\mu = c$ leads to:

$$(\boldsymbol{S}'\boldsymbol{M}\boldsymbol{S} - \lambda \boldsymbol{I})\mu = 0 \tag{19}$$

with:

$$\boldsymbol{M} \triangleq \boldsymbol{I} - \boldsymbol{T}(\boldsymbol{T}'\boldsymbol{T})^{-1}\boldsymbol{T}' \tag{20}$$

a (2×2) – in general $\left((k+1)\times(k+1)\right)$ – eigenvalue problem, the eigenvector μ belonging to the smallest non-zero eigenvalue $\lambda_{\min}$ (***M***, the well-known symmetric,

---

[4] Taken from Paelinck (1985, p. 246).

singular and idempotent projection matrix, being positive semi-definite). The coefficient of z_i should be scaled up to 1, to determine the absolute value of $a^*\sigma$, and substitution into (17) allows of computing $a^*\rho$. Coefficient a then follows from equation (5).

Generalization to the multiple exogenous variable case should follow the same lines combined with equations (10) and (11). Here, however, iteration is needed, not to compute the a_k's (they are defined from $a_k^* = a_k\rho_k\sigma^{-1}$), but to generate the E_{ik}'s and hence the E_i's.

Three types of biases can appear in applying the exercise, which leads to partial reconsideration of the method. First of all, a bias might be introduced by using a linear approximation as was the case in equation (4). Second, a bias might be generated by using so-called 'arc-elasticities', i.e., elasticities computed by logarithmic differences. It is well-known that they are nearly equivalent to 'point-elasticities' for the constant elasticity case. Finally, attention must be paid to some constraints imposed by Box-Cox specifications. These points will now be broken down in succession.

6.2.1 Linearization Bias

In order to find out whether (4) is fully appropriate for the four test cases to be presented in detail in section three (see equations (29) - (32)), the exactly computed E_i's and the linear approximations to them are compared in absolute and relative terms, the latter being quite important (see Table 6.2). Remark that, for the computation of the E_i's and their linear expansions, the *observed* values of the x_i's and y_i's are taken.

6.2.2 Discretization Bias

Another biasing approximation might be the substitution, for the computation of the E_i's, of the differential quotient by the difference quotient. Table 6.3 compares the results of both approaches. Again large differences are observable, though the means are just slightly affected, except for equation (31). Again observed values of the x_i's and y_i's are taken.

This leads to the idea of reducing the intervals between the variables (later on the x_i's, for instance, have been reduced by a factor 100 and the y_i's have been adapted correspondingly). As will be seen below, the results improve considerably.

Table 6.2: Linearization bias for the models given in (29) - (32).

i	E_i (computed)	Linear expansion	Difference	Relative difference (%)
Equation (29)				
1	0.6061	-0.3878	-0.9939	-164
2	0.7692	0.0890	-0.6803	-88
3	0.8696	0.3342	-0.5354	-62
4	0.9412	0.4925	-0.4487	-48
5	0.9259	0.4598	-0.4661	-50
6	0.9160	0.4383	-0.4777	-52
7	0.8805	0.3592	-0.5213	-59
8	0.8791	0.3560	-0.5231	-60
9	1.0000	0.6137	-0.3863	-39
10	0.8889	0.3781	-0.5107	-57
11	0.9167	0.4397	-0.4770	-52
12	0.8889	0.3781	-0.5107	-57
13	1.0400	0.6921	-0.3479	-33
14	1.0219	0.6570	-0.3649	-36
mean	0.8960	0.3786	-0.5174	-58
Equation (30)				
1	0.3871	-0.1577	-0.5448	-141
2	0.4748	0.0873	-0.3875	-82
3	0.4767	0.0923	-0.3845	-81
4	0.4841	0.1106	-0.3735	-77
5	0.4737	0.0847	-0.3890	-82
6	0.5360	0.2329	-0.3031	-57
7	0.5056	0.1629	-0.3428	-68
8	0.5314	0.2224	-0.3089	-58
9	0.5486	0.2609	-0.2878	-52
10	0.5158	0.1868	-0.3290	-64
11	0.5468	0.2568	-0.2900	-53
12	0.5286	0.2161	-0.3125	-59
13	0.5525	0.2693	-0.2832	-51
14	0.5652	0.2965	-0.2687	-48
mean	0.5091	0.1658	-0.3432	-67

Table 6.2: Continued.

i	E_i (computed)	Linear expansion	Difference	Relative difference (%)
Equation (31)				
1	0.7160	-0.0544	-0.7704	-108
2	0.5320	-0.6485	-1.1805	-222
3	0.3727	-1.3604	-1.7331	-465
4	0.3445	-1.5176	-1.8621	-541
5	0.2624	-2.0621	-2.3245	-886
6	0.2223	-2.3934	-2.6158	-1177
7	0.2203	-2.4120	-2.6322	-1195
8	0.2180	-2.4328	-2.6508	-1216
9	0.1909	-2.6984	-2.8893	-1514
10	0.1552	-3.1124	-3.2676	-2105
11	0.1625	-3.0199	-3.1824	-1958
12	0.1487	-3.1984	-3.3471	-2251
13	0.1363	-3.3720	-3.5083	-2574
14	0.1319	-3.4384	-3.5702	-2708
mean	0.2724	-2.2658	-2.5382	-932
Equation (32)				
1	1.9820	1.9820	-0.0001	-0
2	1.9658	1.9655	-0.0003	-0
3	1.9572	1.9568	-0.0005	-0
4	1.9542	1.9537	-0.0005	-0
5	1.9478	1.9471	-0.0007	-0
6	1.9442	1.9434	-0.0008	-0
7	1.9436	1.9428	-0.0008	-0
8	1.9385	1.9375	-0.0010	-0
9	1.9374	1.9364	-0.0010	-0
10	1.9354	1.9343	-0.0011	-0
11	1.9330	1.9318	-0.0011	-0
12	1.9309	1.9296	-0.0012	-0
13	1.9296	1.9283	-0.0013	-0
14	1.9276	1.9263	-0.0013	-0
mean	1.9448	1.9440	-0.0008	-0

Table 6.3: Discretization bias for the models given in (29) - (32).

i	E_i (computed)	Linear expansion	Difference	Relative difference (%)
Equation (29)				
1	0.6061	0.6560	0.0500	8
2	0.7692	0.6976	-0.0716	-9
3	0.8696	0.7249	-0.1447	-17
4	0.9412	1.0732	0.1320	14
5	0.9259	1.0589	0.1330	14
6	0.9160	1.2566	0.3406	37
7	0.8805	1.0118	0.1313	15
8	0.8791	-0.0938	-0.9729	-111
9	1.0000	2.1179	1.1179	112
10	0.8889	0.6771	-0.2117	-24
11	0.9167	1.3537	0.4370	48
12	0.8889	-0.9615	-1.8504	-208
13	1.0400	1.2369	0.1969	19
14	1.0219	1.3140	0.2921	29
mean	0.8960	0.8660	-0.0300	-3
Equation (30)				
1	0.3871	0.3055	-0.0816	-21
2	0.4748	0.5898	0.1150	24
3	0.4767	0.5468	0.0701	15
4	0.4841	0.6968	0.2127	44
5	0.4737	-0.0772	-0.5510	-116
6	0.5360	0.9785	0.4425	83
7	0.5056	0.2282	-0.2774	-55
8	0.5314	0.3283	-0.2031	-38
9	0.5486	1.1860	0.6374	116
10	0.5158	-0.0125	-0.5283	-102
11	0.5468	0.9897	0.4429	81
12	0.5286	0.0458	-0.4828	-91
13	0.5525	0.2941	-0.2584	-47
14	0.5652	0.7935	0.2283	40
mean	0.5091	0.4924	-0.0167	-3

Table 6.3: Continued.

i	E_i (computed)	Linear expansion	Difference	Relative difference (%)
Equation (31)				
1	0.7160	0.5715	-0.1445	-20
2	0.5320	0.1221	-0.4099	-77
3	0.3727	0.7268	0.3541	95
4	0.3445	-0.2200	-0.5645	-164
5	0.2624	0.0912	-0.1712	-65
6	0.2223	0.9400	0.7176	323
7	0.2203	0.9219	0.7016	318
8	0.2180	-0.1275	-0.3455	-158
9	0.1909	-0.9645	-1.1554	-605
10	0.1552	1.4852	1.3300	857
11	0.1625	-0.0257	-0.1883	-116
12	0.1487	-0.0843	-0.2330	-157
13	0.1363	0.5521	0.4158	305
14	0.1319	-1.6621	-1.7940	-1361
mean	0.2724	0.1662	-0.1062	-39
Equation (32)				
1	1.9820	2.1892	0.2072	10
2	1.9658	2.0740	0.1082	6
3	1.9572	1.5399	-0.4173	-21
4	1.9542	2.4583	0.5042	26
5	1.9478	2.0372	0.0893	5
6	1.9442	1.1810	-0.7632	-39
7	1.9436	2.9822	1.0386	53
8	1.9385	1.4626	-0.4758	-25
9	1.9374	2.0156	0.0781	4
10	1.9354	2.3020	0.3666	19
11	1.9330	2.2366	0.3036	16
12	1.9309	1.8342	-0.0967	-5
13	1.9296	2.3569	0.4274	22
14	1.9276	0.5564	-1.3712	-71
mean	1.9448	1.9447	-0.0001	-0

6.2.3 Specification Bias

Consider again equation (1): y_i^σ can only be negative for $\sigma \to 0$ or $|\sigma| \in \mathbb{N}$. In that case also, negative values can be allowed for the x_{ik}'s having $\rho_k \to 0$ or $|\rho_k| \in \mathbb{N}$, the parameters a_k and b can have any sign and an additive error can be conceived of. As soon as one is outside those restrictions, the correct stochastic specification should be, all a_k's and b non-negative:

$$y_i^\sigma = \left(\sum_k a_k x_{ik}^{\rho_k} + b \right) e^{\varepsilon_i} \tag{21}$$

and if $\varepsilon_i \sim N(0,\sigma^{*2})$:

$$E(y_i^\sigma) = \left(\sum_k a_k x_{ik}^{\rho_k} + b \right) e^{\sigma^{*2}/2} \tag{22}$$

which suggests suitable corrections for the estimated a_k's and b, given expectations.
The elasticity, for the one exogenous variable case, is now:

$$E_i = \frac{a\rho}{\sigma} \frac{x_i^\rho}{y_i^\sigma} e^{\varepsilon_i} \tag{23}$$

which, on substituting y_i^σ from (21), is seen to be non-stochastic. One can further specify:

$$E_i^{obs} = \frac{a\rho}{\sigma} \frac{x_i^\rho}{y_i^\sigma} (e^{\eta_i} - e^{\sigma^{**2}/2} + 1) \tag{24}$$

allowing for negative values of E_i^{obs}; one supposes again $\eta \sim N(0,\sigma^{**2})$. Squaring (24) gives:

$$\left(E_i^{obs}\right)^2 = \left(\frac{a\rho}{\sigma}\right)^2 \frac{x_i^{2\rho}}{y_i^{2\sigma}} \left(e^{\eta_i} - e^{\sigma^{**2}} + 1\right)^2 \tag{25}$$

with:

$$\left(e^{\eta_i} - e^{\sigma^{**2}} + 1\right)^2 \equiv e^{\theta_i} \tag{26}$$

In fact, this development is equivalent to taking *absolute* values of the observed elasticities. Taking logarithms leads to:

$$\ln|E_i^{obs}| = \ln|\frac{a\rho}{\sigma}| + \rho \cdot \ln x_i - \sigma \cdot \ln y_i + \theta_i \tag{27}$$

to which the procedure given in (16) and further can be applied, starting from:

$$(\ln|E_i^{obs}| \; \ln y_i)\begin{pmatrix} 1 \\ \sigma \end{pmatrix} = (1 \; \ln x_i)\begin{pmatrix} \ln|\frac{a\rho}{\sigma}| \\ \rho \end{pmatrix} \tag{28}$$

a and *b* can then be obtained from (21), using a procedure leading to non-negative estimators, for instance the SPCE (strictly positive conditional estimator) method [see Paelinck (1990)].

The generalization to the multiple exogenous variable case will be presented in section five.

6.3 The One Exogenous Variable Test

The revised method of sections 6.2.2 and 6.2.3 was applied to the four test cases, already treated in Bailly et al. (1992) quoted in the introduction, and reported hereafter:

$$y_i = 2x_i + 1 \tag{29}$$

$$y_i = 2x_i^{0,6} + 1 \tag{30}$$

$$y_i^{-1} = 2x_i^{-1} + 1 \tag{31}$$

$$\ln y_i = 2\ln x_i + 1 \tag{32}$$

The exact data given in Table 6.4 were used and the ε_i's were chosen at random so as to not exceed, in absolute value, 10 percent of the exact values of the y_i's. Figure 6.1 pictures the apparent stochastic y_i - x_i relations and Table 6.5 shows the corresponding arc-elasticities.

Hereafter (Table 6.6) the results of the exercise follow for $x_i = 1 + (i - 1)/100$ as suggested in section 2.3. The following points should be noted:

1) if σ is exactly known, equation (17) approaches the value of ρ correctly;[5]

---

[5] In fact, when the analytically computed E_i's are used, the program performs identically; this was run as a test on the model.

2) the smallest eigenvalue gives the correct solution, except for test-case (32); the reason for this is that the eigenvalue equations for σ cannot be satisfied for an eigenvalue very close to zero; and
3) as far as experience goes, larger error terms would disturb the estimated values, but this point is now being investigated.

Table 6.4: Test data.

x_i	$2x_i + 1 + \varepsilon_i$	$2x_i^{0.6} + 1 + \varepsilon_i$	$[2x_i^{-1} + 1]^{-1} + \varepsilon_i$	$\exp[2\ln x_i + 1] + \varepsilon_i$
1	3.300	3.100	0.358	2.465
2	5.200	3.831	0.532	11.242
3	6.900	4.866	0.559	26.065
4	8.500	5.695	0.689	40.593
5	10.800	6.653	0.656	70.257
6	13.100	6.560	0.667	101.858
7	15.900	7.628	0.771	122.196
8	18.200	7.864	0.872	181.970
9	18.000	8.174	0.859	216.181
10	22.500	9.262	0.776	267.328
11	24.000	9.251	0.894	332.912
12	27.000	10.083	0.892	404.433
13	25.000	10.120	0.886	468.390
14	27.400	10.343	0.923	557.783
15	30.000	10.925	0.823	579.613

Table 6.5: Arc-elasticities.

	(29)	(30)	(31)	(32)
E_1	0.6560	0.3055	0.5715	2.1892
E_2	0.6976	0.5898	0.1221	2.0740
E_3	0.7249	0.5468	0.7268	1.5399
E_4	1.0732	0.6968	-0.2200	2.4583
E_5	1.0589	-0.0772	0.0912	2.0372
E_6	1.2566	0.9785	0.9400	1.1810
E_7	1.0118	0.2282	0.9219	2.9822
E_8	-0.0938	0.3283	-0.1275	1.4626
E_9	2.1179	1.1860	-0.9645	2.0156
E_{10}	0.6771	-0.0125	1.4852	2.3020
E_{11}	1.3537	0.9897	-0.0257	2.2366
E_{12}	-0.9615	0.0458	-0.0843	1.8342
E_{13}	1.2369	0.2941	0.5521	2.3569
E_{14}	1.3140	0.7935	-1.6621	0.5564
$\bar{E}$	0.8660	0.4924	0.1662	1.9447

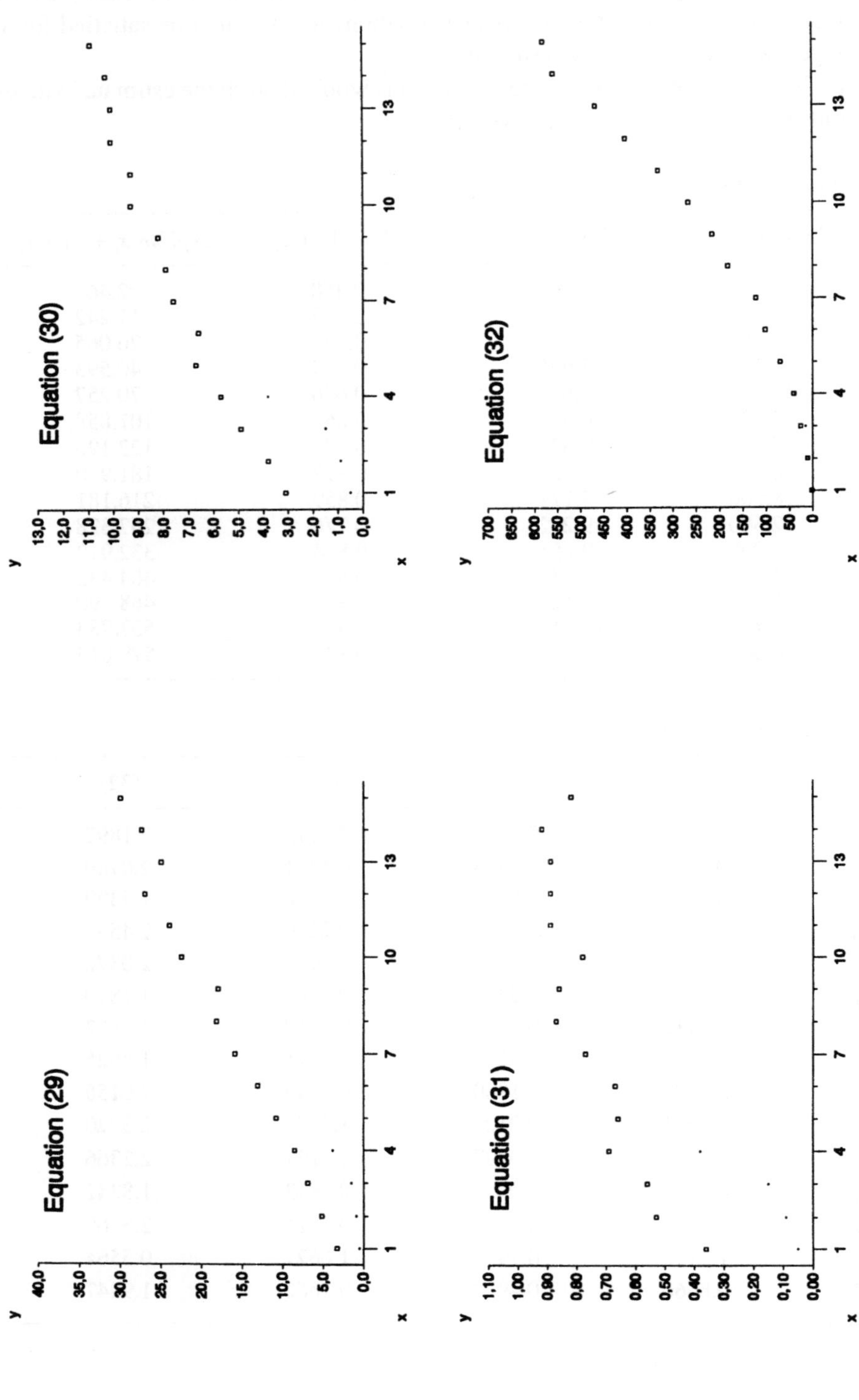

Figure 6.1: The apparent y_i - x_i relations.

Table 6.6: The results of the exercise for $x_i = 1 + (i - 1)/100$.

Equation 29:					
λ	:	1.3213591526E-07			
$\hat{\sigma}$	:	-1.01840		$\hat{\rho}$	: -3.74590
λ	:	1.1933343647E-10			
$\hat{\sigma}$	:	0.98193		$\hat{\rho}$	: 0.98522
σ	:	1.0	then	ρ	: 0.9975
Equation 30:					
λ	:	1.7356493034E-08			
$\hat{\sigma}$	:	-1.02169		$\hat{\rho}$	: -0.21956
λ	:	2.7452325393E-11			
$\hat{\sigma}$	:	0.97877		$\hat{\rho}$	: 0.59012
σ	:	1.0	then	ρ	: 0.5987
Equation 31:					
λ	:	1.4589752504E-07			
$\hat{\sigma}$	:	1.00065		$\hat{\rho}$	: 0.30733
λ	:	1.9430565663E-11			
$\hat{\sigma}$	:	-0.99935		$\hat{\rho}$	: -0.99905
σ	:	-1.0	then	ρ	: -0.9995
Equation 32:					
λ	:	2.5915843492E-09			
$\hat{\sigma}$	:	-0.00929		$\hat{\rho}$	: -0.01856
λ	:	4.2243563926E-13			
$\hat{\sigma}$	:	107.61720		$\hat{\rho}$	: 215.23904
σ	:	0.0001	then	ρ	: 0.0002

Taking averages between succeeding values of the variables y_i and x_i improves the results even further (for small steps of x_i).

6.4 An Application to Spatial Econometrics

In Paelinck and Nijkamp (1975) some work was reported on the spatial correlation properties of economic data. Lebart (1966) initiated this type of analysis using French departmental data. The method exposed above was used to investigate the relationship between regional products in the Dutch provinces as a function of their degree of contiguity. From Figure 6.2 one can easily determine the contiguity matrix of order 1, which has a one in a cell if two provinces have a degree of contiguity of order 1, and zero in all other cases (see Table 6.7). Let it be recalled that two regions have a degree of contiguity 1, if one needs to cross one border to reach the other region.

Table 6.7: Contiguity matrix of order 1 for the Dutch provinces.[a,b]

	Gr	Fr	Dr	Ov	Fp	Gl	Ut	NH	ZH	Zl	NB	Lb
Gr	0	1	1	0	0	0	0	0	0	0	0	0
Fr	1	0	1	1	1	0	0	0	0	0	0	0
Dr	1	1	0	1	0	0	0	0	0	0	0	0
Ov	0	1	1	0	1	1	0	0	0	0	0	0
Fp	0	1	0	1	0	1	1	1	0	0	0	0
Gl	0	0	0	1	1	0	1	0	1	0	1	1
Ut	0	0	0	0	1	1	0	1	1	0	0	0
NH	0	0	0	0	1	0	1	0	1	0	0	0
ZH	0	0	0	0	0	1	1	1	0	1	1	0
Zl	0	0	0	0	0	0	0	0	1	0	1	0
NB	0	0	0	0	0	1	0	0	1	1	0	1
Lb	0	0	0	0	0	1	0	0	0	0	1	0

[a] In constructing the matrix we have ignored the so-called *Afsluitdijk*, the dam between two northern provinces.
[b] See Table 6.9 for the meaning of the symbols.

From this matrix, one can compute the matrix of contiguity degrees (minimum number of borders to be crossed), which is given in Table 6.8.

Figure 6.2: The Dutch provinces.

Table 6.8: Matrix of contiguity degrees for the Dutch provinces.

	Gr	Fr	Dr	Ov	Fp	Gl	Ut	NH	ZH	Zl	NB	Lb
Gr	0	1	1	2	2	3	3	3	4	5	4	4
Fr	1	0	1	1	1	2	2	2	3	4	3	3
Dr	1	1	0	1	2	2	3	3	3	4	3	3
Ov	2	1	1	0	1	1	2	2	2	3	2	2
Fp	2	1	2	1	0	1	1	1	2	3	2	2
Gl	3	2	2	1	1	0	1	2	1	2	1	1
Ut	3	2	3	2	1	1	0	1	1	2	2	2
NH	3	2	3	2	1	2	1	0	1	2	2	3
ZH	4	3	3	2	2	1	1	1	0	1	1	2
Zl	5	4	4	3	3	2	2	2	1	0	1	2
NB	4	3	3	2	2	1	2	2	1	1	0	1
Lb	4	3	3	2	2	1	2	3	2	2	1	0

Table 6.9: Value added (in millions of Dutch guilders) in the Dutch provinces [CBS (1987)].

Groningen	(Gr)	22675	Utrecht	(Ut)	27801
Friesland	(Fr)	13229	Noord-Holland	(NH)	77997
Drente	(Dr)	11187	Zuid-Holland	(ZH)	102864
Overijssel	(Ov)	25448	Zeeland	(Zl)	10908
Flevoland	(Fp)	3703	Noord-Brabant	(NB)	59242
Gelderland	(Gl)	43861	Limburg	(Lb)	29036

Formally the model presents itself as:

$$r_i^{\sigma} = a(r_{ic}^{*})^{\rho} + b \tag{34}$$

where r_i is regional product in province i, and r_{ic}^* the average regional product in provinces contiguous of degree c to province i.[6]

Elasticities, as used in (28), were computed over relevant provinces with contiguity order one, meaning that other cross-terms have not been considered.[7]

[6] In fact, being themselves stochastic, they should have been moved to the left hand side of (28); this correction has not (yet) been performed.

[7] I.e., one only considers E_{ij} for $i > j$ and $C_{ij} = 1$ (C being the contiguity matrix and c the degree of contiguity):

$$E_{ij} \triangleq \frac{\ln r_j - \ln r_i}{\ln r_{jc}^* - \ln r_{ic}^*}$$

Table 6.10 hereafter shows the results of the computations and compares them with the contiguity correlation coefficients c_j given in Paelinck and Nijkamp quoted above (year 1965, regional *per capita* products).

Table 6.10: Results of computations.

Observed elasticities[a]	Order of contiguity	σ	ρ		a	E	c_j
22	1	-4.63	-3.76	(1.77)	- (+)	- (+)	0.92
22	2	-10.83	17.93	(0.02)	- (-)	+ (-)	1.15
20	3	8.89	-8.36	(-1.50)	+ (+)	- (-)	1.10
4	4	-0.71	-2.01	(-4.13)	+ (-)	+ (+)	0.25

[a] This number declines due to the equality of different explanatory variables; for example: $r^*_{Fp,3} = r^*_{Ov,3}$, leading to the impossibility to compute the observed elasticity. Those observations have been skipped.

Using (2) one can easily interpret the results. For the computed σ and ρ the contiguity effects are respectively -, +, -, and + ; for the $\sigma = 1$ constraint (figures between parentheses) they would be +, -, - and +. The c_j's show respectively +, -, - and + effects; they agree with the $\sigma = 1$ case, but it should be remembered that they refer to 1965 and per capita data. In fact the computation of the non-linear model complements that of the spatial autocorrelation coefficients, which only show the sign and strength of the (linear) relationship, but not its eventual specific form.

6.5 The Multiple Exogenous Variable Computation

The method exposed in section three has now to be adapted, as no partial E_{ik}'s are observable. As a starting point one can take:

$$E_{il}^{obs} = \frac{\Delta' \ln y_i - \sum_{k \neq l} a_k^* \Delta' \ln x_{ik}}{\Delta' \ln x_{il}} \tag{35}$$

the a_k*'s being defined in footnote 1. Generalizing (27) one gets:

$$\sum_k \Delta' \ln |E_{ik}^{obs}| = \sum_k \rho_k \cdot \Delta' \ln x_{ik} - K\sigma \cdot \Delta' \ln y_i \tag{36}$$

to be treated according to (16) and further; the a_k's and b can be derived from (21) as specified in section 3.2, and this could lead to an iterative procedure using (2) to compute partial elasticities.

6.6 Conclusions

The research reported on in this chapter has allowed the discovery of relevant aspects of Box-Cox transform estimation, in particular its sensitivity to linearization, discretization and specification.

Though advances have been made with respect to previous trials, various points remain to be further investigated: the multiple regression case, the influence of stochastic terms on the values of the estimators, and a detailed study of the influence of the step length. Work along some of these lines is now in progress.

References

Bailly, A.S., W.J. Coffey, J.H.P. Paelinck and M. Polèse, *The Spatial Econometrics of Services*, Aldershot: Avebury, 1992.

Box, G. and D. Cox, An Analysis of Transformations, *Journal of the Royal Statistical Society*, 16B, 211-243, 1964.

Centraal Bureau voor de Statistiek (CBS), Regionale Economische Jaarcijfers, Den Haag: CBS, 1987.

Cook, R.D. and W. Sandfort, *Residuals and Influence in Regression*, New York: Chapman and Hall, 1982.

Diewert, W.E. and T.J. Wales, Flexible Functional Forms and Global Curvature Conditions, *Econometrica*, 55, 43-68, 1987.

Lebart, L., *Les Variables Socio-Economiques Départementales ou Régionales*, 1, Paris: Méthodes Statistiques, IEDES, 1966.

Paelinck, J.H.P., *Elements d'Analyse Economique Spatiale*, Paris: Economica, 1985.

Paelinck, J.H.P., Quelques Modèles Fonctionnels de Localisation, *Les Cahiers Scientifiques du Transport*, 15/16, 203-212, 1987.

Paelinck, J.H.P., Empirical Results with Some New Estimators in Spatial Econometrics, *Przegląd Statystyczny*, R. XXXVII, zeszyt 4, 257-265, 1990.

Paelinck, J.H.P. and L.H. Klaassen, *Spatial Econometrics*, Farnborough: Saxon House, 1979.

Paelinck, J.H.P. and P. Nijkamp, *Operational Theory and Methods in Regional Economics*, Farnborough: Saxon House, 1975.

Tukey, J.W., On the Comparative Anatomy of Transformations, *Annals of Mathematical Statistics*, 28, 602-632, 1957.

7 Data Problems in Spatial Econometric Modeling

Robert P. Haining
The University of Sheffield, Sheffield, U.K.

7.1 Introduction

Spatial econometrics is a field of study concerned with the development of methods and techniques of data analysis that are appropriate for fitting models arising in regional science [Anselin (1988, p. 10)]. Regional science models relate to events occurring in geographic space and a particular focus of spatial econometrics is to develop methods and techniques that recognize the special problems that arise when fitting models to spatial cross sectional data.

Consider the problem of fitting a regression model to spatial data using the ordinary least squares (OLS) fitting procedure. The regression model is defined:

$$Y = X\beta + u \tag{1}$$

where Y is a vector of length n of the response or dependent variable and X is an $(n \times (k+1))$ array of k predictor or independent variables, where the i-th row corresponding to the i-th case (location or area) is:

$$(1\ X_{i,1}\ X_{i,2} \cdots X_{i,k}) \tag{2}$$

The vector β is of length $(k+1)$ corresponding to $(k+1)$ unknown parameters and u is of length n and is a vector of independent statistical errors drawn from a distribution with a mean of zero and (constant) variance σ^2.

A simple typology of problems that can arise in implementing the fit would distinguish those that arise when the statistical assumptions of OLS are not met from those that arise due to the nature of the data.

There are several statistical assumptions that need to be satisfied if the OLS fitting procedure is to be considered on firm ground. These assumptions include normality of the errors; constant error variance (or homoscedastic errors); independent errors. Considerable effort has been expended in geography, regional science and statistics to identify the consequences of violating these assumptions, develop diagnostics for detecting when such assumptions have been violated and define appropriate remedial action. For recent reviews see, for example Anselin (1988), Haining (1990) and Cressie (1991).

By contrast, rather less attention seems to have been paid to what will be termed here, data problems. These problems include what can be broadly defined as problems that do not violate any statistical assumptions of the model but which, because of the characteristics of the methodology (which some might regard as defects of the methodology), mean that model fitting results ought to be treated with caution. In spite of what may appear a good fit the analyst should consider revising the model or changing the fitting procedure.

The aim of this chapter is to identify various data problems in spatial econometric regression modeling and review progress to date in identifying their consequences, developing diagnostics to detect them (where necessary) and specifying appropriate remedial action.

In an earlier, briefer statement on some of these issues [Haining (1990b, p. 105)], Doreian, in discussion, commented on the need for good database management systems if some of the more serious problems were to be handled efficiently. The point was an important one and in this chapter we will touch on this issue.

Section two describes the nature of spatial data and classifies the main types of spatial econometric data. Section three is a discussion of the main data problems that can arise in spatial econometrics and their consequences. We distinguish between what we term 'generic' and 'technique specific' (in the present case 'regression specific') problems. Section four discusses methodologies for handling data problems including a discussion of diagnostics and appropriate remedial action. The problem of influential cases in regression is described and an example is given. Section five briefly comments on the implementation of the methodologies of section four.

7.2 Data for Spatial Econometric Analysis

Because of its emphasis on quantitative data analysis and model fitting, the conduct of spatial econometrics is greatly influenced by how data are stored in digital data bases, that is, how information about the world is abstracted and discretized into a small enough number of data units for input into computing systems. These processes of abstraction (depiction) and discretization are termed *data modeling*. Goodchild (1992) defines two fundamental geographic data models – the field view and the object view. Briefly, in the field view the world is depicted as a set of layers each defining the continuous or quasi continuous spatial variation of one layer. Discretization results in a set of regular or irregular sample points, contours or a spatial partition. In the object view the world is depicted as an empty space populated by points, lines and areas. The objects are the elements of the discretization. In both cases, attribute values are attached to the point, line and area features that arise from the processes of data modeling.

Both views, or perspectives arise in spatial econometrics. The field view may be appropriate for representing certain features of an urban area, for example land value surfaces or population density surfaces. The set of dwelling units may be depicted as a quasi-continuous surface with attributes attached to small area groupings of the units. The object view arises in, for example, representing the distribution of retail sites of a particular type within a city (points), transport links within a region (lines) and a set of towns in a region (areas).

Spatial econometric modeling is much concerned with the description, explanation and prediction of patterns of spatial variation and modeling may engage variables of the same data type (e.g., unemployment rates as a function of socio-economic characteristics of areas) or of different data types (e.g., models that seek to link patterns of spatial variation and spatial interaction [see Bennett et al. (1985)].

Modeling may be 'vertical' or 'horizontal': vertical in the sense of explaining events at a location in terms of other attributes at the same location; horizontal in the sense of explaining events at a location in terms of other attributes at the same location and other, usually geographically neighboring, locations. The latter type of structure arises where processes of spatial diffusion, interaction or competition occur [Haining (1990a)]. In a regression context we shall refer to the second type of model as a spatial regression model where the information from neighboring locations may be values of the dependent variable (a lagged dependent variable model), one or more of the independent variables (a lagged independent variables model) or the errors (a correlated errors model).

Given this typology of data types and in particular the origins of each data type from a particular representation of geographic reality for purposes of computational analysis, we now consider the data problems that can arise.

7.3 Data Problems in Spatial Econometrics

Data problems may be associated either with attribute values or with the areal system constructed by the processes of data modeling. Since the focus of this chapter is regression modeling we shall distinguish between those problems that arise because of the type of methodology employed in fitting a regression model from those that are of a more generic nature and yet still impact on the interpretation of regression results. We consider generic data problems first; how they arise and the forms they take.

Generic data problems can arise because of the way reality is represented for purposes of storing it in a digital database. They can also arise as a result of the activities of data collection (especially the process of measurement) and data processing (storing, manipulating, retrieving) in computerized systems. We shall term these types of generic data problem: source problems, collection problems and processing problems, respectively.

From the point of view of analysis, such as carrying out a regression analysis, the form the data problem takes is also important. The most extreme form of data problem occurs when values are missing. For recorded values we distinguish between errors, approximations and ambiguities.

We define an error as a departure from a definable true state of affairs. For example, given a definition of unemployment, there is presumably a true number of people unemployed in a district of a city at a given time.[1] We define an approximation as a loss of information that arises from a deficiency in the representation of an event or object. For example, assigning a single unemployment rate to a district when rates are known to vary within the district. The attribute value is not necessarily wrong in any absolute sense, the deficiency stems from attributing the rate to the whole district. Finally, ambiguity arises because of the absence of a unique way of operationalizing a chosen representation of an event or object. In the above example there is no unique districting scheme for a city at a given scale.

Errors generally arise from failures associated with the activities of collection and processing. Approximations and ambiguities, as defined here, arise from the processes of data modeling; that is, abstracting and discretizing that which is real and continuous.

Table 7.1 provides some examples of generic data problems. The problems are classified into those associated with attributes, the areal system or both and further classified by how they arise and the forms they take.

If generic data problems are problems for all forms of data analysis then the next set of data problems discussed are specific to regression modeling. They are data problems only because of the way regression modeling is implemented. These problems do not violate any statistical assumptions and should be seen as quite separate from the concern to identify when statistical assumptions have not been met. Failure to satisfy statistical assumptions often gives rise to well defined consequences. Data problems by contrast may give rise to effects that are sometimes difficult to pin down precisely.

Again, it is useful to distinguish data problems associated with attribute values from those associated with the areal system. We consider the most important types.

- Attribute values may be such that there is strong multicollinearity (or interdependence) amongst the independent variables. In non-experimental sciences this can be a severe problem because the opportunity to collect additional data on the variables and thereby reduce multicollinearity effects is usually not available. The

---

[1] Of course for all practical purposes it may not be possible to know precisely the true value but this does not remove the need to try and assess the scale of any possible over or under counting.

Table 7.1: Examples of generic data problems.[a]

	Problems	Examples
I	*Problems with attributes*	
	Source problems	Areal aggregates or rates may conceal heterogeneity within the area. (I)
	Collection problems	Measurement error. Missing values. Unreliable values. Sampling error where measured values are based on a sample of individuals.
	Processing problems	Keyboard error. Machine rounding error. Error propagation through different manipulations and map operations.
II	*Problems with areal framework*	
	Source problems	Use of line boundaries to delimit subareas in regions where there are no breaks only transitions. (I) Specification of the boundary or edge of the study region. (I or A)
	Collection problems	Errors in laying down or recognizing ground markers used to define the areal system.
	Processing problems	Errors in digitizing boundaries. Errors in recording point locations.
III	*Area/Attribute Interactions*	
	Source problems	Sensitivity of attributes to chosen partition. (A) Variation in area sizes resulting in small number problems in some areas. (I)
	Collection problems	Measurement error arising from failure to identifying correctly which individuals lie inside a spatial unit.
	Processing problems	Wrong attribute value recorded because of the inaccurate location of a point site.

[a] Please note that A indicates ambiguity and I approximation.

effects of multicollinearity are that the precision of estimates falls thus making it difficult to disentangle the contribution of different independent variables. Parameter estimates prove to be highly sensitive to particular data sets and variables may be incorrectly dropped from the model. There is an extensive literature in this area [see, for example Weisberg (1985)].

- Where there are many variables not only may multicollinearity be a problem but in addition it may be difficult to sift variables to select the best fitting regression model. Leamer (1978) discusses different types of specification searches at length.
- Individual data values may have a disproportionate influence on the fit of a model so that if deleted the results of the regression analysis would change. This is often referred to as the problem of influential cases. The influence of a case is a composite measure based on its leverage (a measure in the $\boldsymbol{X}$ domain) and the size of its residual (a measure in the Y domain). We shall consider the problem of influence, which can be an important criterion for assessing the validity of a model, in the next section.
- In considering generic data problems mention was made of missing values and unreliable values (Table 7.1). Such problems have a particularly important effect in regression modeling if they necessitate discarding entire rows in $\boldsymbol{X}$ corresponding to the sites or areas where such missing or unreliable values occur. Apart from reducing the degrees of freedom for fitting the model data deletion of this type may undermine the comparability of regression studies carried out in the same region for two different time periods. If most of the missing or unreliable values are associated with a single variable, deleting that variable when it is part of the correct specification of the model, will lead to biased estimates for the other parameters.

The areal system can also give rise to data problems in regression. The distribution of data points can influence trend surface modeling [Unwin and Wrigley (1987)]. Isolated points on the surface have high leverage and can have a disproportionate effect both on parameter estimates and the order of model fit.

In spatial regression modeling the areal system raises particular problems for the analyst. Spatial relationships between the areas or sites must be defined and the lack of a unique representation for such relationships means that results may be highly sensitive to what may be an ad hoc specification. The effect is felt directly through the analyst's need to define the matrix $\boldsymbol{W}$ [see section four and for an extended discussion Haining (1990a, pp. 69-74)]. Further, results will not only be dependent on the specification of the study region, an issue which arises in fitting the regression model defined by (1), but also on how conditions are defined at the edge or boundary of the study region. The boundary may be a natural one in terms of the process being modeled or it may be an artificial one in the sense that the process extends beyond the limits of the study region [Haining (1990a, pp. 101-110)].

7.4 Methodologies for Handling Data Problems

Detection of some data problems, particularly generic problems often depend on a close detailed knowledge of places or events. A data error may not stand out (in any statistical sense) so that knowing when an attribute value is wrong may ultimately depend on knowing the study region. The same local knowledge may be needed to detect if an areal partition has filtered out important local variations in a variable in parts of a study region.

In the case of data problems associated with regression, many diagnostics have been developed that automate the process of detection, setting criteria for when the data problem is likely to severely undermine the results of the modeling. Weisberg (1985) provides an extensive review of methods for detecting residual outliers, observations with high leverage, influential cases and multicollinearity in the regression model.

It is also important to develop remedial action to deal with data problems. Little and Rubin (1987) review methods of analysis in the presence of missing data. When data values are wrong they will need to be deleted from the data set. If results are sensitive to the areal partition and if there is access to data at smaller scales, the analyst may choose to look at the sensitivity of regression results to alternative spatial frameworks by constructing new regionalizations. In the case of the regression model defined by (1) an extensive literature has evolved for dealing with data problems. When outliers or extreme values are present resistant exploratory tools and resistant estimation and fitting procedures are available for data description while robust methods have been developed for confirmatory analyses [Hoaglin et al. (1985)]. Data deletion methods can be employed to assess the sensitivity of model fitting to specific observations. Li (1985, p. 290) notes that in dealing with these types of problems it is necessary to recognize that 'asking for a *single fit* will often be a mistake'. There is a need to increase the number and diversity of solutions perhaps summarizing results in the form of one 'standard' and one 'robust' fit.

It is not the case that procedures developed to handle data problems arising in the case of fitting a model such as (1) are automatically appropriate in the case of spatial regression models. In the remainder of this section we consider the problem of diagnosing and dealing with influential cases.

7.4.1 Influential Cases in the Standard Regression Model

In the regression model given by (1) where errors are independent Cook's measure of influence for the i-th case, with $p = k + 1$, is defined as [Weisberg (1985, pp. 118-121)]:

$$\hat{\beta} = (X'X)^{-1}X'Y$$
$$\hat{\beta}_{(i)} = (X_{(i)}'X_{(i)})^{-1}X_{(i)}'Y_{(i)} \tag{3}$$

and

$$\hat{\sigma}^2 = \frac{(Y - X\hat{\beta})'(Y - X\hat{\beta})}{(n - p)} \tag{4}$$

$X_{(i)}$ is the $((n\text{-}1) \times (k\text{+}1))$ array with the i-th row deleted and $Y_{(i)}$ is the $(n\text{-}1)$ vector with the i-th case deleted. It can be shown that:

$$D_i = \frac{r_i^2\left(\frac{h_{ii}}{(1 - h_{ii})}\right)}{p} \tag{5}$$

where

$$r_i = \frac{\hat{u}_i}{\hat{\sigma}(1 - h_{ii})} \tag{6}$$

The term h_{ii} is the i-th diagonal entry in the matrix H where:

$$H = X\,(X'X)^{-1}\,X' \tag{7}$$

and r_i is called the internally studentized residual for the i-th case where:

$$\hat{u} = Y - X\hat{\beta} \tag{8}$$

If r_i is sufficiently large the case is called an outlier. In addition h_{ii} is defined as the leverage of the i-th case. It is positive and as $h_{ii} \rightarrow 1$, $\hat{y}_i \rightarrow y_i$ since $\hat{y}_i = \Sigma\, h_{ij}\, y_j$. A leverage is large if it exceeds $2(p/n)$. The influence of the i-th case is therefore seen to be a composite measure based on its leverage and the size of the i-th residual. Rough calibration points are based on the F-distribution with $\alpha = 0.5$ and with $(p, n\text{-}p)$ degrees of freedom [Weisberg (1985, pp. 118-121)].

7.4.2 Influential Cases in a Spatial Regression Model

As noted, there are several types of spatial regression model therefore, in order to keep the discussion focused we shall consider only the regression model with spatially correlated errors which is defined as follows:

$$Y = \boldsymbol{X}\beta + e \tag{9}$$

where Y, $\boldsymbol{X}$ and β are defined as for (1) but:

$$\mathrm{Var}(e) = \sigma^2 \boldsymbol{V} \quad ; \quad \sigma^2 > 0 \tag{10}$$

where $\boldsymbol{V}$ is an $(n \times n)$ non-diagonal matrix. The off-diagonal elements of $\boldsymbol{V}$ describe the structure of spatial correlation in the errors. A permissible form for $\boldsymbol{V}$ is:

$$(\boldsymbol{I} - \rho \boldsymbol{W})^{-1} \tag{11}$$

where $\boldsymbol{I}$ is the identity matrix, $\boldsymbol{W}$ is an $(n \times n)$ weights or connectivity matrix describing the contiguity relations between sites or areas and ρ is an unknown parameter. This is the variance-covariance matrix of the conditional spatial autoregressive model which is discussed at length in, amongst other sources, Haining (1990a).

There are several aspects of the above development of a diagnostic for influential cases that are problematic when transferred to (10).

Aspect (i): Data deletion in the case of (1) reduces the number of degrees of freedom but, since observations are independent, it does not affect the properties of the regression model that is being fit to the (reduced) data set. This is not the case in (10) particularly for a matrix $\boldsymbol{V}$ defined by (12) which is non-stationary. Deleting individual cases changes the variance and covariance properties of the model. In the simple case of a (2×2) lattice let:

$$\boldsymbol{W} = \begin{bmatrix} 0 & 1 & 0 & 1 \\ 1 & 0 & 1 & 0 \\ 0 & 1 & 0 & 1 \\ 1 & 0 & 1 & 0 \end{bmatrix}$$

If the site in the top right position corresponding to the fourth row and column of $\boldsymbol{W}$ is deleted then $\boldsymbol{W}$ turns into $\boldsymbol{W}^*$ given below, and the variance-covariance matrix in the two cases ($\rho = 0.24$) becomes:

$$W^* = \begin{bmatrix} 0 & 1 & 0 \\ 1 & 0 & 1 \\ 0 & 1 & 0 \end{bmatrix}$$

$$\sigma^2 = \begin{bmatrix} 1.15 & 0.31 & 0.15 & 0.31 \\ 0.31 & 1.15 & 0.31 & 0.15 \\ 0.15 & 0.31 & 1.15 & 0.31 \\ 0.31 & 0.15 & 0.31 & 1.15 \end{bmatrix} \quad \text{and} \quad \sigma^2 = \begin{bmatrix} 1.13 & 0.27 & 0.07 \\ 0.27 & 1.13 & 0.27 \\ 0.07 & 0.27 & 1.13 \end{bmatrix}$$

Apart from creating a new boundary, the effect of deleting a case has been to change the model so that parameter shifts, as a result of deleting cases, will compound these two effects. This is likely to be particularly serious for small n but may be significant for larger n where data deletion in the interior creates an awkward internal edge effect. The same problem can be shown to arise for the simultaneous autoregressive model where:

$$V = \left((I - \rho W')(I - \rho W)\right)^{-1} \tag{13}$$

Aspect (ii): In the model defined by (10) suppose V is known. The generalized least squares estimator for β is:

$$\tilde{\beta} = (X' V^{-1} X)^{-1} X' V^{-1} Y \tag{14}$$

and

$$\text{Var}(\tilde{e}) = (V - X(X' V^{-1} X)^{-1} X')\sigma^2 = L\sigma^2 \tag{15}$$

In this case instead of (7) we have:

$$r_i^* = \frac{\tilde{e}_i}{\tilde{\sigma}\, l_{ii}^{1/2}} \tag{16}$$

where

$$\tilde{e} = Y - X\tilde{\beta} \quad ; \quad \tilde{\sigma}^2 = \frac{(\tilde{e}' V^{-1} \tilde{e})}{(n - p)} \tag{17}$$

Usually however, V is not known and its parameters must be estimated. The possibility exists that another kind of outlier exists with respect to the estimation of $\boldsymbol{V}$. If $\boldsymbol{V}$ is given by (13) then the underlying model has the form:

$$e = \rho \boldsymbol{W} e + u \tag{18}$$

where u is as defined in (1). Graphical plots of $\tilde{e}_i$ against $(\boldsymbol{W}\,\tilde{e})_i$ where the row sums of $\boldsymbol{W}$ have been standardized to one, may provide evidence of outliers that have an influence on the estimate of ρ which may in turn influence the estimate of β. The essential point however, is that a thorough assessment of influential cases must include not just the assessment of outliers but *spatial* outliers (that is residual values $(\tilde{e}_i)$ that are markedly different from the local distribution of residual values).

Aspect (iii): In the case of (1) the diagonal elements of the matrix $\boldsymbol{H}$, which is symmetric, define the leverage of the i-th case. Row and column sums of $\boldsymbol{H}$ equal 1 and values $\{h_{ii}\}$ must lie between 0 and 1.

The analogue to $\boldsymbol{H}$ in the case of (10) if $\boldsymbol{V}$ is known is given by:

$$\boldsymbol{K} = \boldsymbol{X}(\boldsymbol{X}'\boldsymbol{V}^{-1}\boldsymbol{X})^{-1}\boldsymbol{X}'\boldsymbol{V}^{-1} \tag{19}$$

Martin (1992) notes that $\boldsymbol{K}$, like $\boldsymbol{H}$ is idempotent but unlike $\boldsymbol{H}$ it is not symmetric and k_{ii} can be negative. Martin (1992) discusses other candidates for measuring the leverage of the i-th case.

Aspect (iv): The analogue to Cook's measure D_i is:

$$D_i^* = \frac{(\tilde{\beta} - \tilde{\beta}_{(i)})(\boldsymbol{X}'\boldsymbol{V}^{-1}\boldsymbol{X})(\tilde{\beta} - \tilde{\beta}_{(i)})}{p\,\tilde{\sigma}^2} \tag{20}$$

where

$$\tilde{\beta}_{(i)} = (\boldsymbol{X}_{(i)}'\boldsymbol{V}_{(i)}^{-1}\boldsymbol{X}_{(i)})^{-1}\boldsymbol{X}_{(i)}'\boldsymbol{V}_{(i)}^{-1}Y_{(i)} \tag{21}$$

and $\boldsymbol{V}_{(i)}$ is obtained by deleting the rows and columns of $\boldsymbol{V}$ associated with i. But as Martin (1992) notes, this measure cannot be expressed in terms of r_i^* nor is there an exact calibration theory.

7.4.3 An Example

The following example illustrates some aspects of the above methodology. For a fuller discussion using a different data set see Haining (1994).

Bayes-adjusted indirect standardized mortality rates for cancer for 87 community medicine areas in Glasgow were regressed on a deprivation index computed for these 87 areas (Table 7.2). The Bayes adjustment procedure follows the gamma model of Clayton and Kaldor (1987) while the deprivation index is a composite measure based on four standardized census variables (percentage in social class four and five, percentage overcrowding, percentage with no car, percentage not owner occupied). The index is similar to the Townsend index [Townsend et al. (1988)] which uses percentage unemployed rather than percentage in social class four and five. Unemployment data for the 87 areas were not available. Details about the study area can be found in Haining (1991).

An ordinary least squares fit gives the following estimates for the regression of cancer mortality (Y) on deprivation (X):

$$\hat{Y} = \underset{(0.70)}{99.60} + \underset{(0.18)}{2.18}\,X$$

with $\hat{\sigma} = 6.65$, $R^2 \times 100 = 61.7\,\%$, a total sum of squares of 9562.7 and an error sum of squares of 3666.3 (values in brackets under the coefficients are standard errors). A standardized weights matrix ($\boldsymbol{W}$ with row sums equal to one) was computed [see Haining (1991)]. The Generalized Moran Coefficient (*GMC*) as a standard normal deviate is 3.05 suggesting significant residual autocorrelation. Table 7.3 gives the extreme values for the diagnostics (6), (7) and (8). Standardized residuals greater than 2.0 in absolute value represent large deviations from the model fit and leverages are deemed large if they exceed 0.046. Sites are influential however if Cooks D exceeds 0.69. The evidence is that none of the cases alone will exert a substantial influence.

A regression model (10) with errors given by (13) was fit to handle the problem of autocorrelated error detected by the *GMC* test. The maximum likelihood fit, where again $\boldsymbol{W}$ is standardized, is given by:

$$\tilde{Y} = \underset{(1.09)}{99.71} + \underset{(0.20)}{2.09}\,X$$

with $\tilde{e} = (\boldsymbol{I} - \rho\,\boldsymbol{W})^{-1}\,\tilde{u}$, $\tilde{\rho} = 0.41$ (standard error: 0.13), $\tilde{\sigma} = 6.21$, a total sum of squares of 9562.7, an error sum of squares for $(\tilde{e})$ of 3675.4, and an error sum of squares for $(\tilde{u})$ of 3281.2.

Table 7.4 records the extreme values for the diagnostics given by (16), (19) and (20) assuming ρ is fixed. In addition, we record the largest percentage changes in the value of the parameter associated with deprivation (β_1) arising from deleting each case in turn (21). Leverage scores given by (19) cannot be assessed in the same way as for the first model [Martin (1992)]. Note the change in the values of the diagnostics for various sites relative to Table 7.3. No site appears to be influential and this is borne out by the evidence of the largest percentage shifts in β_1 arising from deleting individual sites.

Table 7.2: Data for the regression of Bayes-adjusted standardized mortality rates for cancer[a] on deprivation.[b]

Area	Bayes adjusted SMR (cancer)	Deprivation	Area	Bayes adjusted SMR (cancer)	Deprivation
1	104.761	2.148	45	127.818	4.767
2	97.502	0.657	46	101.790	0.880
3	85.772	-1.493	47	96.908	0.638
4	98.195	-0.726	48	104.241	6.407
5	97.592	0.233	50	100.925	3.383
6	90.491	-4.945	49	12.042	3.295
7	91.161	-3.940	51	97.616	1.234
8	87.132	-5.394	52	85.545	-2.379
9	97.940	1.750	53	104.534	4.561
10	101.228	-0.106	54	97.328	2.815
11	100.864	2.278	55	110.332	4.500
12	94.392	4.886	56	112.810	5.605
13	98.215	3.688	57	95.664	-2.086
14	96.224	1.263	58	121.997	4.788
15	98.978	1.716	59	102.720	1.441
16	88.305	1.312	60	89.670	-4.259
17	83.786	-1.358	61	96.757	-1.593
18	88.776	-6.646	62	97.890	0.006
19	88.540	-6.590	63	91.627	-0.513
20	94.290	-4.492	64	97.386	-2.459
21	103.133	2.681	65	107.169	5.397
22	102.983	2.474	66	107.461	4.576
23	106.285	5.340	67	89.512	-5.670
24	100.544	1.634	68	85.476	-6.522
25	118.219	5.801	69	86.668	-5.506
26	102.817	3.034	70	85.756	-6.430
27	118.281	4.245	71	89.753	-5.443
28	113.143	2.477	72	96.559	-5.421
29	104.484	1.617	73	80.709	-7.403
30	84.265	-2.973	74	90.147	-5.605
31	93.986	-3.286	75	84.277	-5.147
32	93.330	-5.189	76	101.507	-1.012
33	87.625	-6.126	77	95.517	-1.963
34	102.927	-3.645	78	113.752	-0.377
35	90.014	-2.768	79	86.475	-0.181
36	97.146	-3.044	80	116.185	3.418
37	94.208	-1.468	81	112.738	3.180
38	101.996	-0.624	82	105.785	3.210
39	90.974	-4.495	83	90.530	-2.377
40	99.884	3.232	84	113.713	3.155
41	110.003	3.116	85	113.403	2.941
42	125.937	4.838	86	114.284	4.144
43	118.963	3.860	87	95.866	-1.139
44	111.668	4.176			

[a] The original (non-adjusted) rates and other details about the spatial system are available in Haining (1991).
[b] The larger the value the greater the (material) deprivation.

Table 7.3: Diagnostics for regression residuals from the standard regression using the least squares fitting procedure.

Site	Standardized residual (7)	Site	Leverage (8)	Site	Cooks *D* (6)
45	2.76	73	0.055	45	0.11
12	-2.45	18	0.046	12	0.09
42	2.44	19	0.046		
78	2.29	68	0.045		
16	-2.17				

Table 7.4: Diagnostics for regression residuals from (10) assuming ρ is fixed.

Site	Standardized residual (16)	Site	Leverage (19)	Site	Cooks *D* (20)	Site	% shift in β_1 (21)
45	2.72	39	1.08	12	0.11	12	4.5
42	2.62	14	1.08			48	3.1
79	-2.37						
12	-2.16						
48	-2.15						
78	2.14						
40	-2.10						
16	-1.50	73	1.04	45	0.06		
		18	0.96				

7.5 Implementing Methodologies

It has been argued elsewhere that Geographic Information Systems may provide a platform for implementing spatial analysis techniques [Goodchild et al. (1992), Haining (1992)]. From the perspective of the issues raised in this chapter a number of facilities within such systems would seem necessary in order to implement procedures for handling data problems.

- The facility to allow different regionalizations (areal aggregates of smaller spatial units) to be constructed so that the sensitivity of results to alternative schemes that are equivalent in scale can be assessed.
- The facility to display data sets and in particular data sets with many variables.

The chances of detecting certain data problems are likely to be better if good visualization tools are developed. Such tools might be based on traditional techniques such as chernoff faces but recently there have been important developments such as those in dynamic graphics [Haslett et al. (1991)] linking maps and statistical graphs. However, both approaches have their shortcomings for the user seeking to develop a feel for the data. The problem is how to present what may be large amounts of data in ways that do not overload the user, either by the amount of data presented at any one time (the chernoff face), or by the sequencing of subsets of the data and the need to make numerous comparisons (dynamic graphics methods).

- The facility to allow the user to fit with some data missing, delete cases and assess the sensitivity of results to individual cases or subsets of cases. Given the methodological problems of deleting cases in assessing spatial regression models another way of assessing model sensitivity may be to perturb rather than delete values. Closely allied to this is the need to be able to restructure the matrix $\boldsymbol{W}$, which describes the structure of spatial relationships between areas. The purpose is to assess the sensitivity of model results to equally plausible forms.

Spatial econometrics has evolved from a concern with testing for spatial pattern to a concern for modeling pattern. That route has led to the emergence of a class of models which have raised special problems if they are to be fit. The problem of fitting not only raises concerns about appropriate model specification and estimation it also raises questions of validation. The concern with data problems enters into all these phases of analysis.

References

Anselin, L., *Spatial Econometrics: Methods and Models*, Dordrecht: Kluwer, 1988.

Bennett, R., R.P. Haining and A. Wilson, Spatial Structure, Spatial Interaction and their Integration: A Review of Alternative Models, *Environment and Planning A*, 17, 625-645, 1985.

Clayton, D. and J. Kaldor, Empirical Bayes Estimates of Age Standardized Relative Risks for Use in Disease Mapping, *Biometrics*, 43, 671-681, 1987.

Cressie, N., *Statistics for Spatial Data*, New York: Wiley, 1991.

Goodchild, M.F., Geographic Data Modeling, *Computers and Geosciences*, 18, 401-408, 1992.

Goodchild, M., R.P. Haining and S. Wise, Integrating GIS and Spatial Data Analysis: Problems and Possibilities, *International Journal of Geographical Information Systems*, 6, 407-424, 1992.

Haining, R.P., *Spatial Data Analysis in the Social and Environmental Sciences* Cambridge: Cambridge University Press, 1990a.

Haining, R.P., Models in Human Geography: Problems in Specifying, Estimating and Validating Models for Spatial Data, in: D.A. Griffith (ed.), *Spatial Statistics: Past, Present and Future*, Ann Arbor: Institute of Mathematical Geography, 1990b (including the discussion by P. Doreian).

Haining, R.P., Bivariate Correlation with Spatial Data, *Geographical Analysis*, 23, 210-227, 1991.

Haining, R.P., *Designing Spatial Data Analysis Modules for Geographical Information Systems*, Proceedings of the NCGIA Initiative 14 Meeting on Spatial Analysis in GIS, San Diego: California, 1992.

Haining, R.P., Diagnostics for Regression Modeling in Spatial Econometrics, *Journal of Regional Science*, 34, 325-341, 1994.

Haslett, J., J. Bradley, P. Craig, G. Wills and A. Unwin, Dynamic Graphics for Exploring Spatial Data with Application to Locating Global and Local Anomalies, *The American Statistician*, 45, 234-242, 1991.

Hoaglin, D., F. Mosteller and J. Turkey, *Understanding Robust and Exploratory Data Analysis*, New York: Wiley, 1985.

Leamer, E.E., *Specification Searches: Ad hoc Inference with Non-Experimental Data*, New York: Wiley, 1978.

Li, G., Robust Regression, in: D. Hoaglin, F. Mosteller and J. Turkey (eds.), *Exploring Data Tables, Trends and Shapes*, New York: Wiley, 1985.

Little, R. and D. Rubin, *Statistical Analysis with Missing Data*, New York: Wiley, 1987.

Martin, R., Leverage, Influence and Residuals in Regression Models when Observations are Correlated, *Communications in Statistics: Theory & Methods*, 21, 1183-1212, 1992.

Townsend P., P. Phillimore and A. Beattie, *Health and Deprivation: Inequality and the North*, London: Croom Helm, 1988.

Unwin, D. and N. Wrigley, Control Point Distribution in Trend Surface Modeling Revisited: An Application of the Concept of Leverage, *Transactions, Institute of British Geographers*, 12, 147-160, 1987.

Weisberg, S., *Applied Linear Regression*, New York: Wiley, 1985.

8 Spatial Filtering in a Regression Framework:
Examples Using Data on Urban Crime, Regional Inequality, and Government Expenditures

Arthur Getis
San Diego State University, San Diego, CA, U.S.A.

> If autocorrelation is found, we suggest that it be corrected by appropriately transforming the model so that in the transformed model there is no autocorrelation [Gujarati (1992, p. 373)].

8.1 Introduction

In a recent paper [Getis (1990)], I develop a rationale for filtering spatially dependent variables into spatially independent variables and demonstrate a technique for changing one to the other. In that paper, the transformation is a multi-step procedure based on Ripley's second order statistic (1981). In this chapter, I will briefly review the argument for the filtering procedure and propose a simplified method based on a spatial statistic developed by Getis and Ord (1992). The chapter is divided into four parts: 1) a short discussion of the rationale for filtering spatially dependent variables into spatially independent variables, 2) a review of a Getis-Ord statistic, 3) an outline of the filtering procedure, and 4) three examples taken from the literature on urban crime, regional inequality, and government expenditures.

8.2 Rationale for a Spatial Filter

One of the most difficult problems facing those who develop regression models of spatial series is finding ways to estimate parameters of stochastically autocorrelated variables. A typical stochastically autocorrelated spatial variable is a modified or spatially lagged autocorrelated variable. It is made up of the original autocorrelated variable (y) multiplied by a spatial weight matrix ($\boldsymbol{W}$) and a spatial autocorrelation coefficient (ρ). $\rho \boldsymbol{W} y$ does not fulfill the required fixed-effects linear regression assumption that correlated variables are not to be stochastic. In this case, since ordinary least squares yields biased parameter estimates and R-squared values, other estimation techniques must be considered, such as maximum likelihood estimation.

In addition, any remaining spatial dependence in the regression equation, as may be evident in error terms, must be accounted for. In multiple variable cases, it may be necessary to develop a series of $\boldsymbol{W}$ matrices, thus further complicating the meaning of the various tests on the significance of the parameters.

Because of the complexity of the typical spatial regression formulation, I propose that the spatial dependence within each dependent and independent variable be filtered out before the estimation procedure is adopted. This proposal has the following ramification: one must somehow reintroduce into the regression equation the removed spatial dependence in order to avoid misspecification. No spatial dependence should be evident in the error term since supposedly it has been removed from all of the possible sources. Since each variable on the right hand side is no longer stochastically correlated nor spatially autocorrelated, and there is only the usual spherical error, ordinary least squares can be used for estimation.

This argument for filtering from spatially autocorrelated variables the spatial dependence effects is sound only insofar as there is a way to accomplish the task. Before demonstrating the variable filtering procedure, let me briefly describe a statistic that will act as its foundation.

8.3 The G_i Statistic

This statistic measures the degree of association that results from the concentration of weighted points (or areas represented by weighted points) and all other weighted points included within a radius of distance d from the original weighted point.[1] We are given an area subdivided into n regions, $i = 1, 2, ..., n$, where each region is identified with a point whose Cartesian coordinates are known. Each i has associated with it a value x (a weight) taken from a variable X. The variable has a natural origin and is positive.[2] The statistic is written as:

$$G_i(d) = \frac{\sum_{j=1}^{n} w_{ij}(d)x_j}{\sum_{j=1}^{n} x_j}, \quad j \neq i \tag{1}$$

where w_{ij} is a one/zero spatial weight matrix with ones for all links defined as being within distance d of a given i; all other links are zero. The numerator is, therefore, the sum of all x_j within d of i except when i equals j. The denominator is the sum

[1] For a full discusssion see Getis and Ord (1992).

[2] A more recent version of this statistic in Ord and Getis (1995) avoids these restrictions.

of all x_j except when i equals j. The mean is:

$$E(G_i) = \frac{W_i}{(n-1)} \tag{2}$$

where $W_i = \Sigma_j w_{ij}(d)$.

If we set $\Sigma_j x_j / (n-1) = Y_{i1}$ and $\Sigma_j x_j^2 / (n-1) - Y_{i1}^2 = Y_{i2}$ then:

$$\mathrm{Var}(G_i) = \frac{W_i(n-1-W_i)}{(n-1)^2(n-2)} \left(\frac{Y_{i2}}{Y_{i1}^2}\right) \tag{3}$$

$G_i(d)$ measures the concentration or lack of concentration of the sum of values associated with variable X in the region under study. $G_i(d)$ is to be differentiated from a statistic $G_i^*(d)$ that takes into account the value of x at i, that is, j equal to i. $G_i(d)$ is a proportion of the sum of all x_j values that are within d of i. If, for example, high-valued x_j's are near to the point i, and d includes these high values, so that a large proportion of the sum of all x_j's is within d of i, then $G_i(d)$ is high. Whether the $G_i(d)$ value is statistically significant depends on the statistic's expectation.

For our purposes here, the most important characteristic of the statistic is that it gives the proportion of the summed variable within a specified distance from a particular point i as a part of the entire summed variable. When this value is compared to the statistic's expectation, the difference tells us the degree of clustering of the sum of the x variable in the vicinity of i that is greater or less than chance would have it.

8.4 The Filtering Procedure

The rationale for transforming a spatially dependent variable into a spatially independent variable is that the spatial dependence can be removed from the spatially dependent variable and replaced as a separate independent variable. An easy solution to this problem, but useless, would be to set all values of the spatially dependent variable to the mean. This 'variable' would not be spatially dependent and it would not correlate with any other variable. The solution I outline below attempts to adjust the spatial dependent variables only to the point where spatial dependence is no longer embodied in them. That which is filtered from the original variable becomes a new spatial variable. It may be that the autocorrelation filtered from one variable is highly correlated with that which is filtered from another. In a regression equation, in order to avoid multicollinearity, in the final equation it may be necessary to use only one spatial variable rather than all spatial variables extracted from the original variables.

Suppose that within distance d of a point, x_1, there are two other points with values, x_2, and x_3, which when summed are a greater proportion of all x (minus x_1) than what one should expect in a similar spatial configuration with the same d value when all x values are randomly distributed. This means that $G_1(d)$ is greater than the random expectation, $W_1/(n-1)$. Suppose $G_1(d)$ is the proportion 0.40 and $E[G_1(d)] = 0.30$. Then 40 percent of the sum of all observed x (not counting x_1 itself) is contained within d of x_1, while the expectation is only 30 percent. We then call 30/40 of x_1 the filtered value of x_1. The difference between the original value and the filtered value is that which is filtered out due to the spatial clustering of x values in the vicinity of x_1. In this example, the ratio 30/40 represents the degree to which x_1 is similar to its expectation. The degree of dissimilarity, 10/40, represents that which is due to positive spatial association. Negative spatial association is found in like manner. Thus,

$$x_i^* = \frac{x_i \left(\dfrac{W_i}{n-1} \right)}{G_i(d)} \tag{4}$$

which when solved for all x_i represents the filtered variable X^*. The difference between X and X^* is a new variable, L, that represents the spatial effects embedded in X.

For realistic filtering it is essential to find an appropriate d value. The value should represent the distance within which spatial dependence is maximized. In Getis (1990), d corresponds to the maximum total sum of squared differences between the observed and the expected G_i values. On reflection, however, a different d value seems more appropriate. This is the value that corresponds to the maximum absolute sum of the normal standard variate of the statistic $G_i(d)$ for all i observations of the variable X. This single value is chosen since it represents overall the distance beyond which no further association or nonassociation effects increase the probability that the observed value is different than the expected value. One might argue that beyond this d value there is an overall cessation of spatial effects for the variable in question. For each variable, we use this value in the examples given below. A more detailed approach, but less general, would be to identify a critical d value for each individual point i. Clearly, more research is needed on this subject.

Our approach to spatial filtering can be given credence by showing that the following four conditions hold:

1. There is no spatial autocorrelation embodied in X^*.
2. If X is a variable with spatial dependence embedded in it, then the difference between X and X^* is a spatially autocorrelated variable (L).
3. In any regression model where all variables have been filtered using an appropriate distance d, residuals are not spatially associated.

4. In a regression equation, appropriate variables should be statistically significant after spatial dependence has been removed from them. Of course, appropriateness in this case requires theoretical justification. In this chapter, we will be satisfied if intuitively correct variables are statistically significant after spatial dependence has been removed.

If these conditions are met, one might conclude that a reasonable estimate of spatial autocorrelation has been found and that ordinary least squares is appropriate for regression modeling where this type of filtering has been used. In the next section, these conditions are demonstrated by way of three examples.

8.5 Filtering Variables: Three Examples

8.5.1 Example 1: Urban Crime

Anselin (1988) provides data on three variables (crime, income and housing) by neighborhoods (given as points with *x,y* coordinates) for Columbus, Ohio, for 1980 ($n = 49$). The autocorrelated variable, crime (*CR*), is constructed from the number of burglaries and thefts per thousand households, and income (*IN*) and housing (*HO*) are given by household in thousands of dollars. When $d = 4$ kilometres spatial association is at a maximum for crime, $d = 3$ for housing and income. In the tests that follow, these values for d will be used. The trial OLS model is:

$$\begin{array}{cccc} CR = & 68.62 & -\ 1.597\ IN & -\ 0.274\ HO \\ (t) & (14.49) & (-4.78) & (-2.65) \end{array} \tag{5}$$

The adjusted $R^2 = 0.533$ and the standardized Moran's I on the residuals is 2.765. The criterion for spatial dependence in this and all subsequent tests is the 0.05 level of the normal curve calculated from the I statistic (randomization, distance effect $= 1/d^2$) of Moran [Cliff and Ord (1973)]. This statistic will not be reviewed here. In this case then, the residuals are spatially statistically significant. In addition, it is important to note that two of the variables reveal strong spatial autocorrelation:

Variable	$Z(I)$
CR	7.345
IN	4.168
HO	1.903

Test 1: There is no spatial autocorrelation embodied in X^*.

This test requires that the transformed variables, CR^*, IN^*, and HO^* are not spatially autocorrelated. The results are as expected:

Variable	$Z(I)$
CR^*	-0.152
IN^*	-0.280
HO^*	-0.454

Test 2: If X is a variable with spatial dependence embedded in it, then the difference between X and X^* is a spatially autocorrelated variable (L).

The results of the test (see below) show that all three L variables are spatially autocorrelated. Although it is not required for the housing variable to have a high $Z(I)$ value, it, too, has spatial dependence embedded within it.

Variable	$Z(I)$
L_{CR}	7.659
L_{IN}	2.059
L_{HO}	3.044

Test 3: In any regression model where all variables have been filtered using an appropriate distance d, residuals are not spatially associated.

In this case we use the linear regression:

$$CR^* = a - IN^* - HO^* + e \tag{6}$$

and then test e for spatial association using I. Clearly, this is an inadequate model but it does serve to show how the filtering of variables satisfies our intuition. As expected, the residuals are not spatially associated as $Z(I) = 0.161$. The correlated variables explain 32 percent of the variation in the filtered crime variable. In the unfiltered case shown above, the adjusted R^2 is 0.533. This indicates that by removing the spatial association, the model becomes a weaker predictor of the location of the incidence of crime.

Test 4: In a regression model, appropriate variables should be statistically significant after spatial dependence has been removed from them.

Based on the earlier tests, an intuitively appealing model designed to 'explain' crime is:

$$CR = 60.15 - 0.96\,IN^{*} + 0.94\,L_{IN} - 0.27\,HO^{*} - 0.63\,L_{CR} \qquad (7)$$
$$(t) \quad (14.11) \quad (-3.40) \quad (-2.97) \quad (-3.28) \quad (-6.08)$$

The adjusted R^2 is 0.719, and the residuals are not spatially associated as $Z(I)$ = 1.1618.

Discussion: All 4 tests had satisfactory outcomes. The example tells us much about the crime, income, and housing data of Columbus. Note that the adjusted R^2 value decreases when each of the variables is filtered for spatial association (0.533 to 0.323). When the spatial effects are reintroduced to the filtered equation, the R^2 increases (0.323 to 0.719). The implication of this is that although filtered equations are free of spatial effects, that which is filtered out must be included in the final model in order to account for the spatial effects. In the case of crime in Columbus, it appears that the configuration of the data units has as much to do with explaining crime as does the income and housing variables. The proliferation of small, spatially autocorrelated units in the high crime areas contributed to the inappropriateness of the original unfiltered equation.

8.5.2 Example 2: Regional Inequality

In a study of regional disparities among the 16 regional divisions of Turkey, Atalik (1990) tested the Cobb-Douglas type model:

$$Y = aP^{b}I^{c}S^{d}A^{f} \qquad (8)$$

which can be written as:

$$\log Y = \log a + b \log P + c \log I + d \log S + f \log A \qquad (9)$$

where Y is GDP per capita, P activity rate (share of the country's active population), I infrastructure rate (share of the country's literate population), S industrial employment rate, and A agglomeration rate (the proportion of the people of the region residing in the largest city).

The trial OLS equation is:

$$\log Y = 0.81 \log a - 0.79 \log P + 1.27 \log I + 0.63 \log S + 0.05 \log A \qquad (10)$$
$$(t) \quad (0.23) \quad (-0.47) \quad (2.28) \quad (2.12) \quad (0.38)$$

in which each variable is spatially autocorrelated (see below) but the residuals are not autocorrelated ($Z(I)$ = -0.8061; see Figure 8.1). Log P has the wrong sign, and the log P and log A variables are not significant. Log I and log S are significant predictors of log Y. The adjusted R^2 is 0.797.

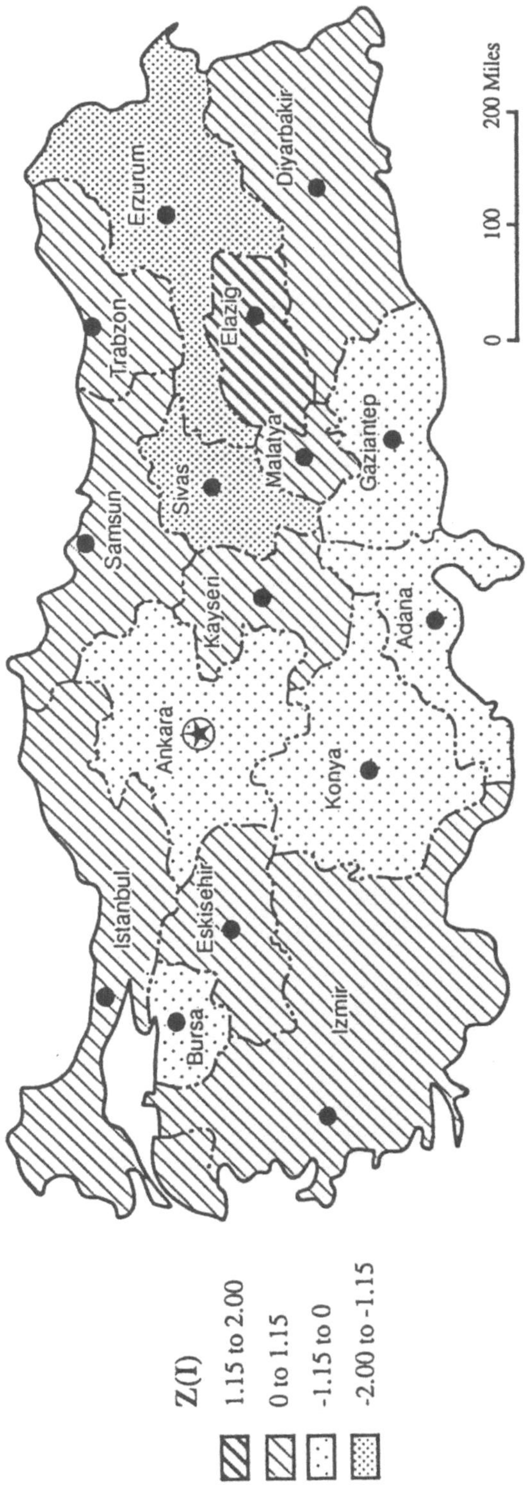

Figure 8.1: *Z(I)* values for 16 regional divisions of Turkey. There is no discernible spatial pattern of the residuals of the trial equation. One might get the impression from this that there is no spatial autocorrelation in the data.

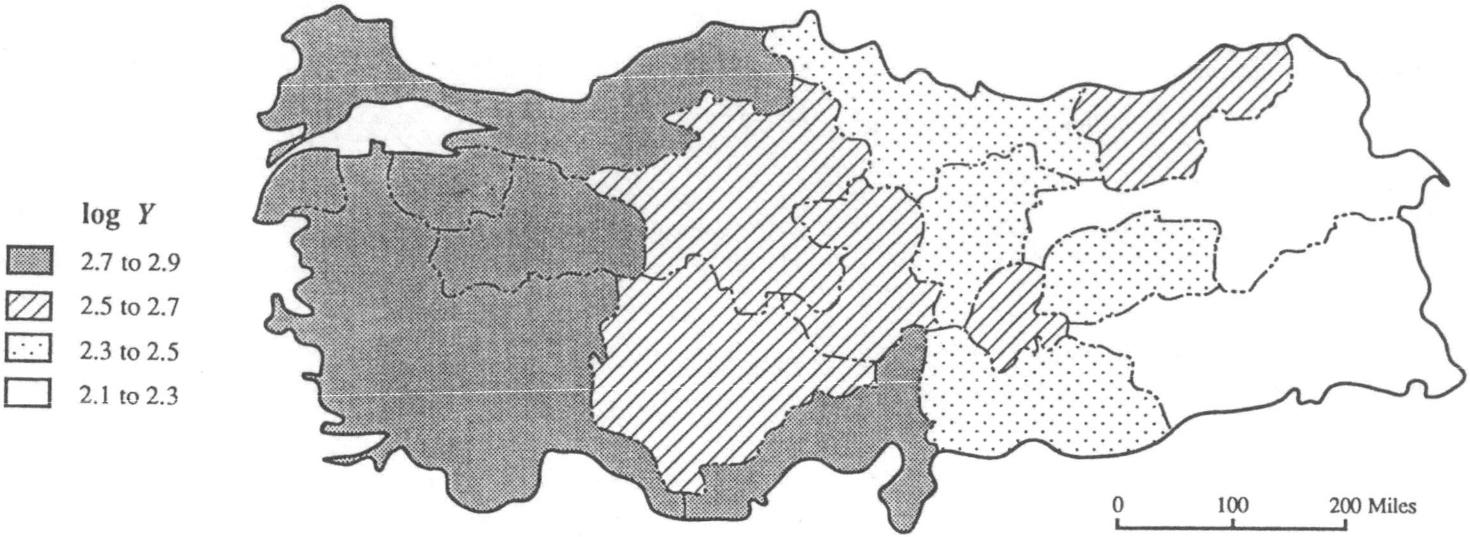

Figure 8.2: The dependent variable, Log Y, for 16 regional divisions of Turkey. It is high in the west and low in the east; all other variables in the trial equation act similarly.

In order to give spatial meaning to these results, coordinates were identified that correspond to the population centroid of each district. Using the procedure described earlier, the critical d value was found to be 187.5 miles for each variable (increments of 18.75 miles were examined starting at 18.75 miles). The $Z(I)$ values shown below indicate that there is strong spatial autocorrelation in the data. There is a clear east-west trend in the data; the more favorable socioeconomic values are in the west (see Figure 8.2).

Variable	$Z(I)$
log Y	5.608
log P	3.777
log I	4.437
log S	5.603
log A	2.790

Test 1: All filtered variables are free of spatial autocorrelation.

Variable	$Z(I)$
log Y^*	-1.203
log P^*	-1.720
log I^*	-0.131
log S^*	-0.282
log A^*	-0.065

Test 2: L variables based on spatially autocorrelated variables are spatially autocorrelated.

Variable	$Z(I)$
$L_{\log Y}$	5.396
$L_{\log P}$	3.292
$L_{\log I}$	4.861
$L_{\log S}$	5.223
$L_{\log A}$	4.486

Test 3: Residuals of the filtered regression equation are not spatially associated. The equation is:

$$\log Y^* = \log a + \log P^* + \log I^* + \log S^* + \log A^* + e \qquad (11)$$

The $Z(I)$ value of 1.8027 for residuals is not significant.

Test 4: In a regression model, appropriate variables should be statistically significant after spatial dependence has been removed from them.

Taking into account the results of the trial OLS model shown above and multicollinearity, an appropriate equation is:

$$\begin{array}{lllll} \log Y = 0.23 \log a & + 0.92 \log I^* & + 0.53 \log S^* & - 4.51 \log L_{\log S} & + e \\ (t) \quad (0.24) & (1.67) & (2.29) & (-6.91) & \end{array} \tag{12}$$

The residuals are not spatially autocorrelated as $Z(I) = -1.5282$. The adjusted R^2 is 0.836, an increase over the 0.797 of the trial model.

Discussion: Again, the four tests show how an inadequate model can be recast in order to account for spatial association in the constituent variables. In this case, not only were we able to show that the east-west trend in the data resulted in a poor trial equation, but that a stronger model can be constructed with fewer variables. As in the first example, the adjusted R^2 value decreases when each of the variables is filtered for spatial association (0.797 to 0.255). When the spatial effects are reintroduced to the filtered equation, the R^2 increases (0.255 to 0.836). In this case, the filtered infrastructure rate (log I^*), i.e., literacy, and filtered industrial employment rate (log S^*) together with the strong west to east diminishing trend in the industrial employment rate satisfactorily explain the level of GDP per capita in Turkey.

8.5.3 Example 3: Government Expenditures

In the well known econometrics text by Pindyck and Rubinfeld (1981, pp. 169-170), an expenditure data set based on Bureau of Census information is given for the states of the United States for 1970. Variables given in the table are transformed by population size by the authors to study the response variable, state and local government expenditures per capita in dollars (*PCEXP* = *EXP/POP*), using the regressors: federal grants to each state per capita in dollars (*PCAID* = *AID/POP*), population in thousands (*POP* = 1/*POP*), and personal income per capita (*PCINC* = *INC/POP*) [Pindyck and Rubinfeld (1981, p. 272)]. The trial OLS equation is:

$$\begin{array}{lllll} PCEXP = & -405.81 & + 1.63\, PCAID & + 25779\, POP & + 0.210\, PCINC \\ (t) & (-4.01) & (6.18) & (1.30) & (10.08) \end{array} \tag{13}$$

and the adjusted R^2 is 0.753.

Population centroids of states were estimated. Using increments of 33.3 miles, the critical d values are 333 miles for the first three variables and 267 miles for *PCINC*. The values shown below indicate that only one variable, *PCINC*, is spatially autocorrelated.

Variable	$Z(I)$
PCEXP	1.861
PCAID	0.291
POP	-0.080
PCINC	5.063

Test 1: All filtered variables are free of spatial autocorrelation.

Variable	$Z(I)$
*PCEXP**	-1.619
*PCAID**	-1.147
*POP**	0.588
*PCINC**	-0.250

Although all filtered variables are free of spatial autocorrelation, the only requirement for this test is that *PCINC** not be spatially autocorrelated. In any case, we always expect filtered variables not to be spatially statistically significant.

Test 2: *L* variables based on spatially autocorrelated variables are spatially autocorrelated.

Variable	$Z(I)$
L_{PCEXP}	3.705
L_{PCAID}	1.238
L_{POP}	1.208
L_{PCINC}	5.274

The requirement that L_{PCINC} be autocorrelated holds. Note that *PCEXP* also contains much embedded spatial dependency.

Test 3: Residuals of the filtered regression equation are not spatially autocorrelated.
For the equation:

$$PCEXP^* = a + PCAID^* + POP^* + PCINC^* + e \tag{14}$$

the $Z(I)$ value for the residuals is -0.666 and the adjusted R^2 is 0.439, which is considerably less than the 0.753 value of the trial OLS equation.

Test 4: In a regression model, appropriate variables should be statistically significant after spatial dependence has been removed from them.
By experimentation, the best model for *PCEXP* is found to be:

$$PCEXP = 369.68 + 1.72\,PCAID^{*} + 0.19\,PCINC^{*} - 2.27\,L_{PCAID} - 0.26\,L_{PCINC} \qquad (15)$$
$$(t) \quad (-3.29) \quad (7.28) \quad (9.42) \quad (-4.06) \quad (-3.84)$$

The adjusted R^2 is 0.747 and the $Z(I)$ of the residuals is -0.819. There is essentially no multicollinearity in this formulation, and all variables are highly significant.

Discussion: The explanation for *PCEXP* is made up of both the non-spatial aspects and the spatial configuration of *PCAID* and *PCINC*. Note that a considerable degree of questionable explanation (0.753) in the trial model exists when spatial association is not taken into account. The final model is slightly less in explained variance (0.747), but it is free of embedded spatial autocorrelation.

8.6 Conclusions

The results of the tests indicate that for the cases presented here, in *every* instance the filtering procedure conforms to our expectation. Clearly, these are but three case studies. Many more examples should be carried out. In the cases shown here, the procedure considerably helps in isolating the spatial dependence embedded within spatial series variables. In addition, the methodology aids in the proper specification of multiple regression relationships. The simplicity and ease of understanding made available by least squares methodology enable us to avoid estimation procedures that rob us of the convenience of R^2 interpretations.

Acknowledgements

I would like to thank Giuseppi Arbia for his suggestions on an earlier version of this chapter, Serge Rey for suggesting the data used in the government expenditures example, and the editors, Luc Anselin and Raymond Florax, for helpful comments.

References

Anselin, L., *Spatial Econometrics: Methods and Models*, Dordrecht: Kluwer Academic Publishers, 1988.

Atalik, G., Some Effects of Regional Differentiation on Integration in the European Community, *Papers of the Regional Science Association*, 69, 11-19, 1990.

Cliff, A.D. and J.K. Ord, *Spatial Autocorrelation*, London: Pion, 1973.

Getis, A., Screening for Spatial Dependence in Regression Analysis, *Papers of the Regional Science Association*, 69, 69-81, 1990.

Getis, A. and J.K. Ord, The Analysis of Spatial Association by Use of Distance Statistics, *Geographical Analysis*, 24, 189-206, 1992.

Gujarati, D., *Essentials of Econometrics*, New York: McGraw-Hill, 1992.

Ord, J.K. and A. Getis, Local Spatial Autocorrelation Statistics: Distributional Issues and an Application, *Geographical Analysis*, 1995 (forthcoming).

Pindyck, R.S. and D. Rubinfeld, *Econometric Models and Economic Forecasts*, New York: McGraw-Hill, 1981 (sec. ed.).

Ripley, B., *Spatial Statistics*, New York: Wiley, 1981.

PART II

Spatial Effects in Limited Dependent Variable Models

9 Spatial Effects in Probit Models: A Monte Carlo Investigation

Daniel P. McMillen
Tulane University, New Orleans, LA, U.S.A.

9.1 Introduction

Heteroscedasticity and autocorrelation typically are assumed to be absent in econometric models. Linear regression models are forgiving if these assumptions fail: ordinary least squares (OLS) estimates remain consistent if errors are not homoscedastic or are autocorrelated. Estimators for models with discrete data are not always as forgiving as OLS. For example, the standard probit estimator continues to provide consistent estimates when error terms are autocorrelated, but the estimates are inconsistent as well as inefficient when errors have non-constant variances. Failure of the homoscedasticity assumption also leads to inconsistent estimates in such common models as tobit and logit. Thus, heteroscedasticity is a serious problem in models with discrete data.

Nevertheless, heteroscedasticity typically is ignored in applied research when models more complicated than OLS are employed. Two reasons likely account for the willingness to employ potentially inconsistent estimators in such situations. First, models such as probit and tobit already involve iterative non-linear estimation techniques, and taking heteroscedasticity into account significantly increases the complexity. High-quality statistical packages include procedures for homoscedastic probit, tobit, logit, etcetera, but additional programming generally is required to estimate a heteroscedastic model.[1] Second, any non-linear model demands a lot of the data. Desirable estimation properties such as consistency and efficiency are proved for large samples, and it is not clear how well complicated estimators per-

[1] LIMDEP is an exception. This package makes it very easy to estimate a heteroscedastic probit model (see Appendix 2 for an example), although the range of heteroscedasticity specifications is limited. This is a recent addition to the LIMDEP probit estimator.

form in common sample sizes. Standard discrete-data estimators indeed may provide inconsistent estimates under heteroscedasticity, but heteroscedasticity-consistent estimators may lead to such large variances in small or moderate sample sizes that they are not worth the programming effort.

This is only speculation, however; little is known about the small-sample performance of discrete-data estimators in situations involving heteroscedasticity. Maddala (1983) discusses several studies that estimate both a standard tobit model and a model employing tobit with a linear heteroscedasticity specification. He concludes that violation of the homoscedasticity assumption may be important: 'Although there were no sign reversals, the differences in the coefficient estimates between the simple tobit and the heteroscedastic tobit were more pronounced than in the case of [OLS]' [Maddala (1983, p. 181)]. This procedure involves comparing OLS to tobit estimates in a situation in which the true model is unknown. The true model may have homoscedasticity but an incorrect functional form or variable list may have been employed. It is not known whether the differences in parameter estimates are caused by heteroscedasticity or another misspecification.

Missing from the literature is a Monte Carlo investigation of discrete-data estimators when the data exhibit autocorrelation and heteroscedasticity. This chapter fills the gap, focusing on probit, the most commonly employed of such estimators. Monte Carlo makes it possible to compare estimators when the true model is known. Sample sizes and degrees of autocorrelation and heteroscedasticity are varied to determine the conditions under which one estimator is preferred to another. The results indicate that the standard probit estimator is preferable to the more complicated heteroscedasticity-consistent estimator when sample sizes are small and heteroscedasticity is not serious. The more complicated estimator significantly out-performs standard probit as sample sizes and the degree of heteroscedasticity increase. In moderate sample sizes (such as 500 observations), standard probit offers little gain over the heteroscedastic probit estimator even when the errors have constant variance. Only in small sample sizes should the researcher hesitate to use heteroscedastic probit because of its possible large variance. Autocorrelation has little or no effect on these results. Thus, the asymptotic prediction that probit estimates are consistent when errors are assumed incorrectly to be independent is borne out in small samples.

The Monte Carlo experiments are also used to estimate the power of standard heteroscedasticity tests. Relative to the OLS case, much larger sample sizes are required before Lagrange multiplier (LM) or likelihood ratio (LR) tests successfully can detect heteroscedasticity in discrete data. The LR test is successful in detecting heteroscedasticity in smaller sample sizes and for smaller amounts of heteroscedasticity than the LM test. The size and power of the tests are unaffected by the presence of autocorrelation: the probability of incorrectly indicating heteroscedasticity or falsely accepting homoscedasticity does not change when autocorrelation is introduced. The results argue strongly for the use of heteroscedastic probit in moderate to large data sets.

9.2 Sources of Heteroscedasticity

Heteroscedasticity long has been recognized as a potential problem in cross-sectional data. The standard textbook example [e.g., Johnston (1984), Judge et al. (1988), and Maddala (1988)] is consumption expenditures: food expenditure by the poor is related closely to income, while expenditure by the rich is more variable and depends on many other factors. The expenditure function has heteroscedastic errors because there is more variation in the errors when income is high than when it is low. Grouped data also lead to heteroscedasticity if the underlying errors have constant variances and the number of observations used to construct the averages varies by group [Johnston (1984)].

Heteroscedasticity is endemic to spatial data sets [Anselin (1988, 1990a), Anselin and Griffith (1988)]. The spatial expansion model of Casetti (1972) is an example: it implies heteroscedasticity unless the expansion is deterministic.[2] In other models, heterogeneity is modeled by assuming that the error terms are spatially autocorrelated. A common assumption is $u = \rho Wu + \varepsilon$, where u is the error vector, W is the 'spatial contiguity matrix', ε is a vector of independently and identically distributed errors, and ρ is a parameter indicating the strength of the spatial dependence. Common specifications of W allow nearby errors to be correlated. For example, a common assumption is $w^*_{ij} = 1$ if two observations are contiguous, $w^*_{ij} = 0$ otherwise, and $W_{ij} = w^*_{ij}/\Sigma_j w^*_{ij}$. This autocorrelation specification implies $u = (I - \rho W)^{-1}\varepsilon$. While this model was developed to account for autocorrelation in the errors, it implies heteroscedasticity in u. A similar problem arises in the spatial autoregressive model [see Haining (1988), McMillen (1992)].

Spatial models begin with the observation that proximity provides information about nearby observations that is not captured directly by the simple explanatory matrix X. The above models were developed to model spatial heterogeneity in variable means, but they also imply non-constant variances. While the heteroscedasticity in these models was derived somewhat mechanically, it is reasonable to expect variances to differ over space. An example is population density: just as expenditures by the poor are correlated more closely with income than expenditures by the rich, population density may be correlated more closely with distance from the city center the farther an observation is from the center. Residences and commercial enterprises are mixed together near city centers, creating more variability in population density than in observations farther from the city center, where land use is more homogeneous. While Griffith (1981) estimates a homoscedastic spatial expansion model and studies such as Anderson (1985) and Lahiri and Numrich (1983)

---

[2] Interestingly, the expansion usually is treated as deterministic. For example, of the papers in the Jones and Casetti (1992) volume, which is a useful introduction to the spatial expansion model, only Anselin (1992) considers the possibility of an error term in the expansion.

account for heteroscedastic errors, a systematic study of the spatial patterns of population density error variances has not been conducted.

9.3 Heteroscedastic Probit

While textbooks devote ample space to heteroscedasticity when discussing the linear regression model and researchers commonly test for heteroscedasticity in cross-sectional data, the problem is all but ignored in models with discrete data.[3] But discrete-data models typically are derived from a standard linear model. For example, the probit model often is assumed to be:

$$y_i^* = X_i \beta + u_i \tag{1}$$

$$y_i = 1 \quad \text{if} \quad y_i^* > 0 \tag{2}$$

$$y_i = 0 \quad \text{if} \quad y_i^* \leq 0 \tag{3}$$

$$u_i \sim N(0,\sigma^2) \tag{4}$$

where $i = 1, ..., n$. In this formulation, y_i^* is an unobserved latent variable while y_i is an observed indicator variable. If y_i^* were observed, the model could be estimated by OLS, regressing y^* on the k variables included in the matrix X. The normality assumption (4) leads to the probit model.

In the linear regression model, the failure of assumption (4) has little effect. As long as u_i has a zero mean, OLS estimates of β are unbiased, even if the errors are heteroscedastic or autocorrelated. The standard probit estimator provides consistent estimates if the errors are autocorrelated, but heteroscedasticity leads to inconsistent estimates. There is no reason to expect the errors to be homoscedastic simply because assumptions (2) and (3) are added to the model. Thus, there is reason to fear that standard probit estimates are inconsistent in many applications.

---

[3] As mentioned in the introduction, Maddala (1983) discusses some applications of heteroscedastic tobit. Davidson and MacKinnon (1984) provide some heteroscedasticity tests for probit and logit models. Case (1992) and McMillen (1992) present estimators for spatial probit models with heteroscedasticity. Autocorrelation has received more attention. Besag (1972) presents a logit model for spatial data with autocorrelation and homoscedastic errors. Avery et al. (1983) and Poirier and Ruud (1988) present covariance matrix estimates for probit models that are consistent when the errors are autocorrelated, but heteroscedasticity causes inconsistent parameter estimates. These approaches virtually have been ignored in applications.

Simple discrete-data models require non-linear estimation techniques even when errors are homoscedastic, and the complexity increases when the errors are heteroscedastic or autocorrelated. Non-linear estimation techniques demand a lot of the data; adding complexity to the model may raise the mean squared error of the estimates by increasing the variance. Yet other complicated models are used routinely if they can be implemented by OLS (e.g., the Heckman (1976) two-stage estimator for selection bias) and there is no Monte Carlo evidence that mean squared error is high for heteroscedastic probit. Thus, it seems reasonable to speculate that apparent estimation complexity has caused heteroscedasticity to be ignored in discrete-data models.

However, the apparent estimation complexity is a misperception because heteroscedastic probit is not inordinately difficult to estimate. Assume that equations (1)-(3) hold, but modify equation (4) as follows:

$$u_i \sim N(0,\sigma_i^2) \tag{5}$$

where $\sigma_i = g(\boldsymbol{Z}_i\gamma)$. In this formulation, the heteroscedasticity is deterministic: g is a known, continuous, twice-differentiable function; $\boldsymbol{Z}$ is an $(n \times q)$ matrix of explanatory variables; and γ is a q-vector of coefficients. This is a quite general functional form, and includes most common specifications. If the error terms are not autocorrelated, the log-likelihood function is:

$$\ln L = \sum_i \left[y_i \ln\left(\Phi\left(\frac{\boldsymbol{X}_i\beta}{\sigma_i}\right)\right) + (1 - y_i)\ln\left(1 - \Phi\left(\frac{\boldsymbol{X}_i\beta}{\sigma_i}\right)\right)\right] \tag{6}$$

Even if the error terms are autocorrelated, consistent estimates of β and γ are obtained by maximizing equation (6). The estimates are also efficient if there is no autocorrelation.

The log-likelihood function (6) is general, and includes most interesting specifications. The spatial expansion model is an example of a model for which this approach is appropriate. Other spatial models may also be handled by this approach. For example, Case (1992) derives a consistent probit estimator for a model in which (using the notation of section two) the autocorrelation structure is $\boldsymbol{u} = (\boldsymbol{I} - \rho\boldsymbol{W})^{-1}\varepsilon$. After transforming the model to produce homoscedastic errors, the model is estimated using standard probit, which is included in equation (6). Note that Case's approach can only be used for data sets in which observations can be grouped into 'districts': errors are assumed to be independent across districts. This is a restrictive assumption, although it may be realistic for many problems. The approach taken here is more general and can be adapted to most econometric models, including those for non-spatial data sets.

Maximization of the log-likelihood function is straightforward. Let ϕ_i and Φ_i be the standard normal density and distribution functions evaluated at $\boldsymbol{X}_i\beta/\sigma_i$. Let $\boldsymbol{R}$ be an $(n \times (k+q))$ matrix whose first k columns consist of a sub-matrix with typical element $(\phi_i/\sigma_i)X_i$, followed by q columns with typical element $(-\sigma_i'\phi_i\sigma_i^{-2}X_i\beta)Z_i$ where $\sigma_i' \equiv d\,\sigma_i / d\,Z_i\gamma$. Let H be the diagonal matrix $\{(\Phi_i(1-\Phi_i))^{-1/2}\}$, and define $e_i \equiv y_i - \Phi_i$ and $\theta \equiv (\beta,\gamma)$. The $(k+q)$-vector of first derivatives is:

$$\frac{\partial \ln L}{\partial \theta} = R'H'He \tag{7}$$

and the expectation of the second-derivative matrix is:

$$\frac{E\,\partial^2 \ln L}{\partial \theta\, \partial \theta'} = -R'H'He \tag{8}$$

The change in estimates between two iterations is:

$$\theta_2 - \theta_1 = (R'H'HR)^{-1}R'H'He \tag{9}$$

This is simply a weighted least squares (WLS) regression of e on R with weights $(\Phi_i(1-\Phi_i))^{-1/2}$, and it is identical to the standard probit algorithm if $\gamma = 0$. Thus, heteroscedastic probit imposes little additional estimation complexity; it merely adds the additional q variables $(-\sigma_i'\phi_i\sigma_i^{-2}X_i\beta)Z_i$ to the WLS regression.[4]

Once the heteroscedastic probit model is estimated, homoscedasticity can be tested easily using either an LR or Wald test. Alternatively, homoscedasticity can be tested using the homoscedastic probit results. The LM test, which originally was proposed by Davidson and MacKinnon (1984), is:

$$e'H'HR(R'H'HR)^{-1}R'H'He \tag{10}$$

This test statistic is the explained sum of squares from a WLS regression of e on R. Under the null of homoscedasticity, the explanatory variables for the heteroscedasticity are simply $-\phi_i X_i\hat{\beta} Z_i$, where $\hat{\beta}$ is obtained from the homoscedastic probit model.

9.4 Monte Carlo Design

While the heteroscedastic probit model is not inordinately difficult to estimate, it does require more programming than standard probit. Is the programming effort worthwhile? Monte Carlo methods provide a convenient and reliable means to

[4] A two-stage estimator, in which β is estimated first and the residuals used to construct estimates of γ, is not available because $(R'H'HR)^{-1}$ is not block-diagonal between β and γ.

compare estimators and statistical tests.[5] The advantage of the Monte Carlo approach is that the true model is known to the investigator, whereas the common approach of assessing the value of an estimator by applying it to an existing data set runs the risk that the results are unique to the data set.

For this study, the true model is:

$$y_i^* = \beta_0 + \beta_1 x_i + u_i \tag{11}$$

$$y_i = 1 \quad \text{if} \quad y_i^* > 12 \tag{12}$$

$$y_i = 0 \quad \text{if} \quad y_i^* \leq 12 \tag{13}$$

$$u_i \sim N(0, \sigma_i^2) \tag{14}$$

where $i = 1, ..., n$. The exogenous explanatory variable (x_i) is drawn from a normal distribution: $x_i \sim N(8,4)$. For all experiments, $\beta_0 = 4$ and $\beta_1 = 1$. The error terms (u_i), which will be discussed in detail next, also are drawn from a normal distribution with a mean of zero and an average standard error of 1. Thus, the expected value of y_i^* is 12, which is the reason this value appears on the right hand side of equations (12) and (13). One-half of the y_i equal 1 on average. This adjusts the intercept in the probit model: since the standard errors of the u_i average 1 for all experiments, the ratio of the intercept to the average standard error is -8. The ratio of β_1 to the standard error remains unity.

The success of the Monte Carlo experiments depends on the specification of the error structure: both autocorrelation and heteroscedasticity must be specified in a simple yet realistic way. The covariance structure used by Dubin (1988, 1992) is a particularly convenient means of introducing spatial autocorrelation to a model because autocorrelation is indexed by a single parameter (λ) and large matrices do not have to be inverted in estimation. I specify a covariance matrix Ω such that the covariance between two errors indexed by i and j is:

$$\Omega_{ij} = e^{\left(\frac{-|x_i - x_j|}{\lambda}\right)} \tag{15}$$

where x is the same as in equation (11). This specification implies homoscedasticity because $\Omega_{ii} = 1$ for all i. This is a fairly general specification, and $|x_i - x_j|$ is not necessarily the distance between sites i and j as in Dubin's study.

Random numbers ξ_i are drawn from a $N(0,1)$ distribution and a new error term $\omega = V\xi$ is formed, where $VV' = \Omega$. The covariance matrix for ω is Ω, so the ω_i are homoscedastic (with unit variance) but autocorrelated. The strength of the

[5] Hendry (1984) provides a useful introduction to Monte Carlo methods. Examples of good Monte Carlo studies are Mizon and Hendry (1980), Engle et al. (1985), Ericsson (1986), Griffiths and Surekha (1986), Anselin and Griffith (1988), and Anselin (1990b).

autocorrelation is indexed by the parameter λ. I let λ range from 0 to 0.01 in increments of 0.0025. There is no autocorrelation when $\lambda = 0$. The correlations reach a high of 0.973 when $\lambda = 0.01$.

To ensure that ample autocorrelation is included in the design, I estimated the asymptotic power of the LR test when the dependent variable is not truncated and the errors are homoscedastic. The log-likelihood function for this model is presented in Dubin (1988). The true parameter values are known in the experiments. Since the information matrix is block-diagonal between λ and the other parameters, the power of the LR test can be estimated consistently by varying only λ. Thus, the restricted model has $\lambda = 0$ and $\beta_0 = 4$, $\beta_1 = 1$, and $\sigma^2 = 1$; the unrestricted model has the same values of the latter three parameters, but sets $\lambda = 0.0025, 0.005, 0.0075$, and 0.01. I calculated LR tests for 1000 draws of errors for each value of λ and for three sample sizes. For $n = 200$, the LR test rejects the incorrect null ($\lambda = 0$) 73.0 percent, 93.2 percent, 98.2 percent, and 99.3 percent of the time when $\lambda = 0.0025, 0.005, 0.0075$, and 0.01. When $\lambda = 0.0025$, the rejection rate is 96.9 percent when $n = 500$ and 99.4 percent when $n = 800$. For both $n = 500$ and $n = 800$, the null is rejected 100 percent of the time when $\lambda > 0.0025$. These are approximate asymptotic results and are only for the continuous dependent variable model, but they suggest that ample autocorrelation is included in the experiments.

The next step is to introduce heteroscedasticity to the model. I specify:

$$\sigma_i = e^{(\gamma_0 z_i + \gamma_1 z_i^2)} \tag{16}$$

where z_i is an exogenous variable drawn from a unit-normal distribution with mean zero, γ_0 and γ_1 are chosen to ensure that the σ_i equal one on average, and u_i in equation (11) is defined as $u_i = \sigma_i \omega_i$. Equation (14) follows for the distribution of u_i because $\omega_i \sim N(0,1)$. The same draws of x and z are used for all experiments. The smallest sample size is 100. Following Davidson and MacKinnon (1984) and Anselin and Griffith (1988), the draws of x and z for $n = 100$ are repeated as necessary to obtain larger sample sizes. This ensures that the average cross-moment matrices for $(1,x)$ and (z,z^2) do not change as the sample size increases.

When $\gamma_0 = \gamma_1 = \lambda = 0$, equations (11)-(16) define a standard homoscedastic probit model, and $u_i \sim N(0,1)$. The expected R^2 is 0.8 for the linear regression implied by equation (11) because $R^2 = \text{var}(\beta_0 + \beta_1 x) / \text{var}(y^*) = 4/5$. This R^2 is on the high side for cross-section continuous dependent variable models. Probit requires a fairly high R^2 in the underlying model to fit the data well, and experimentation revealed that 0.8 is a suitable value for a probit model that successfully predicts in the majority (but not all) of the cases.

Heteroscedasticity is introduced by allowing γ_0 and γ_1 to differ from zero. It may seem reasonable at first glance to simplify the model by specifying $\gamma_0 > 0$ and $\gamma_1 = 0$. The problem with this specification is that the average variance increases as γ_0 increases, so that introducing additional heteroscedasticity leads to a lower R^2. This tends to penalize the heteroscedastic probit estimator because the situations where it has the highest benefits are those where the model fits the data poorly.

Similarly, if both parameters are positive, the average variance of the u_i is larger than 1, and the R^2 is lower than when the errors are homoscedastic. This again penalizes the heteroscedastic probit model. Instead, values of γ_0 and γ_1 are chosen such that $\gamma_0 \geq 0$, $\gamma_1 \leq 0$, and $E(\sigma_i) = 1$. This assures that the R^2 remains near 0.8 when heteroscedasticity is introduced, so that the model fits the data about as well in all experiments.[6] γ_1 is not independent of γ_0, and thus plays no substantive role in the analysis.

The values of β_0, β_1, and the average R^2s are not varied across experiments because the values are representative and the focus of the study is on the effects of heteroscedasticity. γ_0 is allowed to vary from 0 to 1 in increments of 0.05, and γ_1 varies accordingly to keep $E(\sigma_i) = 1$. The errors are homoscedastic when γ_0 and γ_1 equal zero (var $\sigma = 0$), and heteroscedasticity is highest when $\sigma_0 = 1$ (var $\sigma =$ 0.56). Across all experiments, the average variance of σ is 0.32. The sample sizes range from 100 to 800 in increments of 100 (only the results for $n = 200$, 500, and 800 are reported). Probit seldom is used in very small samples; 100 is a suitable starting point.[7] 800 is a compromise between large sample size and estimation cost. Heteroscedastic probit would have little value if it could not match standard probit in mean squared error in sample sizes as large as 800.

400 replications are made of each experiment. This is not a large number; for example, Griffiths and Surekha (1986) use 5000 replications and Anselin and Griffith (1988) use 1000. However, it is not far from other commonly used numbers of replications. Anselin (1990b) uses 500 replications when linear models are estimated, but uses only 300 replications when iterative estimation techniques are used for nonlinear models. Ericsson (1986) uses 500 replications, Mizon and Hendry (1980) use 800, and Engle et al. (1985) use only 21. Monte Carlo estimation is costly, particularly when nonlinear estimation routines are used. 400 replications represent a compromise between statistical precision and estimation cost.

---

[6] The R^2 only is expected to be 0.8 on average. R^2s vary because the ω_i draws vary. Also, the same x_i and z_i are used for all replications of the experiments. Since only one set of x_i and z_i are drawn from the distribution, the variances differ from their expected values. In fact, var(x) = 3.3 and var(z) = 0.77, so the R^2s are lower than would be expected in repeated samples. The R^2s average 0.77 when $\gamma_0 = \gamma_1 = 0$, and the average falls to a low of 0.71 as the degree of heteroscedasticity rises. The R^2 falls as heteroscedasticity increases because the variance of σ increases, although the mean of σ remains 1. For the complete set of experiment results, the average R^2 is 0.74.

[7] Attempts at estimating heteroscedastic probit when $n = 50$ failed. While the estimates would converge for most error draws, convergence could not always be obtained. Failure to converge occurred when the dependent variable values could be predicted perfectly. It is an ironic feature of probit that it cannot be estimated when the model fits the data 'too' well. The weights for WLS are $(\Phi_i(1-\Phi_i))^{-1/2}$; the weights cannot be used when Φ_i equals 0 or 1.

9.5 Tests

Both standard and heteroscedastic probit models are estimated in each replication of the experiments. Standard probit provides consistent estimates of β_0 and β_1 when $\gamma_0 = \gamma_1 = 0$, but is inconsistent otherwise. Standard probit imposes that γ_0 and γ_1 equal zero, so estimates are not available for these coefficients. Heteroscedastic probit provides consistent estimates of β_0, β_1, γ_0, and γ_1 in all experiments. The autocorrelation parameter (λ) is not estimated, and $\lambda = 0$ is imposed. Imposing $\lambda = 0$ does not cause the results to be inconsistent, and one purpose of the experiments is to determine whether this incorrect assumption of the absence of spatial autocorrelation has any effect on the results in small samples.

After each experiment, there are 400 estimates of the coefficients. Let ψ_j be a coefficient estimate for the *j*th replication of an experiment, $j = 1, ..., 400$. Then the Monte Carlo coefficient estimate is simply $\bar{\psi} = (1/400) \sum_j \psi_j$, and the standard error of the estimate is $s = \left((1/399) \sum_j (\psi_j - \bar{\psi})^2\right)^{1/2}$. Since the true coefficient value (ψ^*) is known, the mean squared error (MSE) is estimated easily: MSE = $(\bar{\psi} - \psi^*)^2 + s^2$.

The estimators can be compared easily using the MSE for the β_0 and β_1 estimates. However, this comparison is unfair to heteroscedastic probit. The experiments are designed such that $\sigma = 1$ on average, so the scale of β_0 and β_1 is correct in the homoscedastic probit model despite the imposition of a false restriction on γ_0 and γ_1. This means that homoscedastic probit estimates of β_0/σ_i and β_1/σ_i are incorrect only because of the effects of heteroscedasticity (and possibly autocorrelation), not because the scale of σ_i is changing. β_0 and β_1 should be estimated fairly accurately by standard probit so long as the degree of heteroscedasticity is not severe. Heteroscedastic probit faces the more difficult task of estimating all four coefficients, and is thus at a disadvantage.

Coefficients usually are not interesting in themselves. Rather, the objective of an empirical study is to determine the marginal effect of the explanatory variables. In the probit model defined by (11)-(16), the marginal effect of x_i is:

$$\frac{\partial \Phi_i}{\partial x_i} = \frac{\phi_i \beta_1}{\sigma_i} \tag{17}$$

where ϕ_i and Φ_i are evaluated at $(\beta_0 + \beta_1 x_i) / \sigma_i$. Since σ_i depends on γ_0 and γ_1, equation (17) penalizes a model that includes incorrect values of these parameters. Thus, equation (17) is a better means to assess the relative performance of the estimators. For each replication of the Monte Carlo experiments, the marginal effect of x_i is calculated for each observation, and the statistic ψ is the average of the n values. Thus, $\psi_j = n^{-1} \sum_i \phi_i \beta_1/\sigma_i$, and ψ_j is just another Monte Carlo statistic. The true values of β_0, β_1, γ_0, and γ_1 are known, so the standard and heteroscedastic probit estimates can be compared to the true marginal effects.

Monte Carlo experiments are also used to calculate the power of specification tests. Despite providing inconsistent estimates when errors have non-constant variance, standard probit can be used to test for heteroscedasticity. Alternatively, heteroscedastic probit results can be used to test for heteroscedasticity by means of a Wald test, or the two models' results can be combined and an LR test can be calculated. The three tests are equivalent asymptotically, but may perform differently in small samples. The LM test is the easiest to calculate because it only requires standard probit estimates, but the LR or Wald tests may be preferable if the LM test has low power. The Wald test is sometimes attractive because it does not require estimates of the model under the null of homoscedasticity. But heteroscedastic probit is likely to be estimated only after standard probit, so the Wald test has no advantages and most researchers will not use it in practice. Thus, the analysis focuses on the LM and LR tests.

The explanatory variables for the heteroscedasticity are z_i and z_i^2. The LM test statistic is given by equation (10), and is distributed as χ^2 with two degrees of freedom. The LR test is calculated as $(2 \times (L_u - L_r))$, where L_u is the value of the log-likelihood function evaluated using the heteroscedastic probit estimates and L_r is the value of the function using the standard probit estimates where γ_0 and γ_1 are restricted to zero. A significance level of 0.05 is used in the Monte Carlo experiments, and the critical value for the test statistics is 5.99. In large samples, this critical value will be exceeded in about 5 percent of the replications when γ_0 and γ_1 equal zero. As the degree of heteroscedasticity increases, the proportion of times the null is rejected should approach one.

It also is interesting to assess the power of the probit heteroscedasticity tests relative to the OLS case in which the dependent variable is known. This can be done easily here since y_i is generated from the underlying variable y_i^*. A Breusch-Pagan (1979) test is used to test for heteroscedasticity in the OLS model, with a constant, z_i, and z_i^2 as the explanatory variables for the auxiliary regressions.

Specification tests sometimes have power against alternatives they are not designed to detect. For example, an omitted explanatory variable may cause a heteroscedasticity test to reject the null of homoscedasticity even if errors actually have constant variances. The Monte Carlo experiments allow us to determine whether autocorrelation can be detected by the LR or LM versions of the heteroscedasticity test.

9.6 Monte Carlo Results

While detailed estimation results are provided in Appendix 1, the most important results are summarized conveniently in plots. The MSE for the marginal effect of x_i is summarized in Figures 9.1-9.4.

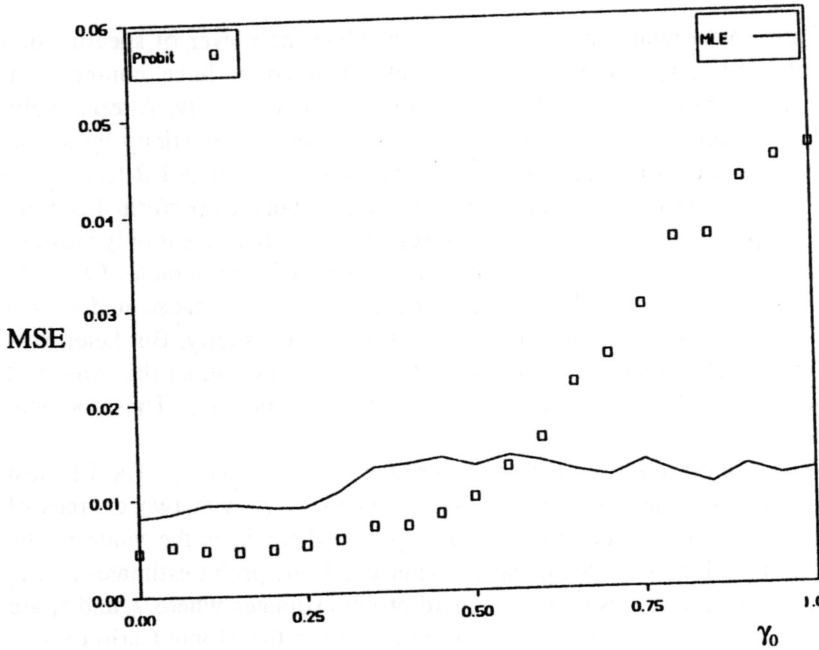

Figure 9.1: The MSE for n = 200 and λ = 0.

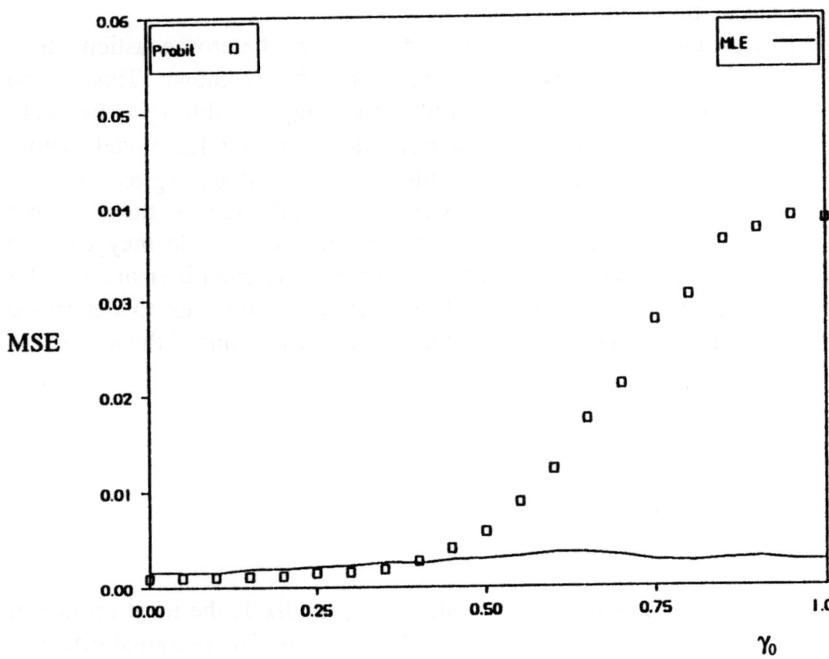

Figure 9.2: The MSE for n = 800 and λ = 0.

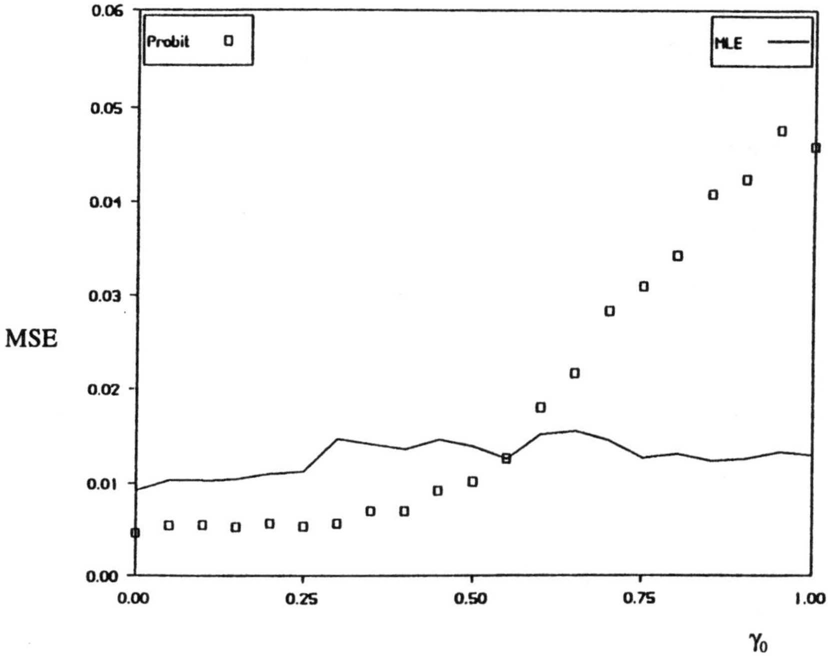

Figure 9.3: The MSE for n = 200 and λ = 0.01.

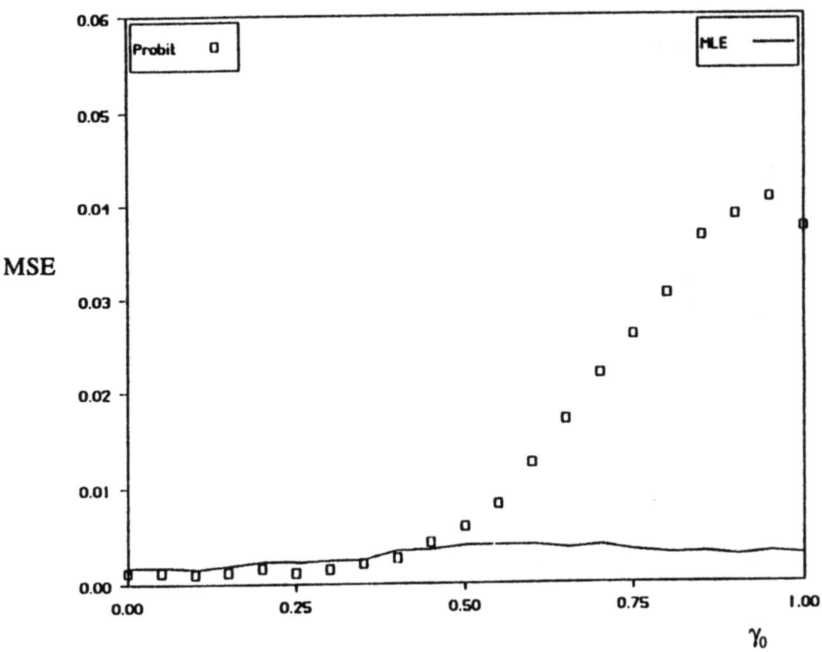

Figure 9.4: The MSE for n = 800 and λ = 0.01.

More detail is provided in Tables 9.A1 and 9.A2 of Appendix 1. Figure 9.1 shows the MSE for the smallest sample size and in the absence of autocorrelation. As expected, the MSE of standard probit is lower than for heteroscedastic probit when there is no heteroscedasticity ($\gamma_0 = 0$). As γ_0 increases, standard probit fares less well. When heteroscedasticity is severe, the MSE of heteroscedastic probit is lower than for standard probit. Thus, even for the smallest sample size (where the variance of heteroscedastic probit estimates is large), the maximum likelihood estimator can be worth the additional estimation complexity if heteroscedasticity is severe.

Figure 9.2 shows that when the sample size increases to 800 standard probit performs poorly relative to heteroscedastic probit for all but small values of γ_0. The heteroscedastic probit MSE is only slightly larger than the standard probit MSE even when errors are homoscedastic. Figures 9.3 and 9.4 duplicate the results of Figures 9.1 and 9.2 for the case of severe autocorrelation ($\lambda = 0.01$). The results are nearly identical to the $\lambda = 0$ results. If the objective of a study is to determine the marginal effect of x_i, there is much to be gained and little to be lost by using heteroscedastic probit to estimate the model except for small sample sizes.

These results, which show the value of the more complicated estimation procedure, hide some important benefits of standard probit. Tables 9.A3-9.A6 of Appendix 1 show that heteroscedastic probit is not as precise as standard probit in estimating individual coefficients. Heteroscedastic probit tends to under-estimate β_0, although this bias tends to disappear in large samples. For all sample sizes, heteroscedastic probit MSE is higher than standard probit MSE when errors are homoscedastic. As the degree of heteroscedasticity increases, the positive bias of standard probit becomes pronounced, and the relative performance of heteroscedastic probit improves. But only for the largest sample size and values of γ_0 does heteroscedastic probit have smaller MSE than the standard estimator. The results for β_1 are similar, except that the bias for standard probit is negative and the heteroscedastic probit bias is positive. The results for both β_0 and β_1 are unaffected by the presence of spatial autocorrelation.

Heteroscedastic probit also provides estimates of γ_0 and γ_1. These estimates, which are summarized in Tables 9.A7 and 9.A8 of Appendix 1, are less accurate than the β_0 and β_1 estimates. Estimates of γ_0 are biased upward, while γ_1 estimates are biased downward. The bias is reduced in large samples, but does not disappear.

These results suggest that heteroscedastic probit should not be relied on to provide accurate estimates of individual coefficients except in large samples. However, even in a sample size as small as 200, heteroscedastic probit provides better estimates of marginal effects than standard probit when heteroscedasticity is severe. In moderate sample sizes ($n = 500$), standard probit may estimate marginal effects substantially worse than the maximum likelihood estimator. Thus, which estimator is preferable depends on the sample size and whether the objective is to measure an individual coefficient or the overall marginal effect of an explanatory variable. Spatial autocorrelation has almost no effect on these results.

The estimation results for the size and power of the probit LM and LR tests are summarized in Figures 9.5-9.8.

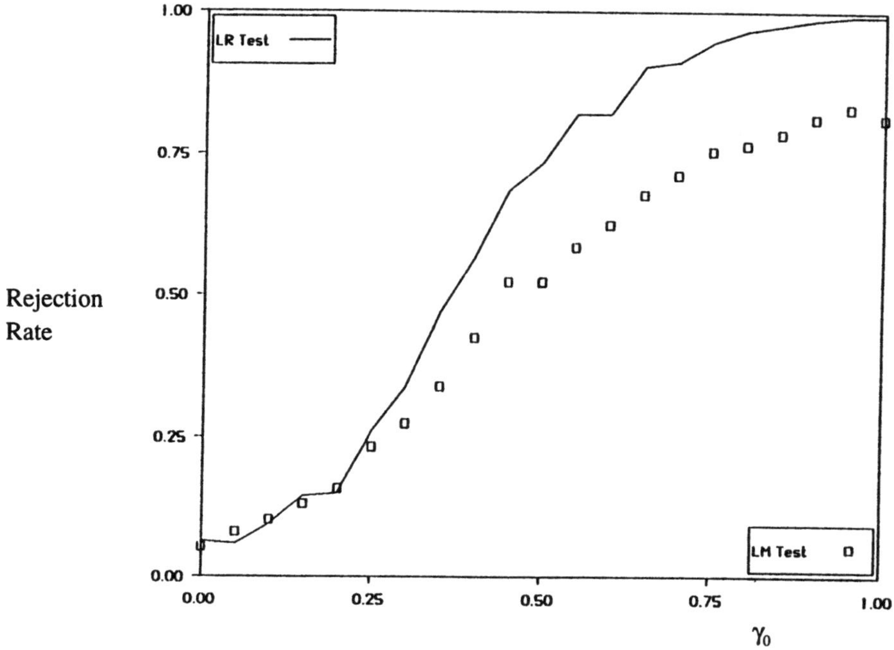

Figure 9.5: Heteroscedasticity test rejection rates for n = 200 and $\lambda = 0$.

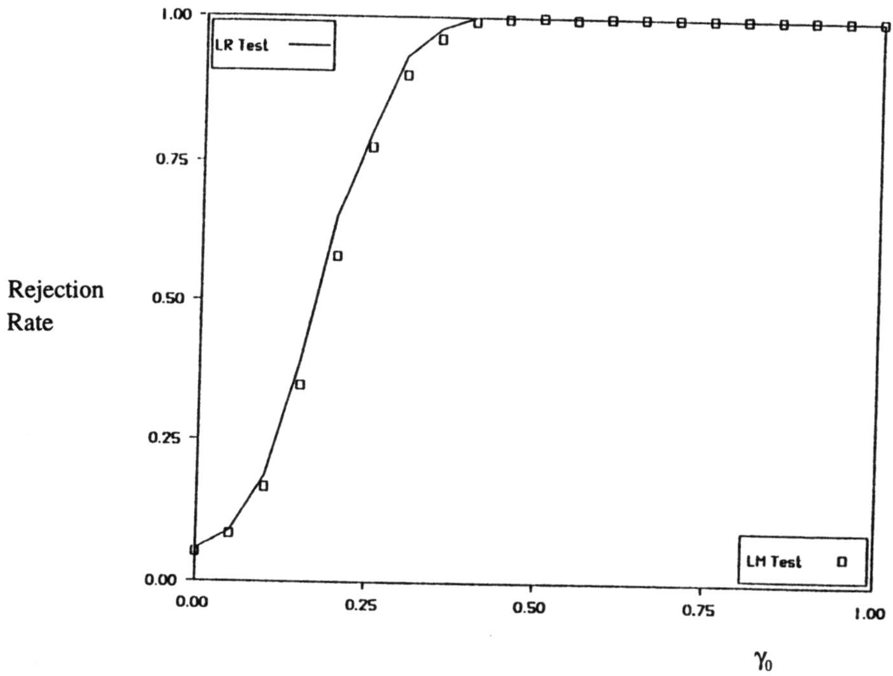

Figure 9.6: Heteroscedasticity test rejection rates for n = 800 and $\lambda = 0$.

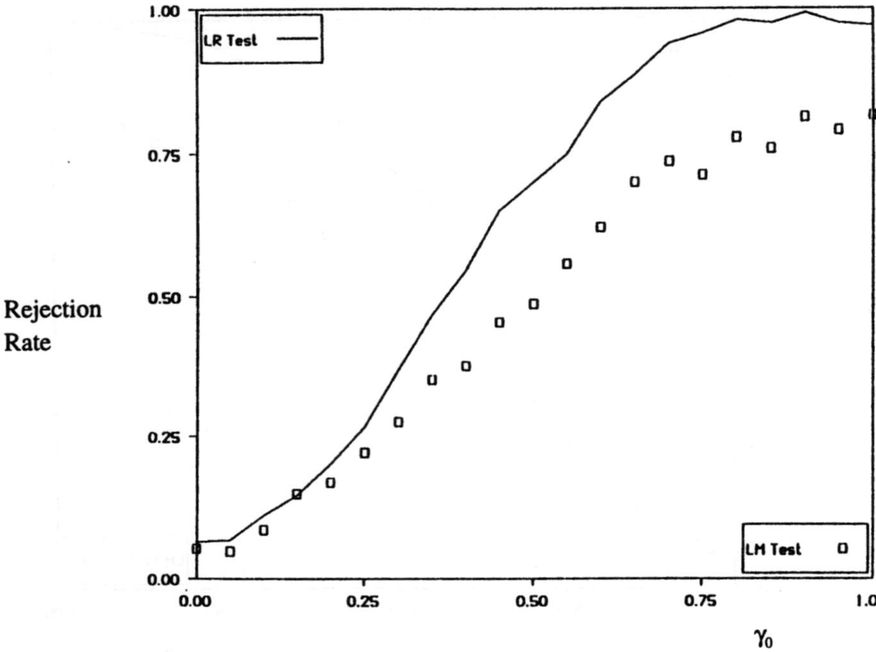

Figure 9.7: Heteroscedasticity test rejection rates for n = 200 and λ = 0.01.

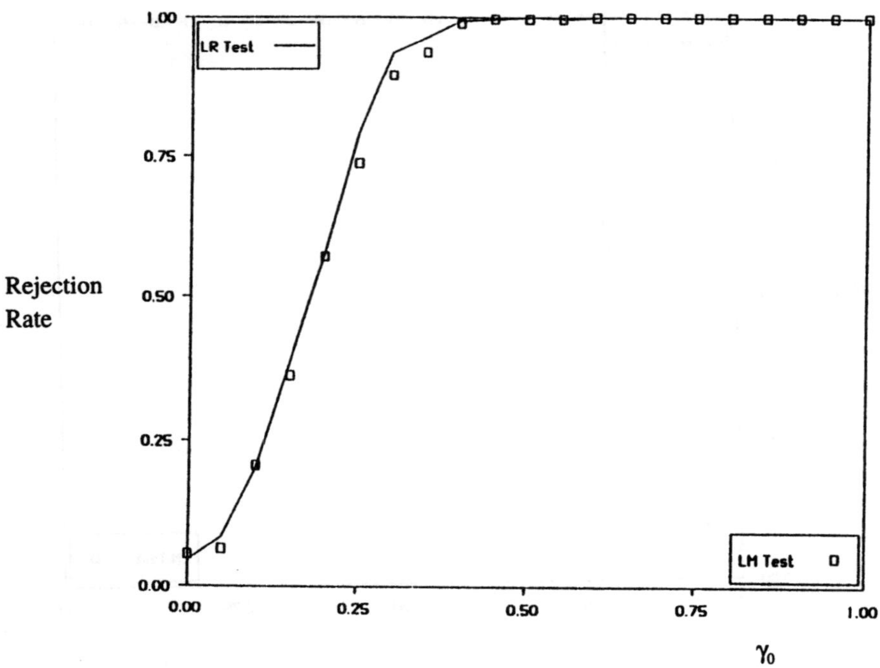

Figure 9.8: Heteroscedasticity test rejection rates for n = 800 and λ = 0.01.

More detail and the results of the Breusch-Pagan (1979) heteroscedasticity test are provided in Table 9.A9 of Appendix 1. The estimated size of both tests is close to the nominal value of 0.05. The LR test has higher power against heteroscedasticity than the LM test in small samples. As the sample size increases, the LM test power approaches that of the LR test.

The low power of the probit LM test in small samples is cause for some concern. The marginal effect of x_i is over-estimated by standard probit when γ_0 is large, but the LM test may not reject homoscedasticity. It is a risky strategy to estimate the standard probit model and test the homoscedasticity assumption using an LM test. The null may be accepted incorrectly, leading to biased estimates of the marginal effect. A safer strategy is to base the heteroscedasticity test on a comparison of the standard and heteroscedastic probit results, the LR test. The superior power of the LR test combined with the accuracy of the heteroscedastic probit model in estimating the marginal effect makes the additional estimation cost worthwhile.

Results for the Breusch-Pagan heteroscedasticity test are presented in Table 9.A9 of Appendix 1. The dependent variable for this test is y_i^* in equation (1), which is the underlying latent variable that is used to construct the dependent variable for the probit model. Not surprisingly, this test has more power than either the probit LM or LR tests. It is surprising, however, that all three tests fail to detect the presence of spatial autocorrelation. This is true for all sample sizes and whether errors have constant variance or are heteroscedastic. In contrast, Anselin and Griffith (1988) find that the power of the Breusch-Pagan test increases when there is severe positive spatial autocorrelation. Anselin (1990b) finds similar results for other specification tests.

The values of λ considered here are large enough to cause severe autocorrelation; why is it that the tests are not affected by λ? Notice that the earlier studies both model spatial autocorrelation as $u = \rho \boldsymbol{W} u + \varepsilon$. As discussed in section two, the errors in the OLS specification become $u = (\boldsymbol{I} - \rho \boldsymbol{W})^{-1}\varepsilon$, which are heteroscedastic even though the ε_i are homoscedastic. Thus, when these studies increase the degree of spatial autocorrelation by increasing ρ, they also increase the amount of heteroscedasticity. In contrast, the specification of spatial autocorrelation used here does not affect the amount of heteroscedasticity. It is reasonable to speculate that this is the source of the difference in the implied effects of spatial autocorrelation on the performance of specification tests.

9.7 Conclusions

In spatial data sets, nearby observations typically exhibit similar characteristics, and one of the goals of spatial econometrics is to model spatial heterogeneity. It is reasonable to expect variances as well as means to differ across space. Indeed, heteroscedasticity is an implicit assumption in standard spatial models. However, heteroscedasticity is ignored in probit models, even when there is reason to expect variances to differ across observations. But neglecting to account for heteroscedasticity produces inconsistent estimates.

The reason for this neglect is that heteroscedastic probit is harder to estimate than the standard model, and the additional estimation complexity may not be worth its cost if the more complicated estimator has high variance. The Monte Carlo results presented in this study suggest that heteroscedastic probit is worth its cost in moderate to large samples. Standard probit estimates of the marginal effects of explanatory variables are biased significantly when there is heteroscedasticity. Even in small samples, heteroscedastic probit estimates have lower MSE when heteroscedasticity is severe. Thus, the Monte Carlo results provide a strong argument for using the heteroscedastic probit estimator. A heteroscedasiticity option has been added to LIMDEP, and the WLS algorithm needed to estimate the model can be implemented easily using other software. Appendix 2 presents LIMDEP, RATS, and TSP programs that can estimate the model analyzed here.

The Monte Carlo results suggest that heteroscedasticity is more of a cause for concern in probit models than spatial autocorrelation. Both standard and heteroscedastic probit estimators are consistent when errors are autocorrelated. This asymptotic result is borne out in small samples: the MSE largely are unaffected by autocorrelation. However, traditional standard error estimates are biased when autocorrelation is ignored, which affects hypothesis testing. Thus, the effects of autocorrelation may not be entirely benign even though parameter estimates are consistent.

Heteroscedasticity causes inconsistent estimates in other standard discrete-data models such as logit and tobit. This study has demonstrated that standard probit has undesirable properties when errors are heteroscedastic, and it may be conjectured that other discrete-data models have similar problems. A profitable area for future research is to develop and apply heteroscedastic versions of other discrete-data model estimators.

References

Anderson, J.E., Estimating Generalized Urban Density Functions, *Journal of Urban Economics* 18, 1-10, 1985.

Anselin, L., *Spatial Econometrics: Methods and Models*, Dordrecht: Kluwer Academic Publishers, 1988.

Anselin, L., Some Robust Approaches to Testing and Estimation in Spatial Regression Analysis, *Regional Science and Urban Economics*, 20, 141-163, 1990a.

Anselin, L., Spatial Dependence and Spatial Structural Instability in Applied Regression Analysis, *Journal of Regional Science*, 30, 185-207, 1990b.

Anselin, L., Spatial Dependence and Spatial Heterogeneity: Model Specification Issues in the Spatial Expansion Paradigm, in: J.P. Jones III and E. Casetti (eds.), *Applications of the Expansion Method*, New York: Routledge, 1992.

Anselin, L. and D.A. Griffith, Do Spatial Effects Really Matter in Regression Analysis?, *Papers of the Regional Science Association*, 65, 11-34, 1988.

Avery, R.B., L.P. Hansen and V.J. Hotz, Multiperiod Probit Models and Orthogonality Condition Estimation, *International Economic Review*, 24, 21-35, 1983.

Besag, J.E., Nearest-Neighbour Systems and the Auto-Logistic Model for Binary Data, *Journal of the Royal Statistical Society*, Series B, 34, 75-83, 1972.

Breusch, T. and A. Pagan, A Simple Test for Heteroscedasticity and Random Coefficient Variation, *Econometrica*, 47, 1287-1294, 1979.

Case, A., Neighborhood Influence and Technological Change, *Regional Science and Urban Economics*, 22, 491-508, 1992.

Casetti, E., Generating Models by the Spatial Expansion Method: Applications to Geographical Research, *Geographical Analysis*, 4, 81-91, 1972.

Davidson, R. and J.G. MacKinnon, Convenient Specification Tests for Logit and Probit Models, *Journal of Econometrics*, 25, 241-262, 1984.

Dubin, R.A., Estimation of Regression Coefficients in the Presence of Spatially Autocorrelated Error Terms, *Review of Economics and Statistics*, 70, 466-474, 1988.

Dubin, R.A., Spatial Autocorrelation and Neighborhood Quality, *Regional Science and Urban Economics*, 22, 433-452, 1992.

Engle, R.F., D.F. Hendry and D. Trumble, Small-Sample Properties of ARCH Estimators and Tests, *Canadian Journal of Economics*, 18, 66-93, 1985.

Ericsson, N.R., Post-Simulation Analysis of Monte Carlo Experiments: Interpreting Pesaran's Study of Non-Nested Hypothesis Test Statistics, *Review of Economic Studies*, 53, 691-707, 1986.

Griffith, D.A., Modelling Urban Population Density in a Multicentered City, *Journal of Urban Economics*, 9, 298-310, 1981.

Griffiths, W.E. and K. Surekha, A Monte Carlo Evaluation of the Power of Some Tests for Heteroscedasticity, *Journal of Econometrics*, 31, 219-231, 1986.

Haining, R., Estimating Spatial Means with an Application to Remotely Sensed Data, *Communications in Statistics*, 17, 573-597, 1988.

Heckman, J.J., The Common Structure of Statistical Models of Truncation, Sample Selection, and Limited Dependent Variables and a Simple Estimator for such Models, *Annals of Economic and Social Measurement*, 5, 475-492, 1976.

Hendry, D.F., Monte Carlo Experimentation in Econometrics, in: Z. Griliches and M.D. Intriligator (eds.), *Handbook of Econometrics*, vol. II, Amsterdam: North-Holland, 1984.

Johnston, J., *Econometric Methods*, New York: McGraw-Hill, 1984.

Jones III, J.P. and E. Casetti, *Applications of the Expansion Method*, New York: Routledge, 1992.

Judge, G.G., R.C. Hill, W.E. Griffiths, H. Lütkepohl and T.C. Lee, *Introduction to the Theory and Practice of Econometrics*, New York: John Wiley & Sons, 1988.

Lahiri, K. and R.P. Numrich, An Econometric Study on the Dynamics of Urban Spatial Structure, *Journal of Urban Economics*, 14, 55-79, 1983.

Maddala, G.S., *Limited-Dependent and Qualitative Variables in Econometrics*, New York: Cambridge University Press, 1983.

Maddala, G.S., *Introduction to Econometrics*, New York: Macmillan Publishing Company, 1988.

McMillen, D.P., Probit with Spatial Autocorrelation, *Journal of Regional Science*, 32, 335-348, 1992.

Mizon, G.E. and D.F. Hendry, An Empirical Application and Monte Carlo Analysis of Tests of Dynamic Specification, *Review of Economic Studies*, 47, 21-45, 1980.

Poirier, D.J. and P.A. Ruud, Probit with Dependent Observations, *Review of Economic Studies*, 55, 593-614, 1988.

Appendix 1: Monte Carlo Results

Table 9.A1: Probit marginal effect of *x*.

	Bias					MSE				
$\gamma_0 \backslash \lambda$	0.0000	0.0025	0.0050	0.0075	0.0500	0.0000	0.0025	0.0050	0.0075	0.0500
					n = 200					
0.0	-0.000	-0.000	0.001	0.000	0.000	0.004	0.005	0.004	0.005	0.005
0.1	-0.000	0.000	-0.000	-0.000	-0.000	0.005	0.004	0.006	0.005	0.005
0.2	0.000	-0.000	0.001	0.000	0.001	0.005	0.005	0.005	0.005	0.006
0.3	0.002	0.002	0.001	0.002	0.001	0.006	0.005	0.005	0.006	0.006
0.4	0.004	0.004	0.004	0.004	0.004	0.007	0.007	0.007	0.007	0.007
0.5	0.007	0.007	0.007	0.007	0.007	0.010	0.010	0.010	0.011	0.010
0.6	0.010	0.010	0.010	0.011	0.011	0.016	0.016	0.017	0.018	0.018
0.7	0.014	0.014	0.014	0.014	0.015	0.024	0.025	0.025	0.026	0.028
0.8	0.018	0.017	0.017	0.017	0.017	0.037	0.035	0.035	0.036	0.034
0.9	0.019	0.019	0.019	0.018	0.019	0.043	0.043	0.042	0.041	0.042
1.0	0.020	0.020	0.019	0.020	0.020	0.046	0.044	0.044	0.045	0.046
					n = 500					
0.0	0.000	0.000	0.000	-0.000	0.000	0.002	0.002	0.002	0.002	0.002
0.1	0.000	0.000	-0.000	-0.000	-0.001	0.002	0.002	0.002	0.002	0.002
0.2	0.001	0.000	0.000	0.000	0.000	0.002	0.002	0.002	0.002	0.002
0.3	0.001	0.002	0.002	0.001	0.002	0.002	0.002	0.002	0.002	0.003
0.4	0.004	0.004	0.003	0.004	0.003	0.003	0.003	0.003	0.004	0.003
0.5	0.007	0.007	0.006	0.007	0.006	0.007	0.007	0.006	0.007	0.006
0.6	0.010	0.010	0.011	0.011	0.010	0.013	0.013	0.014	0.014	0.013
0.7	0.014	0.014	0.014	0.014	0.014	0.022	0.022	0.022	0.022	0.023
0.8	0.018	0.017	0.017	0.018	0.018	0.033	0.032	0.031	0.034	0.034
0.9	0.019	0.019	0.019	0.019	0.019	0.040	0.038	0.038	0.039	0.039
1.0	0.019	0.019	0.020	0.019	0.020	0.040	0.040	0.042	0.039	0.042
					n = 800					
0.0	-0.000	0.000	-0.000	0.000	0.000	0.001	0.001	0.001	0.001	0.001
0.1	0.000	-0.000	0.000	-0.000	-0.000	0.001	0.001	0.001	0.001	0.001
0.2	0.000	0.000	0.000	0.001	0.000	0.001	0.001	0.001	0.001	0.002
0.3	0.002	0.001	0.002	0.001	0.001	0.002	0.002	0.002	0.002	0.002
0.4	0.004	0.003	0.004	0.003	0.003	0.003	0.003	0.003	0.003	0.003
0.5	0.007	0.007	0.006	0.006	0.007	0.006	0.006	0.006	0.006	0.006
0.6	0.010	0.011	0.010	0.010	0.011	0.012	0.012	0.012	0.012	0.013
0.7	0.014	0.014	0.014	0.014	0.014	0.021	0.021	0.022	0.021	0.022
0.8	0.017	0.017	0.018	0.017	0.017	0.030	0.031	0.032	0.031	0.030
0.9	0.019	0.019	0.019	0.019	0.019	0.037	0.037	0.037	0.038	0.039
1.0	0.019	0.019	0.019	0.019	0.019	0.038	0.038	0.039	0.038	0.038

Table 9.A2: MLE marginal effect of *x*.

$\gamma_0 \backslash \lambda$	Bias 0.0000	Bias 0.0025	Bias 0.0050	Bias 0.0075	Bias 0.0500	MSE 0.0000	MSE 0.0025	MSE 0.0050	MSE 0.0075	MSE 0.0500
					$n = 200$					
0.0	-0.001	-0.002	-0.001	-0.000	-0.001	0.008	0.009	0.008	0.009	0.009
0.1	-0.002	-0.001	-0.002	-0.002	-0.002	0.009	0.008	0.010	0.010	0.010
0.2	-0.002	-0.003	-0.003	-0.003	-0.003	0.009	0.011	0.011	0.011	0.011
0.3	-0.002	-0.003	-0.003	-0.003	-0.003	0.010	0.012	0.012	0.012	0.015
0.4	-0.003	-0.003	-0.003	-0.003	-0.003	0.013	0.014	0.012	0.014	0.014
0.5	-0.003	-0.003	-0.003	-0.002	-0.003	0.013	0.015	0.015	0.015	0.014
0.6	-0.002	-0.002	-0.002	-0.001	-0.001	0.013	0.014	0.014	0.014	0.015
0.7	-0.000	0.001	-0.001	-0.000	0.001	0.011	0.011	0.013	0.013	0.015
0.8	0.002	0.001	0.001	0.002	0.000	0.012	0.013	0.012	0.013	0.013
0.9	0.003	0.002	0.002	0.002	0.002	0.012	0.013	0.012	0.013	0.013
1.0	0.002	0.001	0.001	0.001	0.002	0.012	0.011	0.011	0.012	0.013
					$n = 500$					
0.0	-0.001	-0.000	-0.001	-0.001	-0.000	0.003	0.003	0.003	0.003	0.003
0.1	-0.001	-0.001	-0.001	-0.001	-0.001	0.003	0.003	0.003	0.003	0.003
0.2	-0.001	-0.001	-0.001	-0.001	-0.001	0.003	0.003	0.004	0.003	0.004
0.3	-0.001	-0.002	-0.001	-0.002	-0.001	0.004	0.005	0.004	0.004	0.004
0.4	-0.001	-0.002	-0.002	-0.002	-0.002	0.004	0.005	0.005	0.005	0.006
0.5	-0.001	-0.002	-0.001	-0.001	-0.002	0.005	0.005	0.006	0.006	0.006
0.6	-0.001	-0.001	-0.001	-0.001	-0.002	0.005	0.006	0.005	0.007	0.007
0.7	-0.001	-0.001	-0.002	-0.001	-0.000	0.004	0.005	0.005	0.007	0.006
0.8	0.001	0.000	0.000	0.001	0.001	0.004	0.005	0.005	0.005	0.005
0.9	0.002	0.001	0.001	0.001	0.000	0.005	0.004	0.004	0.005	0.005
1.0	0.002	0.001	0.002	0.001	0.002	0.005	0.004	0.004	0.005	0.005
					$n = 800$					
0.0	-0.001	-0.000	-0.001	-0.000	-0.000	0.002	0.002	0.002	0.002	0.002
0.1	-0.000	-0.001	-0.000	-0.000	-0.001	0.002	0.002	0.002	0.002	0.002
0.2	-0.001	-0.001	-0.000	-0.001	-0.001	0.002	0.002	0.002	0.002	0.002
0.3	-0.000	-0.001	-0.001	-0.001	-0.002	0.002	0.002	0.003	0.003	0.002
0.4	-0.001	-0.001	-0.001	-0.001	-0.001	0.003	0.003	0.003	0.003	0.003
0.5	-0.001	-0.001	-0.002	-0.001	-0.001	0.003	0.003	0.003	0.004	0.004
0.6	-0.001	-0.001	-0.001	-0.001	-0.001	0.004	0.004	0.004	0.004	0.004
0.7	-0.000	-0.000	-0.000	-0.001	-0.001	0.003	0.003	0.004	0.004	0.004
0.8	0.000	0.000	0.001	0.000	-0.000	0.003	0.003	0.003	0.003	0.003
0.9	0.001	0.000	0.001	0.001	0.001	0.003	0.003	0.003	0.003	0.003
1.0	0.001	0.001	0.001	0.000	0.001	0.003	0.003	0.003	0.003	0.003

Table 9.A3: Probit β_0 estimates.

$\gamma_0 \backslash \lambda$	Bias					MSE				
	0.0000	0.0025	0.0050	0.0075	0.0500	0.0000	0.0025	0.0050	0.0075	0.0500
					$n = 200$					
0.0	-0.214	-0.171	-0.306	-0.272	-0.316	1.284	1.159	1.352	1.273	1.398
0.1	-0.173	-0.225	-0.258	-0.218	-0.284	1.220	1.271	1.559	1.411	1.779
0.2	-0.180	-0.036	-0.181	-0.114	-0.208	1.293	1.209	1.193	1.341	1.436
0.3	-0.233	-0.121	-0.044	-0.171	-0.057	1.430	1.226	1.259	1.743	1.436
0.4	0.020	0.015	-0.054	-0.039	-0.078	1.307	1.253	1.477	1.580	1.481
0.5	0.073	0.022	0.043	-0.064	0.050	1.148	1.456	1.292	1.559	1.542
0.6	0.102	0.117	0.103	0.013	0.051	1.295	1.275	1.305	1.465	1.715
0.7	0.188	0.204	0.210	0.150	0.018	1.247	1.343	1.335	1.404	1.410
0.8	0.203	0.184	0.210	0.223	0.273	1.221	1.427	1.543	1.411	1.537
0.9	0.205	0.209	0.272	0.304	0.227	1.874	1.607	1.464	1.639	1.639
1.0	0.229	0.279	0.234	0.184	0.179	1.406	1.378	1.522	1.674	1.575
					$n = 500$					
0.0	-0.091	-0.076	-0.110	-0.117	-0.172	0.486	0.390	0.474	0.549	0.587
0.1	-0.083	-0.119	-0.076	-0.048	-0.013	0.395	0.441	0.469	0.510	0.492
0.2	-0.084	-0.018	-0.048	0.019	-0.086	0.487	0.456	0.569	0.510	0.582
0.3	0.109	0.044	0.052	0.039	0.026	0.471	0.514	0.465	0.456	0.561
0.4	0.124	0.151	0.216	0.119	0.160	0.420	0.487	0.499	0.616	0.582
0.5	0.205	0.154	0.270	0.179	0.215	0.540	0.538	0.555	0.603	0.596
0.6	0.259	0.261	0.221	0.235	0.234	0.543	0.586	0.497	0.628	0.643
0.7	0.367	0.291	0.346	0.375	0.290	0.572	0.617	0.675	0.716	0.660
0.8	0.350	0.381	0.401	0.272	0.285	0.576	0.685	0.658	0.663	0.686
0.9	0.370	0.428	0.405	0.431	0.403	0.655	0.654	0.691	0.777	0.713
1.0	0.473	0.443	0.394	0.501	0.408	0.707	0.656	0.673	0.820	0.692
					$n = 800$					
0.0	-0.012	-0.083	-0.035	-0.102	-0.074	0.246	0.272	0.319	0.352	0.306
0.1	-0.063	0.011	-0.073	-0.044	-0.035	0.299	0.298	0.301	0.311	0.283
0.2	0.052	0.019	0.011	-0.008	0.019	0.293	0.279	0.319	0.318	0.352
0.3	0.092	0.151	0.096	0.152	0.168	0.329	0.349	0.304	0.371	0.349
0.4	0.162	0.170	0.122	0.201	0.165	0.331	0.337	0.300	0.393	0.368
0.5	0.268	0.274	0.248	0.268	0.236	0.383	0.384	0.405	0.383	0.394
0.6	0.303	0.297	0.300	0.321	0.288	0.379	0.397	0.372	0.410	0.388
0.7	0.377	0.356	0.344	0.426	0.347	0.477	0.427	0.451	0.492	0.479
0.8	0.426	0.391	0.342	0.382	0.443	0.510	0.475	0.466	0.523	0.537
0.9	0.461	0.463	0.459	0.424	0.413	0.531	0.506	0.520	0.477	0.561
1.0	0.531	0.558	0.540	0.556	0.555	0.592	0.622	0.605	0.606	0.631

Table 9.A4: MLE β_0 estimates.

$\gamma_0\backslash\lambda$	Bias 0.0000	Bias 0.0025	Bias 0.0050	Bias 0.0075	Bias 0.0500	MSE 0.0000	MSE 0.0025	MSE 0.0050	MSE 0.0075	MSE 0.0500
					n = 200					
0.0	-0.181	-0.095	-0.167	-0.208	-0.226	1.908	1.974	1.863	2.055	2.114
0.1	-0.094	-0.135	-0.138	-0.195	-0.220	2.095	1.803	2.086	2.111	2.537
0.2	-0.184	-0.083	-0.154	-0.119	-0.211	2.018	2.232	2.047	2.015	2.300
0.3	-0.279	-0.116	-0.097	-0.212	-0.079	2.007	1.794	2.013	2.219	2.206
0.4	-0.163	-0.157	-0.155	-0.162	-0.187	2.325	2.108	2.344	2.613	2.441
0.5	-0.162	-0.149	-0.223	-0.366	-0.185	1.961	2.187	2.504	2.585	2.514
0.6	-0.185	-0.256	-0.169	-0.354	-0.257	2.073	2.244	2.053	2.876	2.679
0.7	-0.191	-0.241	-0.122	-0.208	-0.363	2.220	2.171	2.244	2.507	2.920
0.8	-0.234	-0.152	-0.185	-0.202	-0.069	2.109	2.190	2.555	2.219	2.562
0.9	-0.327	-0.261	-0.168	-0.218	-0.233	3.041	2.932	2.475	2.487	2.493
1.0	-0.181	-0.089	-0.104	-0.220	-0.280	2.100	2.030	1.999	2.669	2.338
					n = 500					
0.0	0.037	0.001	-0.054	-0.070	-0.075	0.687	0.626	0.746	0.827	0.797
0.1	-0.033	-0.066	-0.015	-0.067	0.043	0.562	0.665	0.695	0.710	0.698
0.2	-0.105	-0.079	-0.140	-0.024	-0.124	0.681	0.714	0.885	0.737	0.779
0.3	-0.055	-0.029	-0.099	-0.037	-0.107	0.820	0.854	0.642	0.653	0.853
0.4	-0.072	-0.099	0.013	-0.077	-0.051	0.669	0.819	0.703	0.888	0.945
0.5	-0.064	-0.084	-0.048	-0.071	-0.037	0.704	0.818	0.763	0.850	0.865
0.6	-0.113	-0.088	-0.097	-0.095	-0.132	0.880	0.809	0.679	0.812	0.844
0.7	0.001	-0.074	0.066	-0.061	-0.105	0.740	0.925	0.912	1.008	0.859
0.8	-0.057	-0.007	-0.039	-0.146	-0.137	0.673	0.793	0.974	0.973	1.002
0.9	-0.152	-0.011	-0.050	-0.079	0.002	0.837	0.724	0.798	0.972	0.845
1.0	-0.115	-0.064	-0.098	-0.053	-0.108	0.831	0.777	0.776	1.115	0.922
					n = 800					
0.0	0.078	-0.022	-0.006	-0.037	-0.013	0.440	0.409	0.482	0.468	0.429
0.1	-0.061	0.041	-0.072	-0.072	-0.048	0.428	0.428	0.499	0.496	0.416
0.2	-0.005	-0.008	-0.071	-0.084	-0.059	0.466	0.379	0.460	0.459	0.486
0.3	-0.075	-0.016	-0.055	0.004	0.048	0.495	0.453	0.477	0.484	0.456
0.4	-0.022	-0.036	-0.049	-0.006	-0.040	0.446	0.441	0.418	0.425	0.532
0.5	0.000	-0.020	0.013	-0.007	-0.027	0.421	0.475	0.407	0.432	0.486
0.6	-0.068	-0.024	-0.032	-0.046	-0.056	0.457	0.443	0.455	0.492	0.437
0.7	-0.072	-0.050	-0.056	0.018	-0.051	0.497	0.499	0.505	0.506	0.530
0.8	-0.071	-0.011	-0.088	-0.087	0.013	0.502	0.502	0.539	0.542	0.503
0.9	-0.103	-0.056	-0.028	-0.060	-0.077	0.535	0.552	0.528	0.466	0.542
1.0	-0.027	-0.039	0.002	0.002	-0.004	0.485	0.523	0.492	0.493	0.511

Table 9.A5: Probit β_1 estimates.

$\gamma_0 \backslash \lambda$	Bias					MSE				
	0.0000	0.0025	0.0050	0.0075	0.0500	0.0000	0.0025	0.0050	0.0075	0.0500
					n = 200					
0.0	0.027	0.021	0.039	0.036	0.040	2.041	1.863	2.138	2.040	2.196
0.1	0.022	0.030	0.031	0.028	0.035	1.974	2.033	2.476	2.204	2.811
0.2	0.022	0.005	0.023	0.015	0.026	2.043	1.917	1.896	2.141	2.292
0.3	0.028	0.015	0.005	0.021	0.007	2.294	1.929	1.987	2.718	2.249
0.4	-0.003	-0.001	0.008	0.004	0.010	2.142	2.019	2.388	2.466	2.310
0.5	-0.010	-0.004	-0.005	0.006	-0.007	1.821	2.281	2.031	2.471	2.416
0.6	-0.014	-0.015	-0.014	-0.002	-0.006	2.028	2.004	2.100	2.296	2.682
0.7	-0.025	-0.026	-0.027	-0.019	-0.002	1.937	2.091	2.077	2.220	2.205
0.8	-0.025	-0.025	-0.027	-0.029	-0.035	1.950	2.279	2.419	2.226	2.409
0.9	-0.027	-0.028	-0.035	-0.040	-0.029	2.931	2.549	2.303	2.581	2.555
1.0	-0.029	-0.036	-0.032	-0.025	-0.024	2.224	2.215	2.377	2.613	2.521
					n = 500					
0.0	0.012	0.010	0.013	0.014	0.022	0.774	0.621	0.752	0.857	0.916
0.1	0.011	0.015	0.009	0.006	0.001	0.626	0.700	0.744	0.790	0.769
0.2	0.011	0.002	0.006	-0.003	0.010	0.752	0.711	0.895	0.792	0.920
0.3	-0.014	-0.005	-0.006	-0.006	-0.003	0.744	0.805	0.739	0.715	0.880
0.4	-0.016	-0.020	-0.027	-0.015	-0.021	0.669	0.785	0.791	0.969	0.921
0.5	-0.026	-0.020	-0.035	-0.023	-0.028	0.857	0.860	0.887	0.958	0.938
0.6	-0.033	-0.033	-0.028	-0.029	-0.031	0.858	0.923	0.795	0.999	1.017
0.7	-0.047	-0.038	-0.045	-0.047	-0.037	0.908	0.976	1.059	1.133	1.037
0.8	-0.044	-0.049	-0.051	-0.035	-0.037	0.897	1.089	1.042	1.056	1.081
0.9	-0.047	-0.055	-0.052	-0.054	-0.052	1.018	1.053	1.087	1.215	1.137
1.0	-0.060	-0.057	-0.049	-0.063	-0.051	1.130	1.050	1.056	1.268	1.069
					n = 800					
0.0	0.002	0.011	0.004	0.013	0.009	0.387	0.431	0.505	0.549	0.481
0.1	0.008	-0.001	0.009	0.005	0.004	0.469	0.471	0.475	0.487	0.437
0.2	-0.006	-0.003	-0.001	0.001	-0.002	0.464	0.440	0.506	0.506	0.570
0.3	-0.012	-0.019	-0.012	-0.019	-0.022	0.525	0.569	0.484	0.590	0.555
0.4	-0.021	-0.023	-0.016	-0.026	-0.022	0.529	0.541	0.475	0.608	0.577
0.5	-0.034	-0.035	-0.032	-0.034	-0.030	0.613	0.606	0.638	0.604	0.628
0.6	-0.039	-0.037	-0.038	-0.041	-0.036	0.601	0.629	0.584	0.657	0.610
0.7	-0.048	-0.045	-0.044	-0.054	-0.044	0.760	0.675	0.721	0.783	0.750
0.8	-0.054	-0.050	-0.044	-0.049	-0.056	0.805	0.758	0.740	0.827	0.838
0.9	-0.059	-0.059	-0.058	-0.054	-0.052	0.850	0.806	0.824	0.767	0.874
1.0	-0.067	-0.070	-0.067	-0.070	-0.070	0.937	0.992	0.935	0.956	0.992

Table 9.A6: MLE β_1 estimates.

	Bias					MSE				
$\gamma_0\backslash\lambda$	0.0000	0.0025	0.0050	0.0075	0.0500	0.0000	0.0025	0.0050	0.0075	0.0500
					n = 200					
0.0	0.024	0.013	0.022	0.028	0.029	3.082	3.165	2.992	3.301	3.365
0.1	0.014	0.019	0.017	0.026	0.027	3.386	2.888	3.319	3.408	4.037
0.2	0.024	0.012	0.020	0.017	0.027	3.200	3.625	3.248	3.219	3.705
0.3	0.034	0.015	0.013	0.028	0.011	3.202	2.855	3.211	3.579	3.497
0.4	0.022	0.022	0.021	0.020	0.024	3.822	3.408	3.749	4.125	3.835
0.5	0.021	0.018	0.029	0.046	0.024	3.177	3.420	4.002	4.102	3.925
0.6	0.023	0.033	0.021	0.046	0.033	3.270	3.552	3.286	4.565	4.234
0.7	0.023	0.032	0.015	0.027	0.046	3.480	3.447	3.605	3.981	4.665
0.8	0.031	0.019	0.023	0.026	0.009	3.358	3.479	3.977	3.533	4.066
0.9	0.041	0.032	0.021	0.027	0.029	4.830	4.655	3.955	3.953	3.951
1.0	0.023	0.011	0.011	0.026	0.034	3.353	3.253	3.156	4.206	3.724
					n = 500					
0.0	-0.004	0.001	0.007	0.009	0.010	1.101	1.015	1.194	1.305	1.268
0.1	0.005	0.009	0.002	0.009	-0.006	0.891	1.066	1.101	1.106	1.102
0.2	0.014	0.010	0.018	0.003	0.015	1.052	1.130	1.411	1.148	1.246
0.3	0.007	0.005	0.013	0.004	0.015	1.304	1.361	1.039	1.031	1.357
0.4	0.010	0.013	-0.000	0.010	0.007	1.054	1.292	1.123	1.412	1.502
0.5	0.009	0.011	0.006	0.010	0.004	1.115	1.307	1.220	1.356	1.356
0.6	0.015	0.012	0.013	0.014	0.016	1.383	1.274	1.081	1.304	1.341
0.7	-0.000	0.009	-0.009	0.008	0.014	1.168	1.448	1.421	1.618	1.375
0.8	0.008	0.001	0.005	0.018	0.017	1.059	1.251	1.534	1.527	1.580
0.9	0.019	0.001	0.005	0.010	-0.001	1.309	1.155	1.262	1.538	1.352
1.0	0.015	0.007	0.013	0.007	0.014	1.326	1.233	1.229	1.748	1.469
					n = 800					
0.0	-0.009	0.003	0.000	0.005	0.001	0.700	0.651	0.775	0.742	0.690
0.1	0.008	-0.005	0.009	0.009	0.006	0.677	0.684	0.799	0.784	0.653
0.2	0.001	0.001	0.009	0.011	0.008	0.741	0.610	0.732	0.729	0.781
0.3	0.010	0.003	0.007	-0.000	-0.006	0.793	0.730	0.766	0.779	0.721
0.4	0.003	0.004	0.007	0.001	0.005	0.705	0.705	0.669	0.656	0.839
0.5	0.001	0.003	-0.001	0.002	0.003	0.675	0.748	0.640	0.682	0.765
0.6	0.009	0.004	0.004	0.006	0.007	0.728	0.702	0.713	0.784	0.688
0.7	0.009	0.007	0.007	-0.002	0.007	0.786	0.784	0.816	0.796	0.842
0.8	0.009	0.001	0.011	0.011	-0.001	0.791	0.792	0.849	0.853	0.794
0.9	0.013	0.007	0.003	0.007	0.010	0.856	0.873	0.839	0.747	0.846
1.0	0.004	0.005	0.000	-0.001	0.001	0.771	0.833	0.780	0.778	0.810

Table 9.A7: MLE γ_0 estimates.

	Bias					MSE				
$\gamma_0 \backslash \lambda$	0.0000	0.0025	0.0050	0.0075	0.0500	0.0000	0.0025	0.0050	0.0075	0.0500
					n = 200					
0.0	0.004	0.020	0.007	0.001	0.000	4.219	4.827	3.845	4.240	4.042
0.1	0.018	0.012	0.018	0.015	0.008	4.491	4.051	4.567	4.453	4.971
0.2	0.006	0.025	0.029	0.027	0.024	4.041	4.229	4.853	5.165	4.932
0.3	0.021	0.049	0.033	0.033	0.036	4.250	3.781	5.937	4.762	5.740
0.4	0.057	0.053	0.049	0.052	0.054	5.363	5.528	5.589	6.424	5.876
0.5	0.077	0.075	0.080	0.045	0.061	6.249	6.517	6.568	5.612	5.958
0.6	0.065	0.100	0.072	0.084	0.065	7.094	7.091	7.086	7.205	6.675
0.7	0.074	0.076	0.075	0.076	0.070	6.585	6.621	6.354	5.925	6.439
0.8	0.053	0.068	0.055	0.052	0.055	5.723	6.433	5.761	6.218	5.379
0.9	0.030	0.015	0.042	0.029	0.022	5.511	5.449	5.073	5.982	4.468
1.0	-0.003	-0.034	-0.050	-0.024	-0.034	5.105	4.223	5.572	5.269	5.509
					n = 500					
0.0	-0.005	-0.002	0.003	0.001	-0.001	1.219	1.190	1.219	1.297	1.366
0.1	0.008	-0.001	0.003	0.001	0.004	1.185	1.259	1.221	1.178	1.098
0.2	0.012	-0.002	0.002	0.011	0.005	1.300	1.170	1.209	1.322	1.590
0.3	0.016	0.022	0.017	0.018	0.007	1.346	1.463	1.539	1.566	1.340
0.4	0.016	0.028	0.022	0.026	0.022	1.499	1.726	1.616	1.658	1.712
0.5	0.027	0.028	0.023	0.021	0.025	1.860	1.973	1.965	1.896	2.034
0.6	0.030	0.041	0.032	0.052	0.047	2.266	2.323	2.232	3.020	2.659
0.7	0.045	0.043	0.050	0.059	0.049	2.781	2.685	2.978	3.479	3.108
0.8	0.060	0.072	0.053	0.046	0.053	3.767	3.963	3.221	3.327	3.613
0.9	0.043	0.032	0.040	0.053	0.052	3.410	2.968	3.333	3.358	3.522
1.0	0.026	0.021	0.007	0.028	0.007	2.932	2.734	2.883	2.853	2.723
					n = 800					
0.0	0.004	0.001	-0.001	0.003	0.003	0.678	0.670	0.764	0.678	0.764
0.1	0.003	-0.001	0.002	0.001	0.000	0.648	0.636	0.787	0.692	0.707
0.2	0.011	0.012	-0.001	0.010	0.007	0.769	0.803	0.692	0.899	0.815
0.3	0.002	0.007	0.005	0.009	0.015	0.743	0.833	0.836	0.955	0.765
0.4	0.013	0.013	0.015	0.016	0.015	0.768	0.954	1.092	1.057	0.960
0.5	0.011	0.019	0.021	0.023	0.018	0.946	1.093	1.075	1.284	1.055
0.6	0.030	0.036	0.019	0.024	0.026	1.497	1.598	1.296	1.158	1.612
0.7	0.034	0.026	0.030	0.044	0.045	1.638	1.402	1.975	1.835	2.055
0.8	0.058	0.039	0.031	0.043	0.040	2.497	2.155	2.082	2.352	2.038
0.9	0.050	0.043	0.044	0.039	0.033	2.687	2.214	2.294	2.403	2.495
1.0	0.035	0.041	0.039	0.030	0.038	2.404	2.244	2.342	2.186	2.386

Table 9.A8: MLE γ_1 estimates.

	Bias					MSE				
$\gamma_0\backslash\lambda$	0.0000	0.0025	0.0050	0.0075	0.0500	0.0000	0.0025	0.0050	0.0075	0.0500
					n = 200					
0.0	-0.073	-0.085	-0.088	-0.076	-0.084	3.798	4.196	3.879	3.178	4.125
0.1	-0.089	-0.080	-0.089	-0.076	-0.085	3.968	3.464	3.491	3.893	4.207
0.2	-0.070	-0.074	-0.093	-0.078	-0.083	3.265	3.496	4.268	3.801	4.249
0.3	-0.079	-0.094	-0.086	-0.089	-0.093	3.831	4.193	4.465	4.975	4.969
0.4	-0.088	-0.084	-0.092	-0.094	-0.098	4.517	4.971	4.512	5.310	5.068
0.5	-0.096	-0.108	-0.101	-0.075	-0.091	4.645	5.886	5.211	4.754	4.686
0.6	-0.097	-0.102	-0.102	-0.100	-0.091	5.764	5.241	5.590	5.127	5.318
0.7	-0.093	-0.080	-0.100	-0.096	-0.100	5.027	4.425	5.019	4.494	5.610
0.8	-0.080	-0.098	-0.089	-0.080	-0.095	4.483	4.646	4.746	4.020	5.414
0.9	-0.061	-0.059	-0.076	-0.057	-0.059	4.533	4.338	4.456	4.188	4.041
1.0	-0.052	-0.040	-0.035	-0.047	-0.031	3.056	3.114	3.732	3.496	3.770
					n = 500					
0.0	-0.048	-0.038	-0.035	-0.034	-0.041	1.080	1.000	0.958	1.018	1.003
0.1	-0.036	-0.035	-0.039	-0.023	-0.035	0.964	1.063	1.136	0.914	0.980
0.2	-0.033	-0.023	-0.021	-0.030	-0.032	1.018	0.842	0.909	0.935	1.084
0.3	-0.026	-0.040	-0.024	-0.037	-0.026	1.099	1.108	0.841	1.111	1.044
0.4	-0.030	-0.029	-0.031	-0.034	-0.032	1.107	1.090	1.034	1.108	1.443
0.5	-0.033	-0.039	-0.024	-0.033	-0.035	1.263	1.263	1.229	1.179	1.305
0.6	-0.034	-0.041	-0.040	-0.051	-0.043	1.540	1.619	1.494	1.931	1.547
0.7	-0.050	-0.052	-0.066	-0.051	-0.049	1.854	2.136	2.208	1.901	1.909
0.8	-0.064	-0.074	-0.056	-0.058	-0.060	2.270	2.465	2.319	2.271	2.446
0.9	-0.046	-0.049	-0.054	-0.054	-0.067	2.143	1.982	2.066	2.161	2.504
1.0	-0.032	-0.041	-0.033	-0.037	-0.030	1.820	1.946	1.841	1.970	1.654
					n = 800					
0.0	-0.032	-0.026	-0.021	-0.025	-0.026	0.576	0.522	0.522	0.496	0.515
0.1	-0.018	-0.022	-0.020	-0.014	-0.016	0.517	0.586	0.591	0.487	0.470
0.2	-0.021	-0.025	-0.013	-0.018	-0.016	0.574	0.579	0.481	0.572	0.507
0.3	-0.011	-0.013	-0.015	-0.015	-0.022	0.525	0.551	0.554	0.519	0.530
0.4	-0.022	-0.019	-0.025	-0.021	-0.022	0.581	0.609	0.620	0.591	0.701
0.5	-0.018	-0.020	-0.029	-0.024	-0.024	0.555	0.682	0.659	0.683	0.705
0.6	-0.025	-0.036	-0.024	-0.023	-0.027	0.758	1.082	0.800	0.686	0.926
0.7	-0.027	-0.028	-0.030	-0.038	-0.040	0.921	0.898	1.085	1.158	1.291
0.8	-0.045	-0.044	-0.038	-0.039	-0.044	1.371	1.353	1.343	1.459	1.438
0.9	-0.040	-0.040	-0.046	-0.043	-0.040	1.515	1.305	1.543	1.489	1.627
1.0	-0.036	-0.036	-0.041	-0.033	-0.040	1.253	1.437	1.435	1.413	1.348

Table 9.A9: Rejection rates for heteroscedasticity tests.

	$n = 200$			$n = 500$			$n = 800$		
γ_0	OLS	Probit	LR	OLS	Probit	LR	OLS	Probit	LR
				$\lambda = 0.0$					
0.00	0.050	0.052	0.062	0.040	0.060	0.050	0.043	0.052	0.058
0.10	0.220	0.102	0.095	0.600	0.130	0.145	0.860	0.168	0.188
0.20	0.765	0.160	0.152	1.000	0.380	0.403	1.000	0.580	0.652
0.30	0.983	0.273	0.335	1.000	0.708	0.770	1.000	0.900	0.932
0.40	1.000	0.425	0.565	1.000	0.892	0.945	1.000	0.993	1.000
0.50	1.000	0.522	0.735	1.000	0.978	0.990	1.000	1.000	1.000
0.60	1.000	0.623	0.820	1.000	0.990	1.000	1.000	1.000	1.000
0.70	1.000	0.713	0.912	1.000	1.000	1.000	1.000	1.000	1.000
0.80	1.000	0.765	0.968	1.000	1.000	1.000	1.000	1.000	1.000
0.90	1.000	0.812	0.988	1.000	1.000	1.000	1.000	1.000	1.000
1.00	1.000	0.812	0.993	1.000	1.000	1.000	1.000	1.000	1.000
				$\lambda = 0.0025$					
0.00	0.052	0.052	0.068	0.048	0.050	0.055	0.040	0.045	0.055
0.10	0.260	0.080	0.098	0.615	0.122	0.142	0.840	0.188	0.182
0.20	0.772	0.185	0.198	0.993	0.378	0.388	1.000	0.560	0.630
0.30	0.975	0.265	0.357	1.000	0.632	0.733	1.000	0.875	0.910
0.40	1.000	0.405	0.568	1.000	0.885	0.938	1.000	0.993	1.000
0.50	1.000	0.532	0.750	1.000	0.960	0.985	1.000	1.000	1.000
0.60	1.000	0.640	0.863	1.000	0.995	1.000	1.000	1.000	1.000
0.70	1.000	0.710	0.915	1.000	1.000	1.000	1.000	1.000	1.000
0.80	1.000	0.760	0.955	1.000	1.000	1.000	1.000	1.000	1.000
0.90	1.000	0.825	0.988	1.000	1.000	1.000	1.000	1.000	1.000
1.00	1.000	0.782	0.998	1.000	1.000	1.000	1.000	1.000	1.000
				$\lambda = 0.0050$					
0.00	0.037	0.050	0.055	0.030	0.065	0.072	0.060	0.050	0.058
0.10	0.268	0.070	0.092	0.593	0.122	0.138	0.855	0.198	0.215
0.20	0.805	0.188	0.230	0.993	0.355	0.400	1.000	0.545	0.595
0.30	0.985	0.278	0.338	1.000	0.650	0.748	1.000	0.882	0.915
0.40	1.000	0.385	0.555	1.000	0.875	0.925	1.000	0.980	0.993
0.50	1.000	0.535	0.723	1.000	0.965	0.985	1.000	1.000	1.000
0.60	1.000	0.635	0.853	1.000	0.998	1.000	1.000	1.000	1.000
0.70	1.000	0.675	0.943	1.000	1.000	1.000	1.000	1.000	1.000
0.80	1.000	0.760	0.973	1.000	1.000	1.000	1.000	1.000	1.000
0.90	1.000	0.838	0.990	1.000	1.000	1.000	1.000	1.000	1.000
1.00	1.000	0.802	0.990	1.000	1.000	1.000	1.000	1.000	1.000

Table 9.A9: continued.

	n = 200			n = 500			n = 800		
γ_0	OLS	Probit	LR	OLS	Probit	LR	OLS	Probit	LR
				$\lambda = 0.0075$					
0.00	0.068	0.045	0.050	0.060	0.043	0.060	0.030	0.048	0.035
0.10	0.235	0.102	0.107	0.652	0.138	0.138	0.855	0.210	0.207
0.20	0.775	0.182	0.217	0.998	0.375	0.438	1.000	0.595	0.647
0.30	0.978	0.285	0.365	1.000	0.647	0.733	1.000	0.868	0.917
0.40	1.000	0.390	0.522	1.000	0.887	0.940	1.000	0.985	0.995
0.50	1.000	0.515	0.725	1.000	0.973	0.990	1.000	1.000	1.000
0.60	1.000	0.637	0.875	1.000	1.000	1.000	1.000	1.000	1.000
0.70	1.000	0.733	0.920	1.000	0.993	1.000	1.000	1.000	1.000
0.80	1.000	0.760	0.960	1.000	1.000	1.000	1.000	1.000	1.000
0.90	1.000	0.825	0.988	1.000	1.000	1.000	1.000	1.000	1.000
1.00	1.000	0.800	0.985	1.000	1.000	1.000	1.000	1.000	1.000
				$\lambda = 0.010$					
0.00	0.040	0.052	0.065	0.055	0.045	0.072	0.048	0.055	0.045
0.10	0.253	0.085	0.110	0.640	0.102	0.107	0.825	0.207	0.205
0.20	0.765	0.168	0.200	0.993	0.400	0.412	1.000	0.573	0.575
0.30	0.980	0.275	0.367	1.000	0.635	0.750	1.000	0.895	0.938
0.40	1.000	0.375	0.545	1.000	0.902	0.955	1.000	0.990	0.995
0.50	1.000	0.487	0.703	1.000	0.970	0.993	1.000	0.998	1.000
0.60	1.000	0.623	0.838	1.000	0.998	1.000	1.000	1.000	1.000
0.70	1.000	0.738	0.940	1.000	1.000	1.000	1.000	1.000	1.000
0.80	1.000	0.777	0.980	1.000	1.000	1.000	1.000	1.000	1.000
0.90	1.000	0.812	0.993	1.000	1.000	1.000	1.000	1.000	1.000
1.00	1.000	0.815	0.970	1.000	1.000	1.000	1.000	1.000	1.000

Appendix 2: Heteroscedastic Probit Computer Programs

This appendix shows how to estimate heteroscedastic probit for the model used in the text. I1 is the dependent variable, D1 is the explanatory variable, and $\sigma = \exp(\gamma_1 * G1 + \gamma_2 * G2)$. In LIMDEP, the heteroscedastic probit model can be estimated with the command: `PROBIT; LHS=I1; RHS=ONE,D1; RH2=G1,G2; HETERO $.`
RATS and TSP involve some additional programming. These are common statistical packages that are easy to use and provide flexible programming capability. Both packages have generic maximization procedures (MAX in RATS and ML in TSP) that provide convenient ways to estimate complicated models. However, while the maximization procedures are easy to use, they also are slow. When data sets are large, significant time savings may be available when nonlinear weighted least squares is used to estimate the probit model.

The programs which follow estimate the heteroscedastic probit model. The first two programs estimate the model using the programs' own maximization procedures, while the third program shows how to estimate the model using iterative WLS in RATS. The parameters to be estimated are called B0, B1, GAM1, and GAM2; and there are 500 observations stored in the file HSPROBIT.DAT.

1. TSP Maximization Procedure.

```
SMPL 1 500 ;
LOAD(FILE='HSPROBIT.DAT') D1 G1 G2 I1 ;
PARAM B0 B1 GAM1 GAM2 ;
FRML XB (B0 + B1*D1)/EXP(GAM1*G1+GAM2*G2) ;
FRML LNL I1*LCNORM(XB) + (1-I1)*LCNORM(-XB) ;
EQSUB LNL XB ;
PROBIT I1 C D1 ;
UNMAKE @COEF B0 B1 ;
SET GAM1 = 0. ;
SET GAM2 = 0. ;
ML(HITER=B,HCOV=N) LNL ;
```

2. RATS Maximization Procedure.

```
ALL 0 500
OPEN DATA HSPROBIT.DAT
DATA(ORG=OBS) / D1 G1 G2 I1
NONLIN B0 B1 GAM1 GAM2
FRML XB = (B0 + B1*D1)/EXP(GAM1*G1+GAM2*G2)
FRML I1L = %CDF(XB)
FRML I0L = 1-%CDF(XB)
FRML LNL = %IF(I1,LOG(I1L),LOG(I0L))
```

```
PRB I1
# CONSTANT D1
COM B0 = %BETA(1), B1 = %BETA(2), GAM1 = 0., GAM2 = 0.
MAX(METHOD=BHHH) LNL
```

3. RATS Nonlinear Weighted Least Squares.

```
ALL 0 500
OPEN DATA HSPROBIT.DAT
DATA(FORMAT=FREE,ORG=OBS) / D1 G1 G2 I1
DEC VECT BETAGAM(4) SIGMAT(4)
PRB I1
# CONSTANT D1
COM BETAGAM = ||%BETA(1),%BETA(2),0.,0.||
******
PROC VARS
  SET SIG = EXP(BETAGAM(3)*G1 + BETAGAM(4)*G2)
  SET XB = BETAGAM(1)+BETAGAM(2)*D1
  SET PHI = %DENSITY(XB/SIG)
  SET CPHI = %CDF(XB/SIG)
  SET ERR = I1-CPHI
  SET WGHT = CPHI*(1.-CPHI)
  SET X1 = PHI/SIG
  SET X2 = D1*X1
  SET Z1 = -XB*G1*X1
  SET Z2 = -XB*G2*X1
END VARS
****
COM CHK = 1.
UNTIL CHK.LT.0.00001
{
  EXEC VARS
  LIN(SPREAD=WGHT,NOPRINT) ERR
  # X1 X2 Z1 Z2
  MAT BETAGAM = BETAGAM + %BETA
  EWISE SIGMAT(I) = ABS(%BETA(I))
  COM CHK = %MAXVALUE(SIGMAT)
  DIS CHK
  }
END UNTIL
******
WRITE BETAGAM
EWISE SIGMAT(I) = SQRT(%XX(I,I)); WRITE SIGMAT
EWISE SIGMAT(I) = BETAGAM(I)/SIGMAT(I); WRITE SIGMAT
```

Appendix 3: Monte Carlo Computer Programs

This appendix presents computer code that generates results similar to those reported in the text. The reported results were obtained using RATS version 3.10. This RATS version recently has been replaced by version 4.0, which is a substantial revision: programs run under version 3.10 will no longer work. Thus, the following code is for the more recent version.

The programs are divided into two parts. The first program (MAKE.COM) generates the exogenous variables x_i and z_i, which are referred to as D1 and G1. The second part of the estimation procedure performs the Monte Carlo simulations. The code for the simulations is presented in MONTE.COM. This program is set up to display on the screen the progress being made, while the results are written to a file. The programs are time consuming, so I found it preferable to break them into parts. I did this in part by setting the programs up to be run separately for each sample size. This can be altered by changing the value for N1, which is 500 in the version presented here. The results are written to the file MONTE500.DAT; the number in the file name also should be changed when running a new program. The do-loops DO IGAMO=0,100,5 and DO ILAM=0,100,25 control the values of γ and λ. I found it desirable to divide these into parts (such as IGAMO = 0,50,5 and IGAMO = 55,100,5) when estimating the models.

1. MAKE.COM

```
ALL 0 800
SET D1 = %RAN(2.)
SET D1 1 100 = 8. + D1(T)
SET D1 101 200 = D1(T-100)
SET D1 201 300 = D1(T-200)
SET D1 301 400 = D1(T-300)
SET D1 401 500 = D1(T-400)
SET D1 501 600 = D1(T-500)
SET D1 601 700 = D1(T-600)
SET D1 701 800 = D1(T-700)
******
SET G1 = %RAN(1.)
SET G1 101 200 = G1(T-100)
SET G1 201 300 = G1(T-200)
SET G1 301 400 = G1(T-300)
SET G1 401 500 = G1(T-400)
SET G1 501 600 = G1(T-500)
SET G1 601 700 = G1(T-600)
SET G1 701 800 = G1(T-700)
OPEN COPY D1G1.DAT
COPY(FORMAT=FREE,ORG=OBS) / D1 G1
```

```
END
```

2. <u>MONTE.COM</u>

```
COM NUMBER = 1.
COM N1 = 500
OPEN COPY MONTE500.DAT
******
COM NUM = 1., L1 = 1., L2 = 1.
******
ALL 16 N1
EQV 1 TO 16
  D1 G1 G2 SIGVAR Y1 ERR1 ERR2 WGHT CPHI PHI C1 C2 C3 C4 I1 XBHAT
OPEN DATA D1G1.DAT
DATA(FORMAT=FREE,ORG=OBS) / D1 G1
SET G2 = G1**2
STAT(NOPRINT) D1
COM D1BAR = %MEAN
STAT(NOPRINT) G1
COM G1BAR = %MEAN, G2BAR = %MEAN**2
ZEROS TEMP
******
DEC VECT BMAT(2) X1BAR(2) X2BAR(2) TMAT1(2) TMAT2(2)
DEC VECT X3BAR(2) X4BAR(2) X5BAR(4) X6BAR(4) HMAT1(4) HMAT2(4)
DEC VECT SMAT(41) PMAT1(6) PMAT2(6)
DEC SYMM V1(100,100)
DEC RECT V2(100,100) EMAT(100,1)
COM TMAT1 = ||-8., 1.||
COM GAM0 = 0., GAM1 = 0., CHK = 2.
******
PROC FINDGAM1
  COM INCGAM1 = .1
DO I=1,15
  COM CHK = 2.
  UNTIL CHK.LT.1.
  {
    SET SIGVAR = EXP(GAM0*G1+GAM1*G2)
    STAT(NOPRINT) SIGVAR
    COM CHK = %MEAN
    COM GAM1 = GAM1-INCGAM1
  }
  COM GAM1 = GAM1+2.*INCGAM1
  COM INCGAM1 = INCGAM1/10.
```

```
END DO
END FINDGAM1
******
COM LAM = 0.
PROC ERRMAT
  MAT V1 = %IDENTITY(100)
  IF LAM.NE.0.
  BEGIN
    DO I=1,100
    DO J=1,I
      IF I.NE.J
        COM V1(I,J) = EXP(-ABS(D1(I)-D1(J))/LAM)
    END DO J
    END DO I
  END
  MAT V2 = %DECOMP(V1)
END ERRMAT
******
PROC MAKEY
  COM A1 = 0
  DO I=1,N1,100
    MAT EMAT = %RAN(1.)
    MAT EMAT = V2*EMAT
    SET Y1 I I+99 = 4. + D1 + SIGVAR*EMAT(T-A1*100,1)
    COM A1 = A1+1
  END DO I
  SET I1 = Y1.GE.12.
END MAKEY
******
PROC CRMODEL
  DIM BMAT(2)
  LIN(NOPRINT) Y1 / ERR1
  # CONSTANT D1
  COM SMAT(9) = SMAT(9)+%RSQUARED/NUMBER
  MAT BMAT = %BETA
  MAT X1BAR = X1BAR + %SCALAR(1./NUMBER)*BMAT
  EWISE BMAT(I) = %BETA(I)**2
  MAT X2BAR = X2BAR + %SCALAR(1./NUMBER)*BMAT
  SET ERR2 = N1*ERR1*ERR1/%RSS - 1.
  LIN(NOPRINT) ERR2
  # CONSTANT G1 G2
  PRJ ERR2
  SET ERR2 = ERR2**2
  ACC ERR2 / ERR2 1
```

```
  COM LM1 = ERR2(N1)/2.
  IF LM1.GE.5.9914644
    COM SMAT(10) = SMAT(10)+1.
END CRMODEL
******
PROC PROBIT
  DIM BMAT(2)
  PRB(NOPRINT,INITIAL=TMAT1) I1
  # CONSTANT D1
  COM L1 = %LOGL
  COM L2 = 2.*(L2-L1)
  IF L2.GE.5.9914644
    COM SMAT(26) = SMAT(26)+1.
  PRJ(CDF=CPHI,DENSITY=PHI) XBHAT
  MAT BMAT = %BETA
  MAT X3BAR = X3BAR + %SCALAR(1./NUMBER)*BMAT
  EWISE BMAT(I) = %BETA(I)**2
  MAT X4BAR = X4BAR + %SCALAR(1./NUMBER)*BMAT
  SET C1 = (I1.EQ.1.).AND.(CPHI.GT.0.5)
  SET C2 = (I1.EQ.0.).AND.(CPHI.LE.0.5)
  SET C1 = (C1.EQ.1.).OR.(C2.EQ.1.)
  ACC C1 / C1 1
  COM PMAT1(1) = PMAT1(1) + C1(N1)/NUMBER
  COM PMAT1(2) = PMAT1(2) + (C1(N1)**2)/NUMBER
  SET ERR1 = PHI*%BETA(2)
  STAT(NOPRINT) ERR1
  COM PMAT1(3) = PMAT1(3) + %MEAN/NUMBER
  COM PMAT1(5) = PMAT1(5) + (%MEAN**2)/NUMBER
  COM R2 = %DENSITY(%BETA(1)+%BETA(2)*D1BAR)*%BETA(2)
  COM PMAT1(4) = PMAT1(4) + R2/NUMBER
  COM PMAT1(6) = PMAT1(6) + (R2**2)/NUMBER
  SET WGHT = CPHI*(1.-CPHI)
  SQRT WGHT
  SET ERR1 = I1-CPHI
  SET C1 = PHI
  MULT D1 PHI / C2 1
  SET C3 = -XBHAT*G1*PHI
  SET C4 = -XBHAT*G2*PHI
  DOFOR TEMP=ERR1 C1 C2 C3 C4
    DIV TEMP WGHT / TEMP 1
  END DO
  LIN(NOPRINT) ERR1
  # C1 C2 C3 C4
  PRJ ERR1
```

```
  MULT ERR1 ERR1 / ERR1 1
  ACC ERR1 / ERR1 1
  COM LM1 = ERR1(N1)
  IF LM1.GE.5.9914644
    COM SMAT(15) = SMAT(15)+1.
END PROBIT
******
PROC FINDLNL
  SET ERR1 = EXP(HMAT2(3)*G1+HMAT2(4)*G2)
  SET C1 = 1./ERR1
  SET C2 = D1/ERR1
  PRB(NOPRINT,ITER=10,INITIAL=TMAT2) I1
  # C1 C2
  MAT TMAT2 = %BETA
  COM CHK2 = %CONVCRIT
  PRJ(CDF=CPHI,DENSITY=PHI)
  SET C3 = (I1.EQ.1.).AND.(CPHI.GT.0.5)
  SET C4 = (I1.EQ.0.).AND.(CPHI.LE.0.5)
  SET C3 = (C3.EQ.1.).OR.(C4.EQ.1.)
  ACC C3 / C3 1
  COM CHK3 = C3(N1), CHK4 = %ITER
END FINDLNL
******
PROC LMTEST
  PRJ(DENSITY=PHI,CDF=CPHI) XBHAT
  MULT C1 PHI / C1 1; MULT C2 PHI / C2 1
  SET C3 = -PHI*XBHAT*G1; SET C4 = -PHI*XBHAT*G2
  SET ERR2 = I1-CPHI
  SET WGHT = SQRT(CPHI*(1.-CPHI))
  DOFOR TEMP=C1 C2 C3 C4 ERR2
    DIV TEMP WGHT / TEMP 1
  END DO
  LIN(NOPRINT) ERR2
  # C1 C2 C3 C4
IF (ABS(%BETA(3)).GT.ABS(BMAT(3))).AND.(ABS(%BETA(4)).GT.ABS(BMAT(4)))
    COM CHK1 = 1.
  MAT BMAT = %BETA
  PRJ ERR2
  MULT ERR2 ERR2 / ERR2 1
  ACC ERR2 / ERR2 1
  COM LM = ERR2(N1)
END LMTEST
******
PROC HSMODEL
```

```
  MAT TMAT2 = TMAT1
  DIM BMAT(4)
  COM HMAT1 = ||-8., 1., GAM0, GAM1||
  MAT HMAT2 = HMAT1
  EXEC FINDLNL
  COM L2 = %LOGL
  EXEC LMTEST
  COM CHK = 0., CHK1 = 0.
  IF CHK2.LE.0.001
  BEGIN
UNTIL (LM.LT.0.001).OR.(CHK.GT.10.).OR. $
  (CHK1.EQ.1.).OR.(CHK2.GT..001).OR.(CHK3.GE.(.96*N1)).OR.(CHK4.LE.1.)
  {
    COM SQ = 1., CHK = CHK+1.
 UNTIL (%LOGL.GT.L2).OR.(CHK2.GT.0.001).OR.(CHK3.GE.(.96*N1)).OR. $
  (CHK.GT.10.).OR.(CHK4.LE.1.)
    {
      MAT HMAT2 = HMAT1 + %SCALAR(SQ)*BMAT
      EXEC FINDLNL
      COM SQ = SQ/2.
    }
    COM MAXB1 = %BETA(1), MAXB2 = %BETA(2), L2 = %LOGL
    MAT HMAT1 = HMAT2
  IF (CHK2.LE.0.001).AND.(CHK3.LT.(.96*N1)).AND.(CHK4.GT.1.)
    EXEC LMTEST
  }
 END
  IF (CHK.LE.10.).AND.(CHK1.EQ.0.).AND.(CHK2.LE..001) $
  .AND.(CHK3.LT.(.96*N1)).AND.(CHK4.GT.1.)
  BEGIN
    COM NUM = NUM+1., HMAT1(1) = MAXB1, HMAT1(2) = MAXB2
    MAT HMAT2 = HMAT1
    MAT X5BAR = X5BAR + %SCALAR(1./NUMBER)*HMAT1
    EWISE HMAT1(I) = HMAT1(I)**2
    MAT X6BAR = X6BAR + %SCALAR(1./NUMBER)*HMAT1
    SET C1 = (I1.EQ.1.).AND.(CPHI.GT.0.5)
    SET C2 = (I1.EQ.0.).AND.(CPHI.LE.0.5)
    SET C1 = (C1.EQ.1.).OR.(C2.EQ.1.)
    ACC C1 / C1 1
    COM PMAT2(1) = PMAT2(1) + C1(N1)/NUMBER
    COM PMAT2(2) = PMAT2(2) + (C1(N1)**2)/NUMBER
    SET ERR2 = PHI*HMAT2(2)/ERR1
    STAT(NOPRINT) ERR2
    COM PMAT2(3) = PMAT2(3) + %MEAN/NUMBER
```

```
    COM PMAT2(5) = PMAT2(5) + (%MEAN**2)/NUMBER
    COM E2 = EXP(HMAT2(3)*G1BAR + HMAT2(4)*G2BAR)
    COM R2 = (HMAT2(1)+HMAT2(2)*D1BAR)/E2
    COM P2 = %DENSITY(R2)
    COM R2 = P2*HMAT2(2)/E2
    COM PMAT2(4) = PMAT2(4) + R2/NUMBER
    COM PMAT2(6) = PMAT2(6) + (R2**2)/NUMBER
    STAT(NOPRINT) ERR1
    COM SMAT(39) = SMAT(39)+%MEAN/NUMBER
    COM SMAT(40) = SMAT(40)+E2/NUMBER
  END
END HSMODEL
******
MAT SMAT = %CONST(0.)
DO IGAM0=0,100,5
  COM GAM0 = IGAM0/100.
  IF IGAM0.LT.0.01
    UNITS SIGVAR
  IF IGAM0.GT.0.01
    EXEC FINDGAM1
DO ILAM=0,100,25
  COM LAM = ILAM/10000.
  MAT X1BAR = %CONST(0.), X2BAR = %CONST(0.), X3BAR = %CONST(0.)
  MAT X4BAR = %CONST(0.), X5BAR = %CONST(0.), X6BAR = %CONST(0.)
  MAT PMAT1 = %CONST(0.), PMAT2 = %CONST(0.), SMAT = %CONST(0.)
  COM NUM = 0.
  COM SMAT(41) = LAM, SMAT(1) = GAM0, SMAT(2) = GAM1
  STAT(NOPRINT) SIGVAR
  COM SMAT(3) = %MEAN, SMAT(4) = SQRT(%VARIANCE)
  SET XBHAT = (-8. + D1)/EXP(GAM0*G1+GAM1*G2)
  SET XBHAT = %DENSITY(XBHAT)/EXP(GAM0*G1+GAM1*G2)
  STAT(NOPRINT) XBHAT
  COM SMAT(29) = %MEAN
  COM SMAT(30) = %DENSITY((-8.+D1BAR)/EXP(GAM0*G1BAR+GAM1*G2BAR))
  COM SMAT(30) = SMAT(30)/EXP(GAM0*G1BAR+GAM1*G2BAR)
UNTIL NUM.GE.NUMBER
{
  EXEC ERRMAT
  EXEC MAKEY
  EXEC HSMODEL
  IF (CHK.LE.10.).AND.(CHK1.EQ.0.).AND.(CHK2.LE.0.001) $
  .AND.(CHK3.LT.(.96*N1)).AND.(CHK4.GT.1.)
    EXEC CRMODEL
  IF (CHK.LE.10.).AND.(CHK1.EQ.0.).AND.(CHK2.LE.0.001) $
```

```
  .AND.(CHK3.LT.(.96*N1)).AND.(CHK4.GT.1.)
    EXEC PROBIT
DISPLAY N1 GAMO LAM NUM
}
  COM SMAT = ||SMAT(1),SMAT(2),SMAT(3),SMAT(4),X1BAR(1),X1BAR(2), $
  X2BAR(1),X2BAR(2),SMAT(9),SMAT(10),X3BAR(1),X3BAR(2),X4BAR(1), $
  X4BAR(2),SMAT(15),PMAT1(1),PMAT1(2),X5BAR(1),X5BAR(2),X5BAR(3), $
  X5BAR(4),X6BAR(1),X6BAR(2),X6BAR(3),X6BAR(4),SMAT(26),PMAT2(1), $
  PMAT2(2),SMAT(29),SMAT(30),PMAT1(3),PMAT1(4),PMAT1(5),PMAT1(6), $
  PMAT2(3),PMAT2(4),PMAT2(5),PMAT2(6),SMAT(39),SMAT(40),SMAT(41)||
  WRITE(NOSKIP) SMAT
  WRITE(UNIT=COPY,NOSKIP) SMAT
END DO LAM
END DO GAMO
END
```

10 Estimating Logit Models with Spatial Dependence

Robin Dubin
Case Western Reserve University, Cleveland, OH, U.S.A.

10.1 Introduction

Much research has been done on estimating models of spatial dependence with continuous random variables [for example, see Ripley (1981), Anselin (1988), Cressie (1991)]. However, relatively little work has been done on incorporating spatial dependencies into models with qualitative dependent variables. Boots and Kanaroglou (1988) have incorporated spatial considerations into a migration model and Anne Case (1992) has done likewise for a model of technology adoption.

Despite the limited amount of research, there are a number of applications for these types of models. For example, an individual's decision of whether to vote for the Republican or Democratic candidate might depend on how his neighbors vote. Or whether or not a county grows in population might depend in part on its proximity to fast growing counties. Or whether or not a firm adopts a technological innovation might depend on its proximity to adopters. The last is the application that will be used to illustrate the proposed model.

The diffusion of innovations has been studied extensively. Diffusion is typically thought of as following a contagion model. Just as a disease spreads by contact with infected individuals, an innovation becomes adopted as more firms become familiar with it, by observing prior adopters. Thus a few firms adopt the innovation initially; other firms observe and adopt in a later period. As the number of adopters increases, more knowledge becomes available, until finally, the entire population of potential users is saturated. This type of model produces an *S*-shaped, or logistic, diffusion curve.[1]

---

[1] Griliches (1957), Mansfield (1968) and Nabseth and Ray (1974) have all found *S*-shaped diffusion curves.

The rate of diffusion is thought to depend on characteristics of the innovation, the industry, and the firm. In terms of innovation characteristics, the higher the profitability and the lower the required investment, the higher the rate of diffusion. The degree of competition, usually measured by the number of firms, is an important characteristic of the industry. Important firm characteristics include: firm size, management attitudes, age of existing capital stock, and the level of information regarding the innovation.[2]

Although the majority of interfirm diffusion studies have been conducted at the industry level, it is possible to formulate a logit model of the firm's adoption decision. Here a firm will adopt if expected profits (unobserved) are positive. Expected profit is a function of the firm's characteristics and its level of information.

A model formulated along these lines has been proposed by Case (1992). Case posits that a farmer's expected profit from adopting a new technology is a function of the own characteristics and the neighbors' expected profits. Since the neighbors' expected profits are a function of their characteristics, any farmer's expected profit becomes a function of their own and their neighbors' characteristics. The farmer adopts the new technology if the expected profit is positive.

While Case's model makes an important contribution, her model is strictly cross section. That is, a farmer examines their own characteristics and those of their immediate neighbors and, based on this information, decides whether or not to adopt the technology. However, the diffusion of a technological innovation is a dynamic process: the innovation is first adopted by a few individuals and gradually spreads to the remainder of the population.

In what follows, I present a dynamic logit model of diffusion. I discuss the model and present the derivatives necessary for maximizing the likelihood function and for constructing the information matrix. Simulated data is used to illustrate the estimation procedure.

10.1.1 Model

A firm's unobserved utility (expected profit) from adopting an innovation is a function of its own characteristics plus its distance from previous adopters. Formally:

---

[2] Studies which have explored the effects of these variables include: profitability and number of firms in the industry [Mansfield (1977)], profitablility and firm size [Nabseth and Ray (1974)] and management attitudes [Carter and Williams (1957)].

$$Y^*_{it} = X_{it}\beta + \sum_{j=1}^{N} \rho_{ij} y_{j,\,t-1} + u_{it} \tag{1}$$

where Y^*_{it} is the unobserved expected profit from the innovation for firm i at time t, with $i = (1, ..., N_t)$. N is the total number of firms while N_t is the number of firms that have not adopted prior to time t, X_{it} (a row vector) are the characteristics of firm i at time t, and u_{it} is the error term, assumed here to have a logistic distribution.[3] β is a vector of coefficients (to be estimated) which represent the influence of the firm's own characteristics on the profitability of the innovation. ρ is a matrix of coefficients, each element of which represents the influence firm j has on firm i and is modeled as a negative exponential function of the distance separating the two firms:

$$\rho_{ij} = b_1 e^{\left(\frac{-D_{ij}}{b_2}\right)} \tag{2}$$

where b_1 and b_2 are parameters to be estimated and D_{ij} is the distance separating the two firms. Thus, a firm's expected profit from an innovation is larger if it is near to current users of the innovation. This dependency occurs because nonusers learn from adopters. Distance as used here is intended to measure the interactions between firms. In this illustration I use geographic distance; however, distance in other metrics might be more appropriate. For example, distance in information space, as proxied by attendance at association meetings or the number of boards on which the directors of the company sit, might be better measures than geographic space.

The parameters b_1 and b_2 allow the researcher to estimate the influence of prior adopters. The parameter b_1 determines the effect that adjacent firms (small D_{ij}) have on each other, while b_2 determines the rate that influence attenuates with distance. If b_1 and b_2 are jointly insignificant, there is no spatial interaction.

The dynamic nature of the model is clear from equation (1). As time progresses, more firms will adopt the technology; thus the probability that hold out-firms will also adopt increases as nearby firms become adopters. After enough time has passed, virtually all firms will adopt the technology, giving rise to an *S*-shaped diffusion curve. However, the model is flexible, in that some firms' characteristics may so dispose them against the innovation that they never adopt.

The expected profit of the firm is not observed; instead we observe whether the firm adopts the innovation in time period t, y_{it}. The firm will adopt the innovation if expected profits (Y^*) are positive. Thus,

[3] The error terms are assumed to be independent. That is the only spatial dependencies are those modeled by the spatial interaction term, $\Sigma_j\, \rho_{ij} y_{i,\,t-1}$.

$$\begin{cases} y_{it} = 1 & \text{if} \quad Y_{it}^* > 0 \\ y_{it} = 0 & \text{otherwise} \end{cases} \tag{3}$$

and

$$\begin{aligned} P(y_{it}=1) &= P\left(u_{it} > -\left(X_{it}\beta + \sum_{j=1}^{N} \rho_{ij} y_{j,\,t-1}\right)\right) \\ &= P\left(u_{it} > -(X_{it}\beta + A_i)\right) \\ &= 1 - F\left(-(X_{it}\beta + A_i)\right) \\ &= \frac{e^{(X_{it}\beta + A_i)}}{1 + e^{(X_{it}\beta + A_i)}} \end{aligned} \tag{4}$$

where F is the cumulative distribution function for u and $A_i = \Sigma_j \rho_{ij} y_{j,\,t-1}$. A_i represents the influence of prior adopters. The last equality in (4) is based on the assumption that the error terms come from a logistic distribution.

The log-likelihood function is given by:

$$\begin{aligned} \ln(L) &= \sum_{i=1}^{N_t}\left[y_{it}\ln\left(\frac{e^{(X_{it}\beta + A_i)}}{1 + e^{(X_{it}\beta + A_i)}}\right) + (1 - y_{it})\ln\left(\frac{1}{1 + e^{(X_{it}\beta + A_i)}}\right)\right] \\ &= \beta\sum_{i=1}^{N_t} y_{it}X_{it} + \sum_{i=1}^{N_t} y_{it}A_i - \sum_{i=1}^{N_t}\ln\left(1 + e^{(X_{it}\beta + A_i)}\right) \end{aligned} \tag{5}$$

Equation (5) is maximized with respect to b and β to obtain maximum likelihood (ML) estimates of these parameters. First and second partial derivatives of the log-likelihood function with respect to the parameters are given below.

$$\frac{\partial \ln(L)}{\partial b} = \sum_{i=1}^{N_t} y_{it}\sum_{j=1}^{N}\frac{\partial \rho_{ij}}{\partial b}y_{j,\,t-1} - \sum_{i=1}^{N_t}\frac{e^{(X_{it}\beta + A_i)}\sum_{j=1}^{N}\frac{\partial \rho_{ij}}{\rho b}y_{j,\,t-1}}{1 + e^{(X_{it}\beta + A_i)}} \tag{6.a}$$

$$\frac{\partial \ln(L)}{\partial \beta} = \sum_{i=1}^{N_t} y_{it}X_{it} - \sum_{i=1}^{N_t}\frac{e^{(X_{it}\beta + A_i)}X_{it}}{1 + e^{(X_{it}\beta + A_i)}} \tag{6.b}$$

$$\frac{\partial^2 \ln(L)}{\partial b^2} = \sum_{i=1}^{N_t} y_{it} \sum_{j=1}^{N} \frac{\partial^2 \rho_{ij}}{\partial b^2} y_{j,\,t-1}$$

$$-\sum_{i=1}^{N_t} \frac{e^{(X_{it}\beta + A_i)} \left[\left(\sum_j \frac{\partial \rho_{ij}}{\partial b} y_{j,\,t-1} \right)^2 + \sum_j \frac{\partial^2 \rho_{ij}}{\partial b^2} y_{j,\,t-1} \left(1 + e^{(X_{it}\beta + A_i)}\right) \right]}{\left(1 + e^{(X_{it}\beta + A_i)}\right)^2} \tag{7.a}$$

$$\frac{\partial^2 \ln(L)}{\partial \beta^2} = -\sum_{i=1}^{N_t} \frac{e^{(X_{it}\beta + A_i)} X_{it} X_{it}'}{\left(1 + e^{(X_{it}\beta + A_i)}\right)^2} \tag{7.b}$$

$$\frac{\partial^2 \ln(L)}{\partial b \partial \beta} = -\sum_{i=1}^{N_t} \frac{e^{(X_{it}\beta + A_i)} X_{it} \sum_j \frac{\partial \rho_{ij}}{\partial b} y_{j,t-1}}{\left(1 + e^{(X_{it}\beta + A_i)}\right)^2} \tag{7.c}$$

The matrix of second partials is used to form the information matrix. The inverse of the information matrix, evaluated at the ML estimates, gives the asymptotic variances and covariances of the estimated parameters.

10.2 Simulation Example

To illustrate the model and the estimation procedure, I simulated spatially autocorrelated data and used these to estimate the model. These results are purely illustrative: certainly no inferences regarding the properties of the estimators can be made from one simulation. The GAUSS programs used to estimate the parameters and their variances can be found in the appendix.

To capture the dynamic flavor of the diffusion of a technological innovation, I simulated the data in a two stage process. In the first stage, I simulated 100 standard logit observations, using two independent variables and a constant term. I assigned these observations to points on a (10×10) rectangular grid. Thus, taking miles as the distance unit, the data is arranged so that an observation is located every mile in a 100 square mile area. The smallest separation distance is 1 mile, and the largest is approximately 14 miles. In the second stage, I allowed non-adopters to be influenced by their proximity to the stage one adopters, through the influence function, ρ. This process resulted in 27 stage one, and 54 stage two, adopters. The locations of stage one and stage two adopters are shown in Figure 10.1. The parameters of the simulated data are $\beta = (-1, 3, -3)'$ and $b = (0.9, 2)'$. Figure 10.2 shows the true ρ as a function of distance. Examination of Figure 10.2

shows that the influence of adopters falls off fairly rapidly with distance: the influence of a prior adopter is negligible for separation distances exceeding 5 miles.

Using the simulated data, I estimated two models: a standard logit model and the model proposed in this chapter. I used the logit routine in GAUSS to estimate the standard logit model, and GAUSS's optimization routine (OPTMUM) to estimate the spatial model. Only the 73 firms that did not adopt in period one are used to estimate the standard logit. In evaluating the likelihood function for the spatial model (5), N_t is 73 and N is 100. Thus, the likelihood function is evaluated using only the stage one *non*adopters, while the stage one *adopters* are used to calculate A_i.

Note that the likelihood function for the standard logit model is misspecified because the spatial interaction term is excluded. This will cause the error terms to be autocorrelated and the estimators to be inconsistent. Columns three and four of Table 10.1 show the standard logit estimates and asymptotic variances, respectively. All variances shown in Table 10.1 are estimated by taking the inverse of the information matrix. The standard logit estimates of β_1 and β_2 have the right sign but the sign of β_0 is wrong. Only β_1 is significant.

The last two columns of Table 10.1 show the estimated coefficients and asymptotic variances for the spatial model. Again, the variances are estimated using the inverse of the information matrix. The estimate of β_0 now has the correct sign and is significant. Despite the relatively large variances of b_1 and b_2, the spatial model fits the data significantly better than the standard model. The likelihood function is -25.57 for the spatial model and -37.13 for the standard model, giving a likelihood ratio statistic of 23.12, which is clearly significant.

Table 10.1: Estimation results for simulated data.

Coeff	True Value	Standard Logit Estimate	Standard Logit Variance	Spatial Logit Estimate	Spatial Logit Variance
β_0	-1	1.058	0.62	-4.486	3.87
β_1	3	2.600	1.18	3.231	1.71
β_2	-3	-1.876	1.11	-2.235	2.23
b_1	0.9			3.241	7.75
b_2	2			1.345	0.54
$\ln(L)$		-37.13		-25.57	

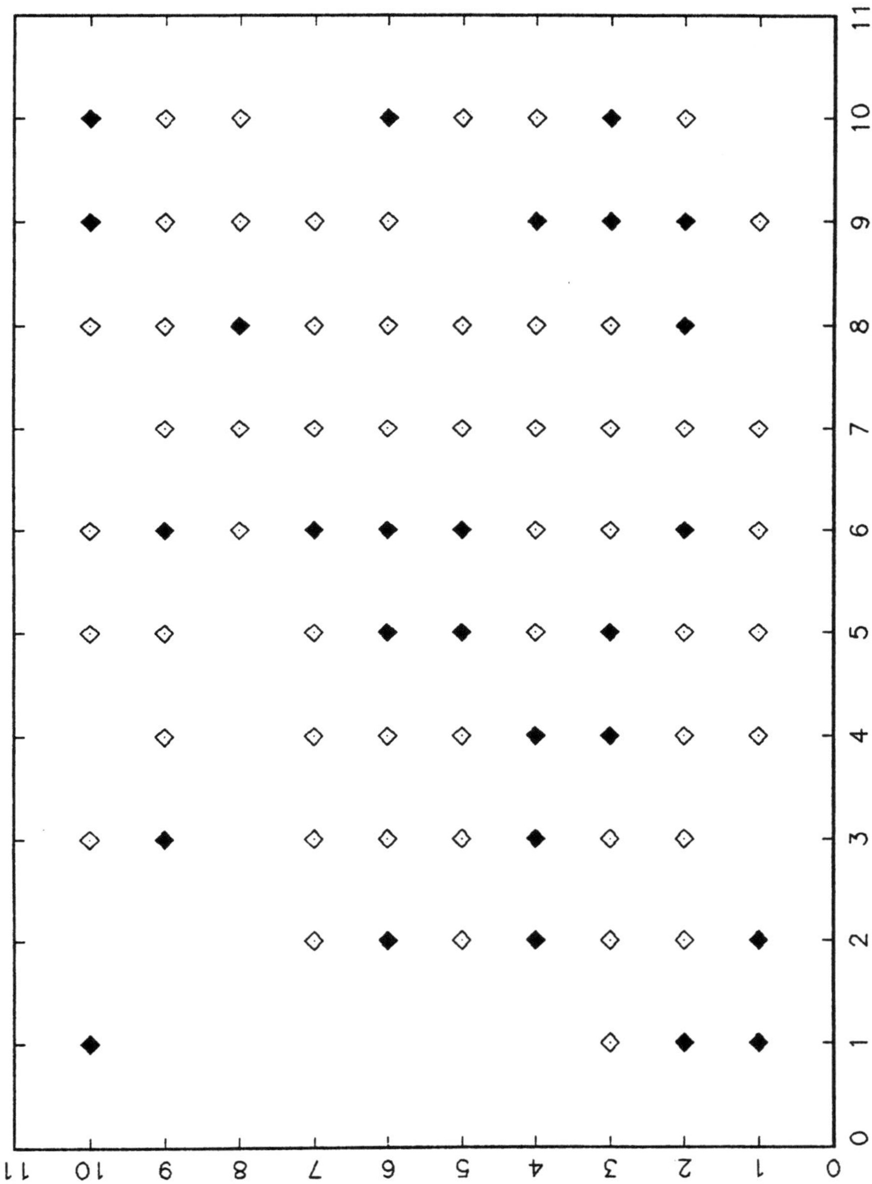

Figure 10.1: Stage one and stage two adopters (◆ stage one adopters, ◇ stage two adopters).

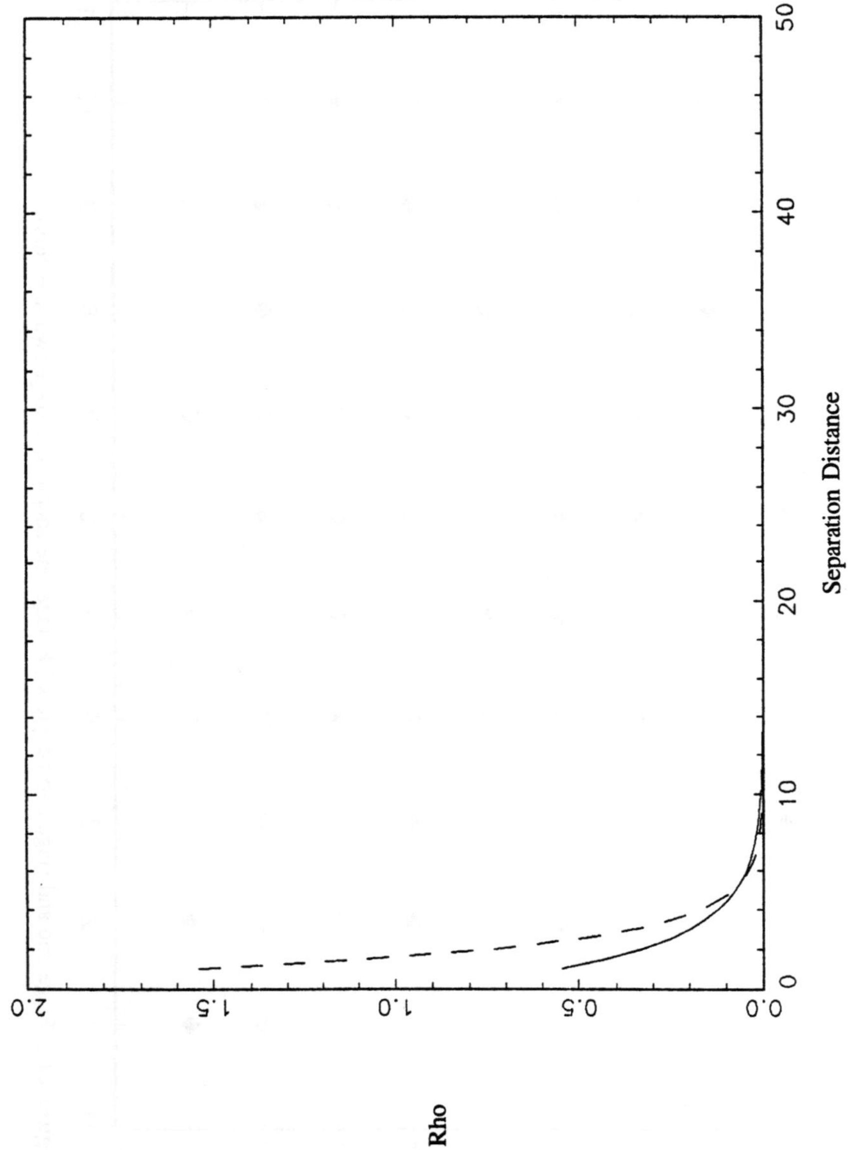

Figure 10.2: True and estimated influence functions (—— true, b = 0.9, 2.0; ---- estimated b = 3.241, 1.345).

While the point estimates of b_1 and b_2 seem to be wide of the mark, jointly they give an estimated influence function not too different from the true one, as shown in Figure 10.2. Although the estimated influence function overstates the influence of nearby firms, both functions give the same influence (negligible) for separation distances greater than 4.

Possible reasons for the overestimation of b_1 include the small sample size and its location in geographic space. Since there are no separation distances less than 1 mile, there is no information with which to estimate the first part of the influence function. The use of maximum likelihood estimation makes the estimators consistent, however the small sample properties are unknown. A Monte Carlo study might usefully explore the small sample properties of the estimators.

10.3 Conclusions

In this chapter I have presented a dynamic logit model of the diffusion of a technological innovation in which the probability of adoption varies with the firm's own characteristics and its interactions with previous adopters. The number of interactions are modeled here as a diminishing function of geographic distance, however, other metrics could be used. Simulated data is used to illustrate the estimation procedure as well as properties of the interaction function.

Diffusion of technology has been used to illustrate this model. However, the model could be applied to a number of other applications. For example, since neighborhood quality affects rents and since neighborhood quality is in part affected by maintenance decisions, a landlord's decision to invest in maintenance might be affected by prior maintenance expenditures of nearby landlords. Other possible applications include modeling voting patterns and county growth. In short, many models with qualitative dependent variables might be subject to spatial dependencies which, if uncorrected, will cause econometric problems. In this chapter, I have proposed a method for modeling these dependencies.

References

Anselin, L., *Spatial Econometrics: Methods and Models*, Dordrecht: Kluwer Academic Publishers, 1988.

Boots, B. N. and P.S. Kanaroglou, Incorporating the Effect of Spatial Structure in Discrete Choice Models of Migration, *Journal of Regional Science*, 28, 495-507, 1988.

Carter, C.F. and B.R. Williams, *Industry and Technical Progress: Factors governing the Speed of Application of Science*, London: Oxford University Press, 1957.

Case, A., Neighborhood Influence and Technological Change, *Regional Science and Urban Economics*, 22, 491-508, 1992.

Cressie, N., *Statistics for Spatial Data*, New York: John Wiley and Sons, 1991.

Griliches, Z., Hybrid Corn: An Exploration in the Economics of Technical Change, *Econometrica*, 501-522, 1957.

Mansfield, E., *Industrial Research and Technical Innovation*, New York: W.W. Norton, 1968.

Mansfield, E., *The Production and Application of New Industrial Technology*, New York: W.W. Norton, 1977.

Nabseth, L. and G.F. Ray, *The Diffusion of New Industrial Processes: An International Study*, London: Cambridge University Press, 1974.

Ripley, B., *Spatial Statistics*, New York: John Wiley and Sons, 1981.

Appendix 1: Gauss Program for Finding ML Estimates

```
/* logit3e.gss MAXIMIZES LIKELIHOOD FUNCTION FOR SPATIAL LOGIT */
library optmum;
#include optmum.ext;
  output file=logit3.out reset;
closeall;
  @ :::::::::::::::::::: options/defaults ::::::::::::::::::::: @

@set parameter values@
@_opshess=1;@
  _opstmth="steep one";
  _opmdmth="bfgs brent";
  _opmiter=75;
  __output=1;
 @ ::::::::::::::::::::::::::::::::::::::::::::::::::::::::::::: @
 @        Define the objective function to be minimized: OFN(b)
              To maximize the function, reverse its sign.            @
    @ :::::::::::::::::::: function definition(s) :::::::::::: @

@initialize variables and read in variables@
  load dist=dist6;  @dist6 is the distance matrix for the data@
  nobs=rows(dist);
  open data=lsim9a;  @lsim9a is the simulated data@
  d=readr(data,nobs);
@  order of variables: xcoord ycoord ytml yt x1 x2       @
@                         1      2     3   4  5  6        @
@  yt=1 if firm uses innovation in period t              @
@  ytml=1 if firm adopted innovation in previous period  @
  y=d[..4];
  ytml=d[..3];
  x=d[..5 6];
  z=ones(nobs,1)~x;
  k=cols(z);
@start optimization procedure@
  proc ofn(b);
    local vofl,vof,i,j,corr1,corr2,cdiag,sum,xbeta,term1,
    term2,term3,v,indic,n1,sum1,term1b,yb;
    xbeta=z*b[3:k+2,1];
    corr1=b[1,1]*exp(-dist/b[2,1]);  @rho@
    cdiag=zeros(nobs,1);
    corr2=diagrv(corr1,cdiag);
    sum=zeros(nobs,1);
```

```
    sum=corr2*ytm1;
    term1=xbeta+sum;
    term1b=delif(term1,ytm1);
    term2=exp(term1b)./(1+exp(term1b));
    yb=delif(y,ytm1);   @this line deletes stage one adopters so that
                        @ LF only evaluated using stage one nonadopters@
    vof1=yb'*ln(term2)+(1-yb)'*ln(1-term2);
    vof=-vof1;
l1:
    retp(vof);
  endp;

@call optmum@
  let b0=.9 1 -1 3 -4;
  {b,f,g,retcode}=optmum(&ofn,b0);
@write out results@
output on;
  "Autocorrelated logit estimation with simulated data--lsim8a
     --vof based on ytm1=0 only";
?;?;
   format 10,5;"value of objective function   ";;f;
            ?;?;"b           ";;b;
              ?;"gradient   ";;g';
              ?;"number of observations";;nobs-sumc(ytm1);
              ?;"number of variables";;k;
              ?;"return code";;retcode;
              ?;"hessian";_opfhess;
end;
```

Appendix 2: Gauss Program to Estimate Asymptotic Variances of ML Estimates

```
@logit6.gss--calculates information matrix for spatial logit@
closeall;
format /rd 6,2;
output file=logit6.out reset;
load dist=dist6;
nobs=rows(dist);
open data=lsim9a varindxi;
d=readr(data,nobs);
y=d[..,iyt];
ytml=d[..,iytml];
x=ones(nobs,1)~d[..,ix1~ix2];
let beta=-4.4860 3.2312 -2.2355;   @ML estimates@
let b= 3.2411 1.3454;
par=b|beta;
mat4=exp(-dist/b[2,1]);
mat4y=mat4*ytml;
mat1=(dist/b[2,1]^2).*mat4;
mat1y=mat1*ytml;
clear mat1;
mat3=b[1,1]*mat4;
mat3y=mat3*ytml;
mat2=mat3.*((dist^2-2*b[2,1]*dist)/b[2,1]^4);
mat2y=mat2*ytml;
clear mat2, mat3;
mat5=(b[1,1]/b[2,1]^2)*dist.*mat4;
mat5y=mat5*ytml;
clear mat5,mat4;
vec1=exp(x*beta+mat3y);
den1=(ones(nobs,1)+vec1);
den=(ones(nobs,1)+vec1)^2;
vec2=vec1./den;
@delete stage one adopters@
mat1y=delif(mat1y,ytml);
mat2y=delif(mat2y,ytml);
mat3y=delif(mat3y,ytml);
mat4y=delif(mat4y,ytml);
mat5y=delif(mat5y,ytml);
vec1=delif(vec1,ytml);
vec2=delif(vec2,ytml);
den=delif(den,ytml);
```

```
den1=delif(den1,ytm1);
y=delif(y,ytm1);
x=delif(x,ytm1);
vof=y'x*beta+y'*mat3y-sumc(ln(den1));
"vof " vof;
?;
spar=zeros(5,1);  @spar is vector of first derivatives of lf@
spar[1,1]=y'*mat4y-(vec1'*(mat4y./den1));
spar[2,1]=y'*mat5y-(vec1'*(mat5y./den1));
spar[3:5,1]=x'*y-(x'*(vec1./den1));
info=zeros(5,5);
info[1,1]=-vec2'*(mat4y.*mat4y);
info[1,2]=y'*mat1y-vec2'*(mat4y.*mat5y+mat1y.*den1);
info[2,1]=info[1,2];
info[2,2]=y'*mat2y-vec2'*(mat5y.*mat5y+mat2y.*den1);
info[1,3:5]=-vec2'*(mat4y.*x);
info[3:5,1]=info[1,3:5]';
info[2,3:5]=-vec2'*(mat5y.*x);
info[3:5,2]=info[2,3:5]';
info[3:5,3:5]=-(vec2.*x)'*x;
"first derivatives " spar;
?;
"info" info;
?;
invinfo=inv(-info);
"invinfo" invinfo;
?;
"parameter estimates, variances, and t stats";
tstat=par./sqrt(diag(invinfo));
par~diag(invinfo)~tstat;
end;
```

11 Utility Variability within Aggregate Spatial Units and its Relevance to Discrete Models of Destination Choice

Mark R. Ferguson and Pavlos S. Kanaroglou
McMaster University, Hamilton, Canada

11.1 Introduction

In this chapter, we examine a logit model designed specifically for spatial choice among aggregate destinations. Typically, the logit and gravity models are applied to problems of spatial interaction without due consideration for the aggregation scale of the data. There is of course a general awareness that models estimated at different levels of aggregation will yield different parameter estimates but most researchers proceed directly with their analysis at the aggregate level and make no adjustments. Research on aggregation issues has been intertwined with work on the 'modifiable areal unit problem'. This research examines how spatial statistical models and diagnostics are affected by the aggregation scale and spatial configuration of spatial units. Choice processes per se, as are present in all problems of spatial interaction, are not specifically addressed in this body of research. Rather the focus is on the univariate or bivariate spatial processes associated with one or two spatially-referenced attributes [Openshaw (1984) and Arbia (1989)]. Aggregation issues are relevant in both scenarios but as will be illustrated in this chapter, the theoretical framework to accommodate aggregation in a choice context is quite different from that considered in analysis of the modifiable areal unit problem.

From the perspective of geographers and regional scientists, the study of aggregation in the choice context has been confined mostly to work on the gravity model [Webber (1980), Batty and Sikdar (1982a,b,c,d), Schwab and Smith (1985), Putman and Chung (1989), Amrhein and Flowerdew (1992)]. Most research on aggregation in discrete choice modeling has been done by those outside these fields [McFadden (1978), Kitamura (1979)] and is not widely applied by practitioners in the spatial sciences. Given the work by Anas (1983) on the strong similarity between the gravity and logit models, the insights gained from the work outside regional science and geography need to be understood and further developed in the spatial context.

It is interesting that most contemporary work in spatial interaction modeling has tended to focus on spatial substitutability and structure effects as the primary contributors to model misspecification [Griffith and Jones (1980), Fotheringham (1981, 1983), Borgers and Timmermans (1987), Haynes and Fotheringham (1990), Lo (1991)], often not recognizing the fact that the choice process is typically not at the same aggregation level as the data. We suggest that aggregation biases must be addressed before all else in a viable model of spatial choice. Problems of substitutability among alternatives can often be solved with alternate aggregation schemes but aggregation biases cannot be eliminated through accounting for substitutability. Fortunately, the framework described here is capable of addressing each of these issues.

In the practical terms of a choice problem then, interaction data are often available only at the level of aggregate spatial units and in other cases, the interaction data are so disaggregated that the number of spatial units involved would be prohibitively large. In either situation, the researcher is presented with a problem whereby the spatial choice units of the model are more aggregated than those likely evaluated in reality by choice-makers. Considering the spatial aggregation of the data to be exogenous then, and not necessarily optimal, how should a researcher correctly go about representing the utility of aggregates and obtaining appropriate parameter estimates?

The central premise of this chapter is that to succeed in this objective it is necessary to account for utility variability within aggregates. In a study of interregional migration, for example, it is unlikely that choice-makers value each point within a region equally. However, when the ordinary multinomial logit model is applied to a regional choice set, this is exactly what is assumed. All variables employed in the model are some form of regional average implying average regional utilities. Alternatively, it is possible to employ variables measured at the subaggregate level and then represent utility at a finer level of resolution. Using such an approach, a theoretically appealing model can be derived which postulates the utility of an aggregate to equal the point of maximum utility within the aggregate. While the choice data for spatial interaction are often aggregated over space and across decision-makers, there is no need to compound the problem by utilizing spatially aggregated explanatory variables.

It is interesting to note that work in spatial econometrics has not found common ground with that of spatial choice analysis, particularly when aggregation is an issue. Typically, logit and probit models have been favoured with spatial econometricians in studies utilizing a spatially-referenced, binary categorical dependent variable associated with non-spatial alternatives. Furthermore, the models have tended to be purely statistical as opposed to being considered an outcome of choice theory. Such a study is that of McMillen (1992) who employs a probit model in the examination of crime rates as a dependent variable. In a similar vein, Bolduc et al. (1989) examine issues such as spatial autocorrelation and heteroscedasticity but in the context of mode choice where the decisions of travellers are seen to be dependent on location. These examples are to be contrasted with the model in this paper, which is clearly a choice model whose alternatives are multiple spatial units. Just as spatial econometric techniques apply to the work of McMillen and Bolduc et al., they are of interest in the context of spatial choice but it must be recognized that such analysis is made more complex by the fact that the number of alternatives is large, that a heterogeneous population is choosing among these spatial units, and that issues of aggregation are present. In this paper, the model contrasted with the ordinary multinomial logit model (OMNL) addresses the effects of aggregation elegantly but is less effective in accounting for complex spatial dependency.

This chapter then is divided into five main sections. In the second section to follow, we summarize the theoretical basis for the proposed model and provide necessary background information. In the third section, a basic estimation procedure is outlined and differences with the estimation of the OMNL model are stressed. In the fourth section, the specification of the model when data are highly aggregated across choice-makers is contrasted to micro-data specifications. These specifications are then tested in a controlled environment. More specifically, the performance of the average utility model is compared with that of the maximum utility model when the utility heterogeneity of aggregates is systematically altered. In the concluding section, some synthesizing remarks are offered with a view toward future work.

11.2 Theoretical Background

In problems of spatial choice, the study area is typically divided into zones necessitated by the available interaction data or by practical considerations which dictate that the aggregation level must be manageable. The resulting aggregate choice set does not necessarily correspond to the zones perceived by individual choice-makers in the spatial choice process. Usually, decision-makers will 'perceive' a more disaggregate zonal system composed of alternatives which are termed elemental. Although in reality the perception of elemental alternatives and hence the definition of individual choice sets may change with the characteristics and the origin of the decision-maker [Tardiff (1980), Thill (1992)], it is assumed here that all choice-makers perceive the same zonal system on average. In practical models of spatial choice then, the problem of aggregation is always present where zonal systems are imposed. The particular model of spatial choice examined below is the multinomial logit. Theory indicates that the use of aggregate choice units necessitates the introduction of corrective size and heterogeneity terms which modify utility. 'Size' refers to the number of elemental alternatives within an aggregate and 'heterogeneity' is a measure of the variation of elemental utility within that aggregate.

The proposed model in its most basic form dates to the work of Lerman (1975). In a problem of intra-urban housing choice, Lerman showed that the utility of an aggregate spatial unit depends on its size. Simulations illustrated that his model was unsatisfactory when aggregates were heterogeneous, suggesting the incorporation of a variance term in the utility of the aggregate. McFadden (1978) devised the improved theoretical framework which provides the necessary size and variance effects. Other things being equal, a heterogeneous aggregate alternative is more likely to contain a utility peak than a homogeneous aggregate and therefore will be more attractive to choice-makers. The heterogeneity term captures this effect and upgrades the estimated utility of heterogeneous aggregates, unlike the OMNL.

With this background in mind, we now briefly outline the theory behind the model in the context of inter-regional migration. A more comprehensive theoretical discussion is provided in Kanaroglou and Ferguson (1993). The focus below is on means to model the process of destination choice where potential migrants have already made the decision to move. Consider that the study area is divided into M elemental alternatives. These are the actual spatial units which the typical migrant considers as potential destinations. Assume that the M spatial units are combined into L aggregates, each of which contains M^J, $J \in \{1,2, ..., L\}$ elemental units. Thus:

$$\sum_{J=1}^{L} M^J = M \tag{1}$$

Given that individual n has decided to depart origin i, their utility for prospective elemental destination unit j within aggregate J is:

$$U_{ijn}^J = V_{ijn}^J + \varepsilon_{in}^J + \varepsilon_{ijn} \tag{2}$$

where:

- V_{ijn}^J is a deterministic component expressed as a function of attributes which relate to the aggregate alternative J, the elemental alternative $j \in \{1, 2, ..., M^J\}$ and characteristics of the choice-maker. Physical and social distance variables between i and j are also utilized.
- ε_{in}^J is a Gumbel distributed random component that captures unobserved variation relating to the aggregate spatial unit J.
- ε_{ijn} is a Gumbel distributed random component with parameters $(0,\mu^J)$ that captures unobserved variation relating to the elemental spatial unit $j \in \{1, 2, ..., M^J\}$.

The ε_{ijn} terms are independent of each other within and between aggregates and are also independent of the ε_{in}^J terms. The ε_{in}^J terms are themselves assumed independent of each other. Furthermore, it is assumed that $\varepsilon_{ijn} + \varepsilon_{in}^J$ are Gumbel variates with parameters $(0,\mu^*)$. In essence, these assumptions imply that the total error ε_{ijn}^J for an elemental alternative is the sum of two independent component errors. All utilities of elemental units within aggregate J share the same component ε_{in}^{J} implying that the utilities of any two elemental units within the same aggregate are not independent. As is the case with many spatial econometric models, contiguity implies similarity. Of course in this context, contiguity is defined by an elemental alternative having a neighbour within the same aggregate, and the similarity evolves from a spatial process which determines utility. While there is a positive correlation between intra-aggregate elemental utilities, inter-aggregate elemental utilities are uncorrelated in this model.

In addition, the assumptions imply that elemental errors within the same aggregate are homoskedastistic, which is potentially restrictive. Meanwhile $\text{Var}(\varepsilon_{ijn})$ is allowed to vary across aggregates, hence accommodating heteroscedasticity to some extent. In a context other than spatial choice, McMillen (1992) has examined the issue of heteroscedasticity over space in probit models.

Since choice-makers are hypothesized to seek the elemental zone within an aggregate which provides maximum utility, this zone's estimate of utility in effect determines the utility of the aggregate:

$$U_{in}^{J} = \varepsilon_{in}^{J} + \max_{j \in \{1,2,\ldots,M^{J}\}} \left(V_{ijn}^{J} + \varepsilon_{ijn}\right) \tag{3a}$$

where the maximization term is a Gumbel variate with parameters

$$\left(\frac{1}{\mu^{J}} \ln \sum_{j=1}^{M^{J}} \exp(\mu^{J} V_{ijn}^{J}), \mu^{J}\right) \tag{3b}$$

The first expression defines the mode of the distribution (which is interpreted as the aggregate systematic utility) and the second term is a scale parameter which is inversely proportional to the variance of the utility distribution. We can represent the choice probability of an aggregate in the general format of an ordinary multinomial logit:

$$P_{n}(J) = \frac{\exp\left[\frac{\mu^{*}}{\mu^{J}} \ln \sum_{j=1}^{M^{J}} \exp\left(\mu^{J} V_{ijn}^{J}\right)\right]}{\sum_{K=1}^{L} \exp\left[\frac{\mu^{*}}{\mu^{K}} \ln \sum_{k=1}^{M^{K}} \exp\left(\mu^{K} V_{ikn}^{K}\right)\right]} \tag{4}$$

In this expression, μ^{J} is related to $\text{Var}(\varepsilon_{ijn})$ while μ^{*} is related to $\text{Var}(\varepsilon_{ijn} + \varepsilon_{in}^{J})$, the combined uncaptured variation of strictly aggregate and elemental utilities. Note that μ^{J} is allowed to vary across aggregate alternatives, possibly being unique for each aggregate or unique for a subset of aggregates. The model is thus sufficiently general that the distribution of unobserved elemental variation is permitted to differ across aggregates. As a by-product of estimating maximum utilities, the proposed model yields estimates of elemental utilities within the aggregates.

The ratio μ^*/μ^J is critical to the theoretical interpretation of the model and can be viewed as an overall measure of the correlation of elemental utilities within aggregate alternatives. We assume that elemental utilities within the same aggregate are positively correlated but that those between aggregates are uncorrelated. The correlation is assumed to result from common unobserved variation (ε_{in}^J) of the aggregate which they share, not from correlation among ε_{ijn}, $j \in \{1, 2, ..., M^J\}$. A problematic aspect of such assumptions is typified by any occasion when the correlation of utility between two elemental spatial units belonging to different aggregates is high. Similarly, the utility dependency between pairs of elemental alternatives within the same aggregate may vary. Such violations of assumptions may require redefinition of the aggregate units if possible.

Irrespective of the definition of the aggregates, it is possible to show that:

$$\frac{\mu^*}{\mu^J} = \frac{\mathrm{Var}(\varepsilon_{ijn})}{\mathrm{Var}(\varepsilon_{ijn} + \varepsilon_{in}^J)} = \sqrt{1 - \mathrm{corr}\,(U_{ijn}^J, U_{ikn}^J)} \tag{5}$$

One value applies to all pairs of elemental alternatives which belong to the same aggregate. If the correlation tends to become high then $\mu^*/\mu^J \to 0$, while $\mu^*/\mu^J \to 1$ as the correlation becomes low. Estimation procedures allow for the identification of the ratio μ^*/μ^J, $J \in \{1, 2, ..., L\}$, but not the true values of its components. Without loss of generality though, the convention $\mu^* = 1$ is typically adopted. If the unobserved elemental utility variation within aggregate J, is small relative to the unobserved variation shared by all elemental alternatives within J, then μ^J is large relative to μ^*, implying highly correlated total utilities among the elemental alternatives in J. In this case, the estimate of the expected maximum elemental utility corresponds closely with the largest of the elemental utilities predicted by the model. Conversely, if unobserved elemental variation is relatively large implying uncorrelated elemental utilities, then there is a lot of uncertainty with respect to the true levels of strictly elemental utilities. Given the individual expected values of Gumbel variates, the expected value of their maximum will be larger the greater the variance of the individual Gumbel variates. It is this principle which upgrades the estimate of systematic utility for an aggregate which contains a set of elemental alternatives with uncertain utilities.

In the context of the definition of aggregate alternatives, it is desirable that $\mu^*/\mu^J \to 0$ as this implies that there is little uncertainty at the sub-aggregate level and that similar elemental alternatives form the aggregate. The factors affecting the choice of an aggregate are also a good representation of the choice behavior at the sub-aggregate level. If the ratio is close to one, the implication is that the individual elemental utilities cannot be modeled very well within that aggregate when only the aggregate choice behavior is used to estimate the model.

If econometric estimation yields a value of μ^*/μ^J outside the range (0,1) then the result indicates a violation of assumptions or a poorly specified model. In practice, the ratio is often found to be greater than one indicating that elemental utilities are assigned a very large variance in order to upgrade the utility of that aggregate. It makes no theoretical sense however to increase the elemental variance beyond the total variance established by μ^* and thus an improved model specification is desirable. Similar conclusions apply for an estimated negative ratio, although such an event is rare in practice.

Clearly, the behavior of the ratio of scale parameters is critical to the interpretation of the model. The estimates of other explanatory variables as well as the goodness-of-fit of the model is very much related to the estimates of these ratios across aggregates. These issues are sufficiently important that their detailed discussion and analysis are reserved for future work. In this chapter, the focus is on contrasting the performance of OMNL with that of model (4) when μ^*/μ^J is given exogenously.

It is important to understand the composition of the expected maximum utility for an aggregate alternative. Through algebraic manipulation, it is possible to show that the initial model in (4) can be written as:

$$P_n(J) = \frac{\exp\left[V_{in}^{oJ}\right]}{\sum_{K=1}^{L} \exp\left[V_{in}^{oK}\right]} \tag{6}$$

where:

$$V_{in}^{oJ} = \mu^* \bar{V}_{in}^{J} + \frac{\mu^*}{\mu^J} \ln\left[\frac{1}{M^J} \ln \sum_{j=1}^{M^J} \exp\left(\mu^J (V_{ijn} - \bar{V}_{in}^{J})\right)\right] + \frac{\mu^*}{\mu^J} \ln M^J$$

With $\mu^* = 1$, note that the expected maximum elemental utility from the set of elemental alternatives in an aggregate (i.e. the systematic utility of the aggregate) can be broken into a mean utility effect $\bar{V}_{in}^{J}$, a heterogeneity effect:

$$(1/\mu^J)\ln\left((1/M^J)\Sigma_{j=1}^{M^J} \exp(\mu^J(V_{ijn} - \bar{V}_{in}^{J}))\right)$$

and a size effect $(1/\mu^J) \ln M^J$. In contrast, the OMNL utilizes only the mean utility term. The predicted mean utility with the OMNL will be different from that predicted by the proposed model since the parameters estimated will be dependent on the presence of the additional corrective terms. The size effect is constant across decision-makers and is self-explanatory. The heterogeneity effect varies across choice-makers and aggregates and, as mentioned earlier, is a measure of the variability of elemental systematic utilities. Typically, the correction from the size effect will be larger than that of the heterogeneity effect [Lerman (1975)].

In practice, a condensed form of (6) is more convenient:

$$P_n(J) = \frac{\exp\left[\bar{V}_{in}^{J} + \frac{1}{\mu^J}\ln\sum_{j=1}^{M^J}\exp\left(\mu^J(V_{ijn} - \bar{V}_{in}^{J})\right)\right]}{\sum_{k=1}^{L}\exp\left[\bar{V}_{in}^{K} + \frac{1}{\mu^K}\ln\sum_{K=1}^{M^K}\exp\left(\mu^K(V_{ikn} - \bar{V}_{in}^{K})\right)\right]} \tag{7}$$

In this model the convention $\mu^* = 1$ is used. In the remainder of the chapter, this model is referred to as HETRO. The size and heterogeneity terms from (6) are combined to form a term which measures the expected maximum deviation from the mean utility of aggregate alternative J. From the latter term it is easy to extract the heterogeneity and size effects for each choice-maker and aggregate alternative since the size effect does not vary across choice-makers.

It is important to emphasize at this point that the mean utility term in (6) and (7) differs from mean utility as typically represented in applications of the OMNL. The $\bar{V}_{in}^{J}$ term is assumed to be the simple unweighted average of the constituent elemental utilities. By extension, this implies that the attributes of aggregates are taken as the unweighted averages of the attributes measured at the elemental level. In practice, the attributes measured at the aggregate level and applied to represent the utility of aggregates in the OMNL are weighted averages of the contained elemental attributes. Typically, the aggregate averages in this context are weighted by the populations associated with individual elemental alternatives as would be the case with economic variables such as wages. With the exception of a perfectly even population distribution, the unweighted average of attributes relevant to (6) and the weighted average of the OMNL will not be the same. Of course for variables which cannot be related to choice-makers, such as climatic attributes, the concept of a weighted average makes little sense. This discussion highlights another potential difficulty with the OMNL, namely that the elemental alternatives which contribute most to some weighted average of appropriate variable types may not be the ones which provide the likely point of maximum utility for a given choice-maker.

To conclude this section, it should be mentioned that to the knowledge of these authors, no empirical study has utilized the heterogeneity and size effects in the manner advocated above. In the context of a full nested logit model of housing choice, Fischer and Aufhauser (1988) do acknowledge the presence of a heterogeneity effect in the choice of dwelling units within census tracts. They do not acknowledge though that the same idea also applies to the higher levels of their model where the spatial units are more aggregated. For that matter, it does not seem widely known that the theory behind HETRO transcends the field of intra-urban housing choice and that there is a heterogeneity effect to be considered in any spatial or non-spatial choice problem where the alternatives are aggregations of perceived elemental alternatives. The approach seems particularly applicable in problems of inter-regional choice such as human migration or industrial location.

11.3 Estimation of the Maximum Utility Model

In this section the estimation procedure for HETRO is outlined. Also, distinctions with the estimation normally used for the OMNL are highlighted. The procedure established in this section is utilized in section four.

The likelihood function for L aggregate zones and N observed choice-makers is:

$$\mathcal{L}(\beta,\mu) = \prod_{n=1}^{N}\prod_{j=1}^{L} P_n(J)^{Y_{Jn}} \tag{8}$$

where β is a vector of parameters that enter the systematic utility $V_{ijn}{}^{J}$, μ is a vector of parameters representing $(\mu^1, \mu^2, ..., \mu^L)$, and Y_{Jn} is a dummy variable defined as 1 when decision-maker n is observed to choose aggregate J and 0 if not.
The log-likelihood function is:

$$\mathcal{L}^*(\beta,\mu) = \sum_{n=1}^{N}\sum_{J=1}^{L} Y_{Jn} \ln P_n(J) \tag{9}$$

A suitable estimation procedure must identify the values of β and μ that maximize $\mathcal{L}^*(\beta,\mu)$. Endogenous estimation of μ complicates the estimation procedure considerably. Since the focus of this chapter is not on estimation, it is accepted that the values of μ are specified exogenously. Discussion on their endogenous estimation is reserved for a later paper. Bearing this in mind, the log-likelihood function has a maximum where:

$$\frac{\partial \mathcal{L}^*(\beta)}{\partial \beta} = 0 \tag{10}$$

The first derivative of $\mathcal{L}^*(\beta)$ with respect to the k-th parameter is:

$$\frac{\partial \mathcal{L}^*(\beta)}{\partial \beta_k} = \sum_{n=1}^{N}\sum_{J=1}^{L} \frac{Y_n^J}{P_n(J)} \frac{\partial P_n(J)}{\partial \beta_k} \tag{11}$$

In general, the partial derivative of a logit probability with respect to β_k is:

$$\frac{\partial P_n(J)}{\partial \beta_k} = P_n(J)\left[\frac{\partial V_{in}^J}{\partial \beta_k} - \sum_{Q=1}^{L} \frac{\partial V_{in}^Q}{\partial \beta_k} P_n(Q)\right] \tag{12}$$

Now for OMNL, the systematic utility V_{in}^J is given by the mean utility term:

$$V_{in}^J = \sum_{k=1}^{K} \beta_k \bar{X}_{ink}^J \tag{13}$$

which has the partial derivative:

$$\frac{\partial V_{in}^J}{\partial \beta_k} = \bar{X}_{ink}^J \tag{14}$$

Substituting (14) into (12), the partial derivative of the OMNL probability is obtained:

$$\frac{\partial P_n(J)}{\partial \beta_k} = P_n(J)\left[\bar{X}_{ink}^J - \sum_{Q=1}^{L} \bar{X}_{ink}^Q P_n(Q)\right] \tag{15}$$

The equivalent of (14) for HETRO is:

$$\frac{\partial V_{in}^J}{\partial \beta_k} = \bar{X}_{ink}^J + \frac{\sum_{j=1}^{M^J} \left(X_{ijnk} - \bar{X}_{ink}^J\right) \exp\left[\mu^J \sum_{l=1} \beta_l \left(X_{ijnl} - \bar{X}_{inl}^J\right)\right]}{\sum_{j=1}^{M^J} \exp\left[\mu^J \sum_{l=1} \beta_l \left(X_{ijnl} - \bar{X}_{inl}^J\right)\right]} \tag{16}$$

The derivative $\partial P_n(J)/\partial \beta_k$ for HETRO is obtained through the substitution of (16) into (12).

Given the partial derivatives of the HETRO probabilities, the iterative solution is no different from that of OMNL. Cramer (1991) advocates the use of the method of scoring, a gradient technique similar to the Newton-Raphson approach. Like other methods in this family, the gradient vector $q = \partial L^*(\beta)/\partial \beta$, also known as a score vector, and a direction matrix must be calculated at each iteration. The non-linear system of equations defined by setting q equal to zero must be solved iteratively. A Taylor series approximation of $q(\beta)$ around β' in the vicinity of β is utilized:

$$q(\beta) \approx q(\beta') + \boldsymbol{H}(\beta')(\beta - \beta') \tag{17}$$

where $H_{kl} = \partial^2 \mathcal{L}^*(\beta)/(\partial\beta_k \partial\beta_l)$ represents an element in the Hessian matrix $\boldsymbol{H}$ of second derivatives. Since $q(\beta) = 0$, (17) can be rearranged to yield:

$$\beta \approx \beta' - \boldsymbol{H}(\beta')^{-1} q(\beta') \tag{18}$$

When this approximation is applied repeatedly in the vicinity of the optimum, with β being substituted for β', convergence should occur. This forms the basis for the Newton-Raphson algorithm expressed as:

$$\beta_{t+1} \approx \beta_t - \boldsymbol{H}(\beta_t)^{-1} q(\beta_t) \tag{19}$$

With respect to either OMNL or HETRO, the most convenient expression for each element of q is:

$$\frac{\partial \mathcal{L}^*(\beta)}{\partial \beta_k} = \sum_{n=1}^{N} \sum_{J=1}^{L} \left[Y_n^J - P_n(J) \right] \frac{\partial V_{in}^J}{\partial \beta_k} \tag{20}$$

which is obtained when (12) is substituted into (11) and the result simplified.

The method of scoring differs from the Newton-Raphson approach in that the information matrix replaces the Hessian. Since the asymptotic variance-covariance matrix of β is directly related to the information matrix this replacement is useful. The expression for any element of the information matrix is:

$$Q_{kl} = \sum_{n=1}^{N} \sum_{J=1}^{L} \frac{1}{P_n(J)} \frac{\partial P_n(J)}{\partial \beta_k} \frac{\partial P_n(J)}{\partial \beta_l} \tag{21}$$

where (12) is substituted into (21). Since (16) for HETRO is quite different from (14) for OMNL, the results for Q_{kl} will also differ. Because the information matrix is given by $-E[\boldsymbol{H}]$, the negative expectation of the Hessian, the Newton-Raphson sequence of (19) is revised:

$$\beta_{t+1} \approx \beta_t + \left[-E\left(\boldsymbol{H}(\beta_t) \right) \right]^{-1} q(\beta_t) \tag{22}$$

McFadden (1974) has shown that for OMNL, the Hessian matrix will be negative definite under all but extreme circumstances. This implies that the information matrix should be positive definite. After extensive experimentation with simulated data we conclude that the convergence success rate is almost as high for HETRO with μ exogenous as it is for OMNL. As will be shown in future work, this success rate is considerably lower when μ is endogenous. A modified estimation procedure is needed in such circumstances.

An important issue affecting the convergence of HETRO is the possibility of high correlation among the different destination attributes. Such attributes, which are typically used in spatial choice models to specify utilities, and which do not vary across individuals, can potentially create dependencies in the information matrix. In the cases where convergence does not occur, it is likely that the information matrix is near singular. Clearly, it is critical to thoroughly screen destination attributes, and any interaction terms related to them, beforehand. It is suggested that an orthoganilization technique such as principal components analysis be applied in situations where many generic destination variables are used. By utilizing the significant principal components, most of the variance is accounted for without risk of multicollinearity.

11.4 Model Specifications and Simulations

To an extent, the issues revolving around the specification of the maximum utility model depend on whether the interaction data are aggregated or disaggregated across choice-makers. The variables used to specify utilities can be classified as: destination attributes, measures of distance and socio-economic individual characteristics. The latter category is unique to data which are totally disaggregated across choice-makers while the first two categories are relevant to both aggregation levels. In this section, and in the simulations to follow, we consider differences in the models defined by interaction data which are aggregated spatially but may or may not be aggregated across decision-makers

11.4.1 Specification Issues

Aggregate data: Several studies of human migration utilize logit models with interaction data aggregated across both choice-makers and destinations [e.g., Day (1992), Anderson and Papageorgiou (1994)]. While there is no question that the true utilities of choice-makers from a particular destination are highly variable, the use of aggregate data precludes the differentiation of systematic utilities among choice-makers originating from the same zone. In essence, such a model works on the premise that the 'average' individual can be used as representative of all individuals from a given origin. Distance is the one variable type which does introduce some variation in systematic utilities across decision-makers since it can be used to segment the population based on the zone of origin.

A model with aggregate data is primarily specified using destination attributes. As Tardiff (1980) mentions, such variables are a peculiarity of spatial choice models in that they represent no variation across decision-makers. The main specification issue with aggregate data is whether variables are to be specified in a generic way or as alternative-specific. The former specification method implies that the marginal contributions to utility of a variable are constant across all destinations while the latter implies that the marginal importance of a variable in determining utility depends on the destination.

Alternative-specific specifications of distance variables and destination variables can be problematic. With respect to distance, there is the issue of whether it is theoretically valid to allow the friction-of-distance to vary across destinations. Are differences in parameters a product of the fact that distance interacts with destination attributes or does the friction-of-distance vary? Some argue that there is no single valid measure of the friction-of-distance [Lo (1993)]. Alternative-specific specifications of destination attributes can cause problems in that each parameter associated with this variable type implies one fewer alternative-specific constant that can be used. If there are more than L-1 such parameters, a model becomes indeterminate since it is no longer possible to measure utility in relative terms.

With respect to HETRO, distance variables must be measured at the sub-aggregate level. In so doing, the systematic utility of a prospective aggregate destination from a given origin will tend to depend on the closest elemental zone within the aggregate, not the average distance to all elemental zones. On a uniform plain, we would expect this to be the case: people would choose the closest elemental alternative within the aggregate. That people do not do this in reality reflects the fact that other factors are at work which ought not be confused with the friction-of-distance.

Destination attributes can be measured either at the aggregate level or the sub-aggregate level. Variables associated with, for example, climate or employment relate really to sub-aggregate phenomena. Their values fluctuate within aggregates and hence should introduce variability into utilities. Other attributes such as regional tax rates are constant across a given aggregate. Since there is no intra-aggregate utility variability introduced by such variables, they are specified as they would be in OMNL.

Micro-data: Observations in micro-data are at the individual choice-maker level. Such data usually incorporate variables describing the socio-economic characteristics of decision-makers and are thus richer in information content than aggregate data sets. With the help of such variables, it is possible to differentiate the utilities of individual choice-makers or population segments and capture the heterogeneity of tastes in the population. As in the case of aggregate data, however, destination attributes do not vary across decision-makers and can play the role of alternative-specific constants.

The socio-economic individual characteristics available in micro-data are specified in an alternative-specific manner since such variables will typically affect the utilities associated with different destinations in different ways. The marginal contribution to utility of a decision-maker's income, for example, might be positive for one destination but negative in another. However, since utility is a relative concept in the logit model, some spatial unit must act as a zero utility reference against which the utilities of other alternatives are compared. The option is available to specify both OMNL and HETRO in this manner.

A difficulty in using this approach for HETRO is that the socio-economic variables can be used to differentiate utilities only on an aggregate alternative basis. One might reasonably expect that a socio-economic variable will affect the perceived utility of elemental alternatives within the same aggregate quite differently. However, since the choice behavior is unobserved at the elemental level and since the number of elemental alternatives is likely unmanageably large, alternative-specific specifications at the sub-aggregate level are not feasible.

An alternative means of introducing variability into the elemental utilities associated with socio-economic characteristics is to specify these variables interactively with destination attributes. This approach is appealing in the sense that utilities are modeled as a matching of decision-maker characteristics with those of potential destinations. It has been utilized by Liaw and Ledent (1988) and Liaw (1990) in studies which interacted categorical socio-economic variables with destination attributes. There is, however, no practical or theoretical reason that this cannot be done with continuous socio-economic variables. A variable of this type is defined as:

$$Z_{ijn}^{J} = X_{in} X_{j}^{J} \tag{23}$$

where X_{in} is the socio-economic variable score for individual n at origin i, and X_j^J is the value of the destination attribute at elemental alternative j in aggregate J.

An interaction variable such as Z_{ijn}^J can be specified as generic in the context of one aggregate alternative since the same parameter will apply to all elemental alternatives within this aggregate. However it can be specified as alternative-specific across aggregates meaning that elemental alternatives in different aggregates will be associated with different parameters. By allowing this alternative-specific specification of a generic interaction term, one captures essentially the same effects as would be obtained by specifying the socio-economic variable alone in an alternative-specific manner at the aggregate level.

The usefulness of this interaction form, however, is that different choice-makers can be associated with different levels of utility heterogeneity within aggregates. Given a single aggregate and with all things equal, the choice-maker in the sample with the minimum variable score X_{in} will necessarily be associated with less heterogeneous values of Z_{ijn}^J in (23), and therefore $V_{ijn}^{\ J}$, than the choice-maker with the largest variable score. Note that it might be necessary to transform the socio-economic variable in some circumstances since choice-makers associated with the smallest variable score, not the largest, might perceive the greatest amount of utility heterogeneity.

11.4.2 Description of Simulation Method

The purpose of this section is to illustrate through simulations the model specifications discussed in the previous section and to demonstrate that as the heterogeneity of aggregate choice units increases, so does the potential bias of parameters estimated through the OMNL. In parallel, it will be seen that HETRO is not subject to similar problems. To address these points, a Monte Carlo procedure is utilized. Observed choices are generated on the basis of known parameter values. Logit models of both types are fitted to the resulting data and after repeated trials, it is judged whether the models produce the expected parameter estimates. The goal is to systematically alter the variance of a destination attribute within one aggregate while preserving the initial mean value of this attribute within that aggregate. It should become evident that the mean utility of an aggregate can, under some circumstances, become a minor determining factor in how decision-makers select aggregates.

There are four types of simulations under study which can be classified on the basis of two criteria. The first is whether the variables are aggregated across decision-makers or not. The simulations with aggregate data are generated with a distance variable and a destination attribute as is typical in such a model of migration. The simulations with micro-data are generated on the basis of an interaction between a socio-economic individual characteristic, unique to each choice-maker, and a destination attribute. The second classification criterion is based on whether the alternative whose utility variability is altered has the highest level of mean utility or not. It turns out that the biases in OMNL differ if a heterogeneous aggregate has lower mean utility than other aggregates as compared to situations where the heterogeneous aggregate has the highest mean utility among all aggregates.

Some other general points to be noted about the simulations are as follows:

- A bottom-up approach is used. That is, we generate true utilities at the elemental level and assume that the actual utility of an aggregate for that decision-maker is given by the largest of the contained elemental utilities. It is hence of considerable importance to note that the model in (7) is not directly used to generate the true utilities at the aggregate level.
- The effects of mean utility and aggregate alternative size are controlled by maintaining the same set of mean utilities for all levels of heterogeneity and by assigning the same number of elemental alternatives (four) to each aggregate.
- It is assumed that the ratio μ^*/μ' is exogenously set to one for all aggregates. Hence the random error components of the elemental utilities within and between aggregates are independently generated and therefore uncorrelated.
- It is assumed that the attribute value for an aggregate alternative is simply the mean of the constituent elemental attributes.
- Note that even though we generate data which are 'aggregate' across decision-makers in the sense that systematic utilities are not differentiated on the basis of

socio-economic variables, the true utilities which determine the choice shares are generated one decision-maker at a time.

A precise description of the data generation procedure is as follows.

(1) A data set of attribute values is generated for a problem consisting of three aggregate alternatives each containing four elemental alternatives within. The data set is assumed to contain 50 choice-makers positioned at five different origins. None of the five corresponds to any potential destination. The data consist of a destination attribute, a socio-economic individual variable, the interaction between the two and a measure of distance to the given potential destination. The destination attribute values, regardless of the prevailing level of heterogeneity, are adjusted to conform with predefined mean attribute values in the aggregates.

(2) For the appropriate specification, systematic elemental utilities are defined as a linear combination of the attribute values and known parameters.

(3) Actual utilities for a given choice-maker, are obtained with the addition to the systematic utilities of a Gumbel distributed error component having variance parameter $\mu = 1$ and hence a variance of $\pi^2/6$.

(4) In accordance with theory, the chosen elemental alternative for a given individual is taken as the one which provides the maximum utility of all elemental alternatives in the system. Obviously, this defines the chosen aggregate alternative also.

(5) Steps 3 and 4 are repeated N times to correspond with the number of choice-makers in a given trial. The observed aggregate alternative choice behavior and the known variable and attribute values provide the necessary information to estimate the unknown parameters of the two model types (OMNL and HETRO) for that trial.

(6) Steps 3, 4 and 5 are repeated appropriately to generate choice data for 100 trials at a given level of heterogeneity. Preliminary work revealed that little was to be gained by employing a larger number of trials.

(7) In order to alter the utility variance within a single aggregate in relation to the other two aggregates, the elemental destination attribute values in that aggregate are multiplied by some common factor and the result translated to maintain the same mean utilities across aggregates. The appropriate variances are obtained through the identity $\mathrm{Var}(ax) = a^2\mathrm{Var}(x)$ where $\mathrm{Var}(x)$ is the initial variance of the destination attribute in the aggregate alternative and a is the necessary constant used to obtain the new variance. Multiplying the destination attribute by some factor also affects the interaction term. Steps 1 to 6 are then completed for the new level of heterogeneity.

(8) For each new level of heterogeneity (the variances tested ranged from 2 to 50) or new specification, step 7 is repeated.

Table 11.1: Estimation results for aggregate data; variable aggregate does not have highest mean utility.

Variance		β_1	β_2	t_1	t_2	se_1	se_2	ρ^2	shares		
2	HETRO	0.35	-0.10	1.98	-2.48	0.17	0.04	0.12	16.9	23.1	9.9
	OMNL	0.29	-0.08	2.26	-1.85	0.12	0.04	0.11			
5	HETRO	0.31	-0.10	2.10	-2.42	0.14	0.04	0.11	18.5	21.5	9.9
	OMNL	0.26	-0.06	2.08	-1.58	0.12	0.04	0.09			
10	HETRO	0.31	-0.11	2.85	-2.59	0.11	0.04	0.13	20.8	20.1	9.1
	OMNL	0.26	-0.06	2.09	-1.39	0.12	0.04	0.08			
15	HETRO	0.29	-0.11	3.38	-2.54	0.09	0.04	0.13	22.4	18.8	8.8
	OMNL	0.25	-0.04	1.99	-1.08	0.12	0.04	0.07			
25	HETRO	0.30	-0.09	4.67	-2.18	0.06	0.04	0.17	26.8	16.0	7.2
	OMNL	0.26	-0.00	2.03	-0.04	0.13	0.04	0.06			
50	HETRO	0.31	-0.11	6.54	-2.06	0.05	0.05	0.33	34.7	10.4	4.9
	OMNL	0.26	0.06	1.90	1.50	0.13	0.04	0.07			

Table 11.2: Estimation results for aggregate data; variable aggregate has highest mean utility.

Variance		β_1	β_2	t_1	t_2	se_1	se_2	ρ^2	shares		
2	HETRO	0.30	-0.10	2.52	-2.39	0.11	0.04	0.13	23.0	18.5	8.4
	OMNL	0.36	-0.08	2.88	-1.98	0.12	0.04	0.12			
5	HETRO	0.29	-0.10	3.07	-2.27	0.09	0.04	0.15	25.0	16.9	8.1
	OMNL	0.39	-0.07	2.97	-1.76	0.13	0.04	0.13			
10	HETRO	0.30	-0.10	4.04	-2.33	0.07	0.04	0.19	27.3	15.8	6.9
	OMNL	0.46	-0.07	3.28	-1.65	0.14	0.04	0.16			
15	HETRO	0.31	-0.10	4.85	-2.22	0.06	0.05	0.23	30.1	13.6	6.3
	OMNL	0.51	-0.05	3.41	-1.26	0.15	0.04	0.18			
25	HETRO	0.31	-0.11	5.83	-2.17	0.05	0.05	0.31	33.6	11.5	4.8
	OMNL	0.64	-0.04	3.65	-0.98	0.17	0.04	0.24			
50	HETRO	0.31	-0.10	6.70	-1.48	0.05	0.06	0.48	40.3	6.3	3.4
	OMNL	0.91	0.01	3.64	0.15	0.25	0.05	0.37			

Table 11.3: Estimation results for micro-data; variable aggregate does not have highest mean utility.

Variance		β_1	t_1	se_{pred}	se_{true}	ρ^2	shares		
2	HETRO	1.11	2.29	0.46	0.49	0.10	20.6	22.0	7.4
	OMNL	0.86	2.66	0.32	0.33	0.09			
5	HETRO	1.05	2.68	0.38	0.43	0.11	23.8	19.3	6.9
	OMNL	0.76	2.46	0.30	0.26	0.07			
10	HETRO	1.05	4.13	0.25	0.25	0.17	27.7	16.6	5.7
	OMNL	0.73	2.39	0.30	0.21	0.06			
15	HETRO	0.98	4.90	0.20	0.19	0.21	30.4	14.2	5.3
	OMNL	0.67	2.24	0.30	0.22	0.06			
25	HETRO	1.02	6.19	0.16	0.16	0.35	35.9	10.2	3.9
	OMNL	0.59	2.03	0.29	0.16	0.04			
50	HETRO	1.02	6.53	0.16	0.18	0.57	41.9	5.6	2.5
	OMNL	0.52	1.85	0.28	0.09	0.03			

Table 11.4: Estimation results for micro-data; variable aggregate has highest mean utility.

Variance		β_1	t_1	se_{pred}	se_{true}	ρ^2	shares		
2	HETRO	1.03	3.41	0.30	0.31	0.22	30.7	14.5	4.9
	OMNL	1.29	3.65	0.35	0.39	0.21			
5	HETRO	0.99	4.33	0.23	0.19	0.26	33.1	12.5	4.4
	OMNL	1.50	3.86	0.38	0.38	0.25			
10	HETRO	1.03	5.39	0.19	0.21	0.36	36.4	9.7	3.9
	OMNL	1.94	4.06	0.47	0.69	0.32			
15	HETRO	1.02	5.89	0.17	0.17	0.43	38.5	8.1	3.4
	OMNL	2.31	4.21	0.54	0.70	0.38			
25	HETRO	1.03	6.20	0.17	0.17	0.55	41.4	5.9	2.7
	OMNL	3.06	4.37	0.69	0.86	0.48			
50	HETRO	1.06	5.86	0.19	0.22	0.71	44.9	3.3	1.7
	OMNL	5.08	4.36	1.20	2.13	0.64			

11.4.3 Results

Case #1: Aggregate data specification; aggregate with variable heterogeneity does not have largest mean utility. In this case, data are aggregated across choice-makers with the exception of distance which introduces some variability into systematic utilities. The known parameter associated with distance is -0.1 and the parameter for the destination attribute is 0.3. The latter implies that spatial units with high values on that attribute are perceived to have more utility than those with low values. The variance levels for the destination attribute in aggregate 1 range from 2, which is smaller than the corresponding variances in the other aggregates, to 50 which makes aggregate 1 far more heterogeneous than all other aggregates with respect to this attribute. All values in Table 11.1 represent mean results from the 100 trials on each model and at each level of variance. For a given variance level, the top row contains results for HETRO while the bottom row has results for OMNL. The average parameter estimates, *t*-scores, predicted standard errors and rho-squared (ρ^2) goodness-of-fit statistic are displayed. The latter is calculated as:

$$\rho^2 = 1 - \frac{\mathcal{L}^*(\beta)}{\mathcal{L}^*(0)} \qquad (24)$$

where $\mathcal{L}^*(\beta)$ defines the value of the log-likelihood function at convergence and $\mathcal{L}^*(0)$ defines its value with a naive model of zero utilities. The columns labeled 'shares' represent the average number of choice-makers out of the group of fifty, who choose aggregate 1, 2 and 3. The 'true' standard errors, which are given by the parameter standard deviations from 100 trials, are not displayed because they are almost identical, in this case, to the average standard errors derived from the variance-covariance matrix.

Initially, the first aggregate has slightly lower mean utility than aggregate 2 which is also more heterogeneous. As the level of variance in the destination attribute of aggregate 1 increases, with unchanging mean utilities, it becomes progressively more likely that alternative 1 is chosen as reflected in the share column of Table 11.1. It is more likely to contain an attractive elemental alternative with regard to the destination attribute and, therefore, also contain the point of maximum utility in the system. On a mean utility basis the other aggregates are more attractive than 1, but this fact becomes increasingly irrelevant to the utility maximizing choice-maker.

At all but the smallest level of variance, there is a consistent downward bias in the OMNL parameters. At higher levels of variance, the OMNL model is unable to reconcile the patterns implied by the mean utility with that of the choice shares. On a mean utility basis, it makes no sense that aggregate 1 is more often chosen, hence the parameters of variables used to predict the mean utilities become less significant. Even though there are no huge biases in parameters, the fact that OMNL does not include the corrective terms is very damaging to its goodness-of-fit. It is noteworthy also that while the changing variance affects only the destination attribute directly, with all distances being fixed, there are indirect effects which reduce the significance of the distance parameter. In addition, OMNL consistently underestimates the significance of the two variables and, for the destination attribute, is generally less efficient in relation to HETRO.

Case #2: Aggregate data specification; aggregate with variable heterogeneity has largest mean utility. This simulation differs from Case #1 in that aggregate 1 has the largest mean utility with regard to the destination attribute. Hence, for all variance levels, aggregate 1 is more often chosen than the other alternatives. In Table 11.2 we witness the opposite bias in the OMNL. With increasing variance, the OMNL destination attribute parameter begins to display a substantial upward bias. The sampling distributions for the destination attribute are displayed graphically in Figure 11.1 (a-c).

At higher variances, the extra choices which accrue to aggregate 1 are on the basis of an increased heterogeneity effect which in turn increases the likelihood that alternative 1 contains the point of maximum utility. From the perspective of OMNL, however, the prevailing choice shares indicate that the destination attribute under consideration is very important, implying that too much significance is associated with the corresponding parameter and therefore the impact of mean utility. HETRO properly separates the heterogeneity effect from the mean utility effect and consistently produces unbiased and efficient estimates of the two parameters. Apparently, OMNL so overestimates the effect of the destination attribute that the impact of distance is judged to be minimal, resulting in a severe mis-diagnosis of its significance. Again, HETRO results in better fitting models.

Case #3: Micro-data specification; aggregate with variable heterogeneity does not have largest mean utility. This scenario differs from those of Case #1 and #2 since the use of microdata is assumed in which information relating to specific choice-makers is available. It is like Case #1 in that aggregate 1 has lower mean utility than aggregate 2. While destination attributes and distance variables could be included in the specification of Table 11.3, a one variable model in which a destination attribute interacts with a socio-economic characteristic is utilized in order to focus on the behavior of this type of variable. This interaction variable has a true parameter of 1.0. Therefore, a choice-maker who has a large score on the socio-economic variable will perceive more heterogeneity in intra-aggregate utility than someone with a small score. The positive true parameter implies that this interaction adds to utility, regardless of the level of the socio-economic variable, rather than acting as a disutility.

The results in Table 11.3 indicate that OMNL provides biased estimates at any level of variance in aggregate 1 although the performance becomes progressively worse as this variance increases. The bias in the interaction parameter behaves qualitatively the same as did the destination attribute parameter in Table 11.1: when the choice shares are not explainable with the mean utilities, the OMNL parameter will be biased downward.

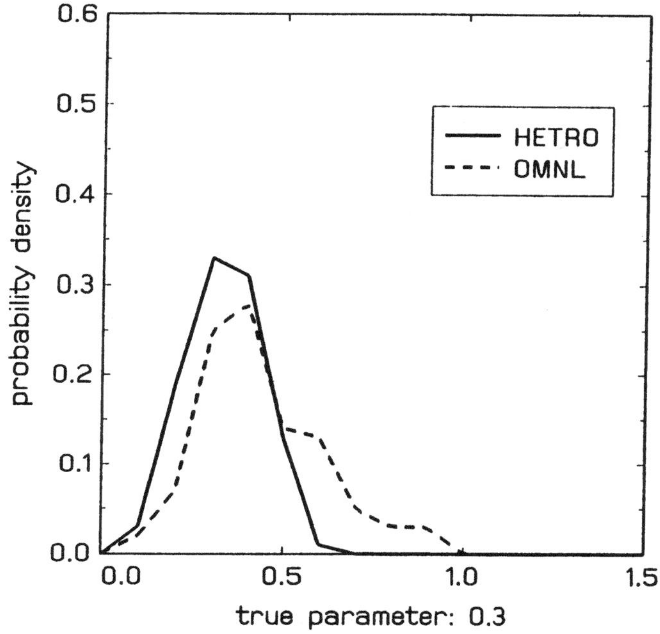

Figure 11.1a: Case 2, variance = 5.0.

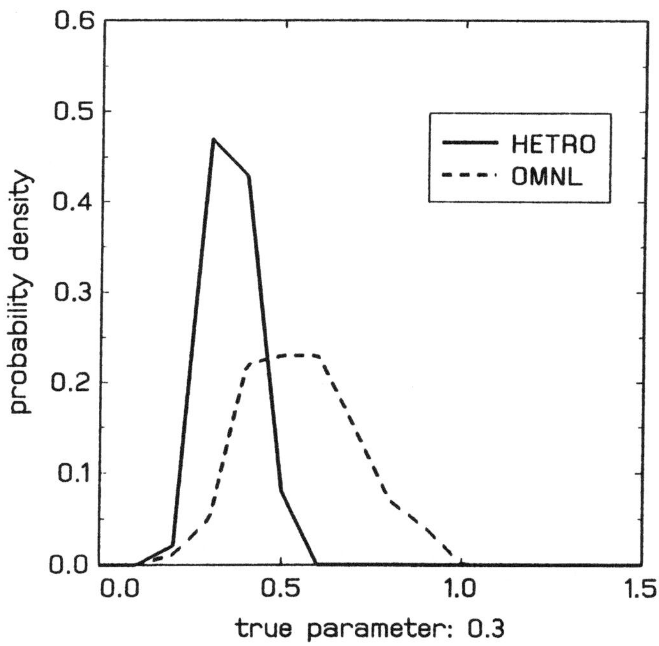

Figure 11.1b: Case 2, variance = 15.0.

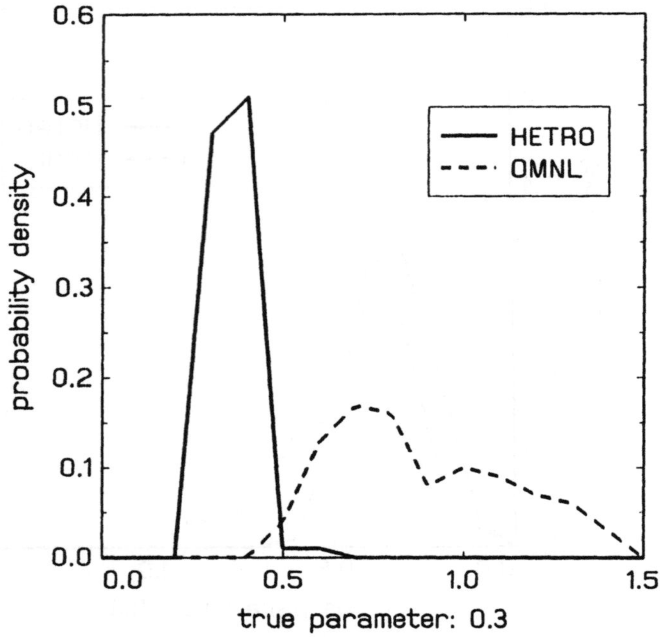

Figure 11.1c: Case 2, variance = 50.0.

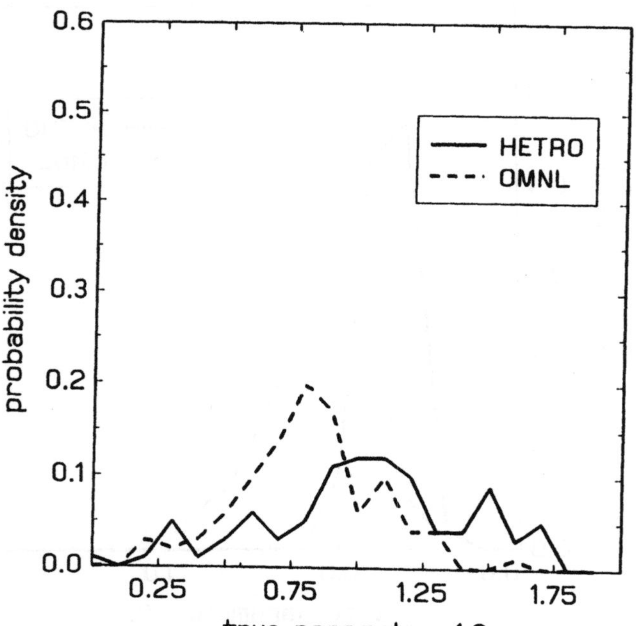

Figure 11.1d: Case 3, variance = 5.0.

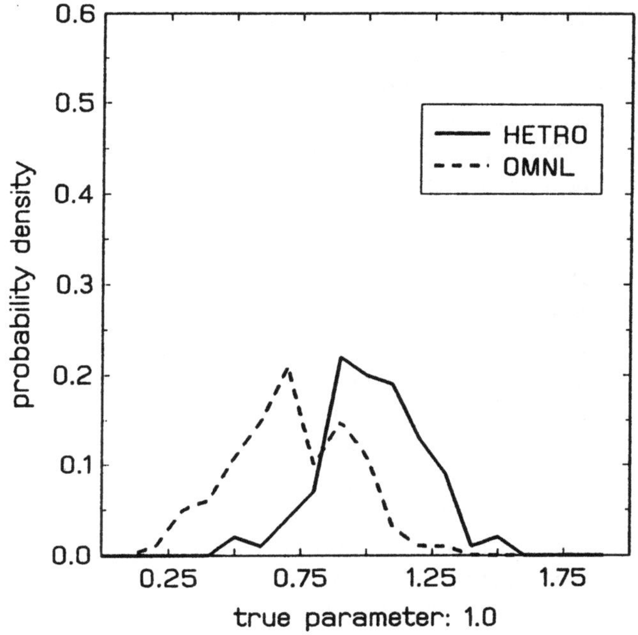

Figure 11.1e: Case 3, variance = 15.0.

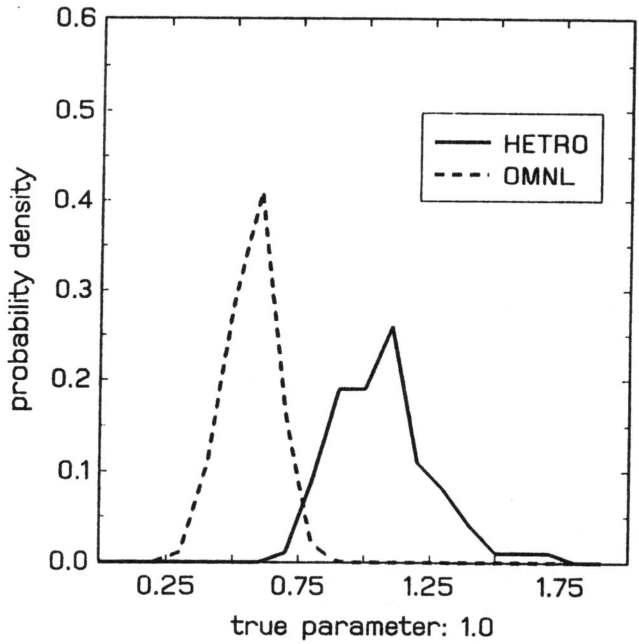

Figure 11.1f: Case 3, variance = 50.0.

In Table 11.3 the true standard errors exhibited in 100 trials are compared with the average of the predicted standard errors because there are some substantial differences. In particular, the predicted standard errors of OMNL are consistently larger than the true ones indicating that there is a tendency to underestimate the significance of this interaction parameter. Interestingly, as can be seen visually in Figure 11.1 (d-f), OMNL generally provides a more efficient estimate than HETRO albeit a substantially biased one. Nevertheless, HETRO by far produces superior fitting models.

Case #4: Micro-data specification; aggregate with variable heterogeneity has largest mean utility. This case employs the same specification as Case #3 but the mean utility in aggregate 1 is always larger than in the other two aggregates. As the share column in Table 11.4 indicates, the number of choices going to aggregate 1 is larger than is the case in Table 11.3 because of the differing mean utility effects. At high levels of variance in aggregate 1, the biases in the OMNL estimate of the interaction parameter become quite drastic. Again, as was the case in Case #2, the bias is upward since OMNL is over-estimating the impact of the mean utility effect while the inflated standard errors are indicative of considerable inefficiency. Moreover, the predicted standard errors for OMNL are biased downward resulting generally in over-estimates of significance. In contrast to Case #3, where the biases were not so severe, the poor parameter estimates do not compromise the overall goodness-of-fit very much. It can be concluded that the general pattern of biases in the interaction parameter estimate is similar to the type occurring with a destination attribute. In comparison, the performance of HETRO is good in all aspects and at all levels of variance.

11.5 Conclusions

The main contribution of this study has been to illustrate the potential importance of a utility heterogeneity effect in properly characterizing the utility of an aggregate spatial unit. The notion that sub-aggregate attribute information should be applied when the available interaction data are at the aggregate level is a new perspective also. While it has long been known that there are size effects which need to be associated with aggregates, this chapter represents an initial attempt at illustrating to regional scientists the usefulness of a heterogeneity term. The proposed theoretical framework also provides the customary size effect but one which does not emerge in the ad hoc manner common to many empirical studies. The theoretical principle under which these corrective terms are derived is simply that decision-makers seek the point within an aggregate which provides them the maximum utility. With the ordinary multinomial logit model, parameters are incorrectly estimated on the basis of some measure of average utility within aggregates rather than through maximum utility.

The simulations here have shown that this theory of choice among aggregates performs very well. The proposed model has excelled in a variety of circumstances whereas with the same data, the ordinary multinomial logit model suffers from potentially severe biases and inefficiencies, especially in a heterogeneous system. In particular, the parameters associated with destination attributes and the interactions of destination attributes with socio-economic variables are prone to downward bias when the most attractive aggregates on a mean utility basis are less heterogeneous than other aggregates. The parameters are biased upward when the heterogeneity of aggregates tends to complement the differences in mean utilities. Also, the goodness-of-fit of the ordinary multinomial logit model is often seriously compromised in relation to the proposed model.

Considerable work lies ahead in this area of research. Much of it relates to the estimation and interpretation of the ratio μ^*/μ^J which is an index of the degree to which elemental alternatives within some aggregate are perceived as similar and as such it can be described as a spatial autocorrelation parameter. The endogenous estimation of this ratio is a considerably more complicated optimization problem than that of the exogenous treatment applied in this chapter. Ultimately, it will be necessary to examine the sampling distributions of μ^*/μ^J under different circumstances to assess its reliability in remaining within the theoretically defined (0,1] interval.

Acknowledgement

Financial support for this research was provided by the Social Sciences and Humanities Research Council of Canada grant 410-91-0901.

References

Amrhein, C.G. and R. Flowerdew, The Effect of Data Aggregation on a Poisson Regression Model of Canadian Migration, *Environment and Planning A*, 1381-1391, 1992.

Anas, A., Discrete Choice Theory, Information Theory and the Multinomial Logit and Gravity Models, *Transportation Research B*, 17, 13-23, 1983.

Anderson, W.P. and Y.Y. Papageorgiou, An Analysis of Migration Flows for the Canadian System 1952-1983: 1. Migration Probabilities, *Geographical Analysis*, 26, 15-36, 1994.

Arbia, G., *Spatial Data Configuration in Statistical Analysis of Regional Economic and Related Problems*, Dordrecht: Kluwer, 1989.

Batty, M. and P.K. Sikdar, Spatial Aggregation in Gravity Models: 1. an Information Theoretic Approach, *Environment and Planning A*, 14, 377-405, 1982a.

Batty, M. and P.K. Sikdar, Spatial Aggregation in Gravity Models: 2. One Dimensional Population Density Models, *Environment and Planning A,* 14, 525-553, 1982b.

Batty, M. and P.K. Sikdar, Spatial Aggregation in Gravity Models: 3. Two-dimensional Trip Distribution and Location Models, *Environment and Planning A*, 14, 629-658, 1982c.

Batty, M. and P.K. Sikdar, Spatial Aggregation in Gravity Models: 4. Generalisations and Large Scale Applications, *Environment and Planning A*, 14, 795-822, 1982d.

Bolduc D., M.G. Dagenais and M.J. Gaudry, Spatially Autocorrelated Errors in Origin-Destination Models: A New Specification Applied to Aggregate Mode Choice, *Transportation Research*, 25B, 1-12, 1989.

Borgers, A. and H. Timmermans, Choice Model Specification, Substitution and Spatial Structure Effects, *Regional Science and Urban Economics*, 17, 29-47, 1987.

Cramer, J.S., *The Logit Model: An Introduction for Economists*, London: Edward Arnold, 1991.

Day, K.M., Inter-Provincial Migration and Local Public Goods, *Canadian Journal of Economics*, 25, 123-142, 1992.

Fischer M.M. and E. Aufhauser, Housing Choice in a Regulated Market, *Geographical Analysis*, 20, 47-69, 1988.

Fotheringham, A.S., Spatial Structure and Distance-Decay Parameters, *Annals of the Association of American Geographers*, 71, 425-436, 1981.

Fotheringham, A.S., A New Set of Spatial Interaction Models: The Theory of Competing Destinations, *Environment and Planning A*, 15, 15-36, 1983.

Griffith, D.A. and K.G. Jones, Explorations Into the Relationship between Spatial Structure and Spatial Interaction, *Environment and Planning A*, 12, 187-201, 1980.

Haynes, K.E. and A.S. Fotheringham, The Impact of Space on the Application of Discrete Choice Models, *The Review of Regional Studies*, 20, 39-49, 1990.

Kitamura, R., L.P. Kostyniuk and K. Ting, Aggregation in Spatial Choice Modeling, *Transportation Science*, 13, 325-342, 1979.

Lerman, S.R., *A Disaggregate Behavioral Model of Urban Mobility Decisions*, PhD dissertation, Cambridge: Department of Civil Engineering, MIT, 1975.

Liaw, K.L. and J. Ledent, Joint Effects of Ecological and Personal Factors on Elderly Interprovincial Migration in Canada, *Canadian Journal of Regional Science*, 11, 77-100, 1988.

Liaw, K.L., Joint Effects of Personal Factors and Ecological Variables on the Interprovincial Migration Patterns of Young Adults in Canada, *Geographical Analysis*, 22, 189-208, 1990.

Lo, L., Substitutability, Spatial Structure and Spatial Interaction, *Geographical Analysis*, 23, 132-146, 1991.

Lo, L., *Unbiased Distance Parameter: Illusion or Solution?*, Paper presented at the June 1993 meetings of the Canadian Association of Geographers, Ottawa, Canada, 1993.

McFadden, D., Conditional Logit Analysis of Qualitative Choice Behavior, in: P. Zarembka (ed.), *Frontiers of Econometrics*, New York: Academic Press, 1974.

McFadden, D., Modelling the Choice of Residential Location, in: A. Karlqvist, L. Lundqvist, F. Snickars and J.W. Weibull (eds.), *Spatial Interaction Theory and Planning Models*, Amsterdam: North Holland, 1978.

McMillen, D.P., Probit with Spatial Autocorrelation, *Journal of Regional Science*, 32, 335-348, 1992.

Openshaw, S., Ecological Fallacies and the Analysis of Areal Census Data, *Environment and Planning A*, 16, 17-31, 1984.

Putman, S.H. and S.H. Chung, Effects of Spatial System Design on Spatial Interaction Models. 1: The Spatial System Definition Problem, *Environment and Planning A*, 21, 27-46, 1989.

Schwab, M.G. and T.R. Smith, Functional Invariance Under Spatial Aggregation from Continuous Spatial Interaction Models, *Geographical Analysis*, 17, 217-230, 1985.

Tardiff, T.J., Definition of Alternatives and Representation of Dynamic Behavior in Spatial Choice Models, *Transportation Research Record*, 775, 25-30, 1980.

Thill, J.C., Choice Set Formation for Destination Choice Modelling, *Progress in Human Geography*, 16, 361-382, 1992.

Webber, M.J., A Theoretical Analysis of Aggregation in Spatial Interaction Models, *Geographical Analysis*, 12, 129-141, 1980.

PART III

Heterogeneity and Dependence in Space-Time Models

12 The General Linear Model and Spatial Autoregressive Models

Daniel A. Griffith
Syracuse University, Syracuse, NY, U.S.A.

12.1 Introduction

Specific instances of the general linear model (GLM) have already been implemented within spatial statistics. Griffith (1978) summarizes how to write the simple one-way ANOVA model in the presence of spatial autocorrelation, more recently extending this to N-way ANOVA and unbalanced designs [Griffith (1992b)]. This recent article highlights the need for more work on mixed and random effects unbalanced design models, which is the case in traditional statistics as well. Griffith (1979) describes how to write the one-way MANOVA model for geo-referenced data, more recently extending this to N-way MANOVA and unbalanced designs [Griffith (1992b)], too. His initial findings in this case are consistent with the perspective promoted by Haining (1991). Griffith (1989) has also outlined spatial statistical two-groups discriminant function analysis and ANCOVA models [see also Anselin (1988), for a spatial econometric implementation of this latter model]. Other attempts along these lines are found in Switzer (1985), who has devised a spatial principal components analysis, Mardia (1988) and Griffith (1988), who have constructed multivariate geo-referenced data models, and Wartenberg (1985), who has formulated a cross-Moran Coefficient (cross-MC). And, Cressie and Hilterbrand (1993) have studied the problem of multivariate geo-statistics. What remains explicitly unaddressed is treatment of canonical correlation, and a spatial statistical k-groups discriminant function analysis model (which should relate to the eigenvectors of the aforementioned spatially adjusted MANOVA model).

The purpose of this chapter is threefold. First, it describes a situation in which spatial autocorrelation latent in variable Y is nearly perfectly accounted for by spatial autocorrelation latent in a single variable X, helping to illuminate the 'missing variables' interpretation of error spatial autocorrelation [see Griffith (1992a)]. Second, it outlines a procedure for implementing the GLM within spatial statistics, emphasizing its utility in conducting canonical correlation analyses. And, third, it furnishes an evaluation of U.S. Census of Agriculture data, in a rare situation for geo-referenced data in which essentially repeated measures are provided by the Commonwealth of Puerto Rico for selected production variables (with special reference here to milk yields).

Two characteristics of spatial autocorrelation are conspicuous for Puerto Rican agricultural yields. First, commonalities arising from soil, climatic, topography, land use practices, and input requisites cause similar densities of a given agricultural commodity to tend to cluster together on the island; hence, milk production should, and does, display positive spatial autocorrelation. Second, since urban population demands encourage milk sheds to form around cities positioned on the coastal lowlands, and a mountain chain traverses the center of the island, from east to west, affiliated yield densities should be at least partially summarized by a second-order, gutter-shaped trend surface model with a more pronounced north-south differentiation. This situation allows trade-offs between spatial autocorrelation and trend surface components to be explored.

The Commonwealth of Puerto Rico (CPR) annually collects yield data on milk production, as well as yields for selected other agricultural commodities. This data collection exercise has been going on for roughly five decades now. Meanwhile, the U.S. Department of Agriculture (USDA) has collected data for these same agricultural commodities in the census years of 1959, 1964, 1969, 1974, 1978, 1982, and 1987. Both agencies report these data by municipio, and have regionalized the island into five groups of municipios (see Figure 12.1). The pairs of municipio variable measures for the seven USDA census years furnish a basis for conducting some illuminating experiments, such as one that addresses the following question: can spatial autocorrelation be completely removed from variable Y with a carefully selected variable X?

Of course, one drawback is that the two agencies have used somewhat different operational definitions of farm, yield, and/or production year interval. Nevertheless, in an ideal situation, a researcher should expect appropriate null hypotheses for a regression of Y on X to be:

$$H_0\text{: } \alpha = 0, \text{ and } H_0\text{: } \beta = 1,$$

where α is the intercept of the linear prediction equation (a constant that refers to the conditional mean of Y given X), and β is the slope in this equation (a ratio of the magnitude of change in Y for a given change in X). Further, classically speaking, the resulting measurement error should behave randomly, and conform to a normal distribution. Perhaps this is one of the few cases in which a statistical application to geo-referenced data comes close to complying with conventional statistical thinking.

Each of these three themes will be treated in detail in this chapter.

Figure 12.1: The Commonwealth of Puerto Rico.

12.2 The GLM

One overview of the GLM appears in Johnson and Wichern (1992). Its regression equation is written as:

$$Y_{n\times Q} = X_{n\times(P+1)}\,\beta_{(P+1)\times q} + \varepsilon_{n\times Q} \tag{1}$$

where $Y_{n\times Q}$ is a matrix comprised of Q > 1 variables, rather than a vector, $X_{n\times(P+1)}$ is the standard matrix of $(P+1)$ predictor variables for regression, $\varepsilon_{n\times Q}$ is a matrix, rather than a vector, and $\beta_{(P+1),g}$ and $\varepsilon_{n,g}$ are the OLS results from regressing $Y_{n,g}$ on $X_{n\times(P+1)}$.

In other words, the GLM results may be obtained by regressing each of the Y_q variables on the same matrix of predictor variables, X, and then pooling the q sets of residuals. The key difference occurs with regard to the assumptions that accompany equation (1).

The generalized assumptions associated with the GLM are:

(1) $E(\varepsilon_g) = 0$ and $COV(\varepsilon_g,\varepsilon_h) = \sigma_{gh}\boldsymbol{I}$,

(2) $COV(\beta_g,\beta_h) = \sigma_{gh}(X'X)^{-1}$, replacing $\sigma^2(X'X)^{-1}$, and

(3) $E(\mathbf{b}) = \beta$, where β is a $(P+1)$-by-Q matrix, and $\hat{\Sigma} = \hat{\varepsilon}'\hat{\varepsilon}/n$, where $\hat{\varepsilon}$ is an n-by-Q matrix (in the classical case, the error variance-covariance matrix Σ is block diagonal).

Assumption (2) allows correlation across measures of each observation (which should occur here, since time series are being analyzed), but not amongst observations (i.e., zero spatial autocorrelation is assumed). One advantage of this specification is that each individual marginal distribution for $\hat{\varepsilon}_g$, itself, need not conform to a univariate normal distribution, or display homoscedastic variance. Multivariate sample variability allows up to a prespecified percentage (α) of the Q variables to lead to univariate null hypothesis rejection. Rather, the Q vectors jointly must conform to multivariate normality (MVN) with constant covariance matrices. This relaxation of assumptions allows more flexibility in satisfying them. Moreover, the univariate test probability (say α) will be equivalent to a considerably smaller joint test probability (which, if the Q vectors are independent, often is written as α^Q for such multiple comparisons). Thus, a second advantage is that the GLM has a greater tolerance of aberrations. A third advantage is that the simultaneous estimation of q regression coefficient vectors enables a researcher to check, statistically, for stability of and changes in these coefficients.

12.3 Data Preprocessing

In terms of agricultural production in Puerto Rico, analysis of the data is facilitated by the Jacobian approximation developed by Griffith see, for instance (1992c), a preprocessing task. Calibration, based upon a systematic sample of size 22 across the feasible parameter space of the spatial autocorrelation parameter, ρ, of this Jacobian approximation term yields the following results for Puerto Rico:

$$\hat{J} = 0.16\ln(1.44) + 0.12\ln(1.04) - 0.16\ln(1.44 + \rho) - 0.12\ln(1.04 - \rho) \qquad (2)$$

The sum-of-squared errors for this calibration is $4.05 \cdot 10^{-4}$; as repeatedly found before, an exceptionally good fit has been attained. Furthermore, the data analysis is enhanced by the existence of a meaningful regionalization of the island, which can serve as a basis for exploratory specification diagnostic purposes. And, it must involve estimation of suppressed USDA values (a standard EM-type of algorithm will be employed here), which need to be imputed before a GLM can be implemented in an optimal way.

12.3.1 Analysis of the 1964 Benchmark Data

The 1964 USDA data are complete, the only set in the space-time series that is. These data will serve as a basis for exploring model specification issues. Analyses of these data are reported in Table 12.1.

Since the USDA and CPR data for 1964 essentially are repeated measures, they should be strongly linearly related, with a regression of the USDA values on the CPR values yielding an intercept statistic that is not significantly different from zero, and a slope that is not significantly different from one. The *raw data* analysis results appearing in Table 12.1 indicate that the slope indeed is not significantly different from one, but that the intercept is significantly different from zero at the 5 percent level. In other words, the somewhat different operational definitions of farm, yield, and/or production year interval used by the two different agencies generates a USDA milk production figure that is, on average, 255169 quarts higher than its respective CPR figure, across the 73 municipios. The accompanying percent of variance accounted for in USDA by CPR suggests that a very good correspondence exists between these two sets of measures. Unfortunately, the reported diagnostics suggest that these data are not well-behaved. The residuals display perceivable heterogeneous variance across the five agricultural administrative regions, and most of the regional frequency distributions do not suggest that those in the population conform to a normal distribution. As is expected, both USDA and CPR values exhibit roughly the same level of positive spatial autocorrelation (it is modest but significant); the regression residuals, however, exhibit no autocorrelation, according to the z-score test statistic based upon MC. One implication here, then, is that the CPR measure accounts for the latent spatial autocorrelation contained in the USDA measure. Another possibility, though, is that model assumption violations are what the diagnostic statistics are detecting. One nuance is alluded to, as well, by the failure of the Geary Ratio to be affected very much by the inclusion of CPR as a predictor of the USDA variable.

This data analysis is a good candidate for the application of variable transformations [see Emerson and Stoto (1983)]. In an attempt to stabilize the residual variance, and induce a greater degree of symmetry in the five regional frequency distributions, both variables were transformed by first dividing by municipio area times 4000, in order to adjust for areal unit size differences and the large magnitude of the numbers, then adding a constant of 6 in order to increase the minimum value from zero, and finally taking logarithms. This transformation was corroborated by fitting the equation:

$$\frac{USDA}{4000 \times area} = -\delta + \alpha\left(\frac{CPR}{4000 \times area} + \delta\right)^{\beta} + \varepsilon \qquad (3)$$

which yielded $\hat{\delta} = 5.92240$. The *transformed data* analysis results also appear in Table 12.1, and now indicate that both the intercept and the slope are significantly different from their respective hypothesized values of 0 and 1, at the 5 percent level.

Table 12.1: Diagnostic statistics for the 1964 data analysis.

Raw data analysis

OLS: a = 255169 [P(H_0: $\alpha = 0$) = 0.0284]
b = 0.999878 [P(H_0: $\beta = 1$) = 0.9950]
$R^2 = 0.9745$

Homogeneity of variance test: $\chi^2 = 9.87805$
[P(H_0: $\sigma_{SJ}^2 = \sigma_A^2 = \sigma_M^2 = \sigma_P^2 = \sigma_C^2$) = 0.0425]

Individual Variables:	USDA	CPR	residuals
Within regions normality test (Wilk-Shapiro):			
San Juan	0.90596*	0.87318**	0.76007***
Arecibo	0.73649***	0.73364***	0.83443**
Mayaguez	0.63857***	0.67063***	0.62992***
Ponce	0.86221**	0.84553**	0.86695**
Caguas	0.83169**	0.78952***	0.92665
Spatial autocorrelation tests			
Moran Coefficient	0.12857**	0.12817**	0.03695
Geary Ratio	0.79128	0.79734	0.80328

Transformed data analysis

OLS: a = 0.16936 [P(H_0: $\alpha = 0$) = 0.0026]
b = 0.96749 [P(H_0: $\beta = 1$) = 0.0406]
$R^2 = 0.9750$

Homogeneity of variance test: $\chi^2 = 6.43412$
[P(H_0: $\sigma_{SJ}^2 = \sigma_A^2 = \sigma_M^2 = \sigma_P^2 = \sigma_C^2$) = 0.1690]

Individual Variables:	ln(USDA/4000area+6)	ln(CPR/4000area+6)	residuals
Within regions normality test (Wilk-Shapiro):			
San Juan	0.90863*	0.91879	0.90354*
Arecibo	0.95986	0.95579	0.93819
Mayaguez	0.82956***	0.86349**	0.77896***
Ponce	0.94157	0.89400*	0.89326*
Caguas	0.97006	0.93562	0.86549**
Spatial autocorrelation tests			
Moran Coefficient	0.33904***	0.33923***	0.01421
Geary Ratio	0.61211	0.59946	0.85809

Here the regression residuals are much better behaved, though: the variance is more homogeneous across the five regions; overall the regional frequency distributions conform more closely to normally distributed ones; and, significant spatial autocorrelation remains absent from them. Spatial autocorrelation in the transformed values is even more pronounced than in their raw data counterparts, while GR quantities are consistent with their respective MC companions.

Because no gross coding errors are conspicuous in the two data sets, unusual or extreme values, called outliers, appear to be attributable to empirical data distribution anomalies. Outlier detection and influential observation statistics for this transformed variables model [see Belsley et al. (1980), Barnett and Lewis (1984), Rousseeuw and Leroy (1987)] highlight the following groups of municipios.

Region	Municipio	StRes	Cook's *D*	H	DFFITS	CovRatio	DfBeta1	DfBeta2
San Juan	Carolina			×				
	Dorado	×	×	×	×		×	×
	Fajardo	×			×	×		
	Luquillo			×				×
	Tao Baja			×		×		
Arecibo	Barceloneta	×						
	Hatillo			×		×		
Mayaguez	Cabo Rojo	×				×		
	Lajas	×				×		
	San German	×			×		×	
Ponce	Santa Isabel	×					×	
Caguas	Aguas Buenas	×				×		
	Gurabo					×		
	Humacao	×				×		
	Juncos			×		×		

Given that each statistic will tend to tune in on various nuances and idiosyncrasies of the data under study, in terms of whether or not an observation has a substantial impact on the OLS regression fit, those municipios uncovered here that should be considered candidates for remedial measures are (1) Dorado being earmarked by virtually all of the statistics (it is a marginal case with Cook's *D*, but a marked extreme with DFFITS and each of the DfBetas), (2) Fajardo and San German being earmarked by three of the statistics, and (3) Hatillo being a markedly extreme leverage point. Having identified these municipios as possibly being unlikely events, coming from a different population than the rest of the municipios, or a failure of the model will help to furnish insights during the ensuing statistical analysis.

The computer code presented in Appendix 1 is for a constant mean situation (equivalent to regressing *Y* on a vector of 1s). The expected value of MC is given by [see Cliff and Ord, 1981]:

$$-\left(\frac{n}{\mathbf{1}'\mathbf{C}\mathbf{1}}\right)\frac{\mathrm{tr}\left((\mathbf{X}'\mathbf{X})^{-1}\mathbf{X}'\mathbf{C}\mathbf{X}\right)}{n - P - 1} \tag{4}$$

where tr is the matrix trace operator, and $\mathbf{1}'\mathbf{C}\mathbf{1} = \Sigma_{i=1}^{n}\Sigma_{j=1}^{n}c_{ij}$. If $\mathbf{X} = \mathbf{1}$ ($P = 0$), then equation (4) reduces to the well-known result of $-1/(n-1)$. If the variables *Y* and *X* are converted to *z*-scores, then regression through the origin is determined, and $\mathbf{X} = z_X$. In this case, equation (4) reduces to:

$$\frac{-\left[1 + \left(\frac{n}{\mathbf{1}'\mathbf{C}\mathbf{1}}\right)\left(z_x'\mathbf{C}z_x\right)\right]}{(n - 1)(n - 2)} \tag{5}$$

where $z_x'\mathbf{C}z_x = \Sigma_{i=1}^{n}\Sigma_{j=1}^{n}c_{ij}z_iz_j$. Accordingly, the diagnostics for the regression residuals reported in Table 12.1 have the expected values of -0.01589 (raw data), and -0.01886 (transformed data).

Although the transformed data do not exhibit perfectly ideal behavior, the diagnostics summarized in this section suggest that the data transformation employed here is quite appropriate. Additional exploratory analyses were undertaken with regard to this issue, following Van Gastel and Paelinck (1995). The entire space-time data set consists of 73 municipio time series, each five years in length. Of these, only 39 series are complete, of which 7 contain at least one zero, and the remaining 34 have missing values. The 32 useful series have been treated simultaneously, invoking the exchangeability principle outlined in Arbia (1989). Their analysis produced no evidence to indicate that the exponents for the USDA and CPR variables are not the same, helping to justify the use of the aforementioned transformation on both variables. Of note is that two of the 32 sets have elasticities that are noticeably, but not markedly, different from the others.

12.3.2 Evaluation of Missing USDA Values Estimation

Clearly, as should be expected, there is a very strong ($R^2 = 0.9750$), positive linear relationship between the transformed USDA and CPR data for 1964. When USDA data are suppressed in subsequent years, and hence their values become missing in the published data sources, this relationship can be exploited in order to provide best estimates of them. Because the residuals of this linear relationship fail to suggest the presence of non-zero spatial autocorrelation, a conventional approach [using an EM type of algorithm; see Little and Rubin (1987)] to this estimation problem can be taken. Otherwise, the approach outlined in Haining et al. (1984), Griffith (1988), or Griffith (1992d) should be followed. Furthermore, since the USDA data contain regional totals, this conventional approach can be modified by constraining the estimates to sum to the difference between these regional totals and the respective sums of known municipio values by region.

Assuming normality, the traditional missing value estimation problem may be stated as follows:

$$\begin{aligned} \text{MAX: } L &= \prod_{i=1}^{n}\left[\frac{1}{\sqrt{(2\pi)}\,\sigma}\exp\left(\frac{-(y_i - x_i\beta)(y_i - x_i\beta)}{2\sigma^2}\right)\right] \\ &= \left(\frac{1}{\sqrt{(2\pi)}}\right)^{n}\sigma^{-n}\exp\left(\frac{-(Y_0 - X_0\beta)'(Y_0 - X_0\beta)}{2\sigma^2}\right) \\ &\quad \times \exp\left(\frac{-(Y_m - X_m\beta)'(Y_m - X_m\beta)}{2\sigma^2}\right) \end{aligned} \tag{6}$$

where the subscript '0' denotes known values, and the subscript '*m*' denotes missing values.

The maximum likelihood (ML) solution to equation (6) involves $\hat{Y}_m = X_m\beta$. Substituting this solution into equation (6) reduces the missing values exponential term to $\exp(0) = 1$, removing this portion of the likelihood function from further analysis. This result can be easily seen in the simple univariate case ($X = 1$), where the estimate of each element of Y_m is μ; in turn the estimate of μ becomes the mean of the known values (i.e., $\bar{y}_0$).

Now if equation (6) is constrained in such a way that $1'Y_m = S$, coupled with $\hat{Y}_m \geq \mathbf{0}$, then the problem can be rewritten as one of estimating the set of terms $m_j/(1 + \Sigma_{j=1}^{M-1} m_j)$, $j = (1, 2, ..., M\text{-}1)$, in the case of M missing values (this specification arbitrarily sets one of the m_js equal to unity, in order to facilitate estimation of the remaining M-1 terms). Here if $X = 1$, then the ML solution results in the unaccounted for part of a regional total being uniformly allocated to the missing values in that region (i.e., $\hat{y}_m = S/M$, $m = 1, 2, ..., M$). When X is a full battery of predictor variables, this uniform allocation is modified in accordance with the multiple regression relationship between the Xs and Y. Because the missing values' sum-of-squares component can be separated from the rest of the likelihood function, their estimation reduces to solving:

$$\text{MIN}: \left[\sum_{k=1}^{M-1} \left(\frac{m_k S}{1 + \sum_{j=1}^{M-1} m_j} - x_k \beta \right)^2 \right] + \left(\frac{S}{1 + \sum_{j=1}^{M-1} m_j} - x_M \beta \right)^2 \tag{7}$$

The analytical derivatives of equation (7) involve each term, since m_j appears in each.

Equation (7) has been used to evaluate estimates based upon equation (6) by artificially suppressing pairs of 1964 USDA values; hence, $M = 2$. Because of some heterogeneity across the five regions, four pairs of municipios have been selected from each. A judgement sample has been resorted to here in an attempt to obtain a representative impression of the estimator. This sample has been constructed so that one pair of selected municipios has (1) the two largest, (2) the two closest to the median, (3) the two smallest, and (4) the two extreme USDA milk production figures in each region. Thus, 20 judiciously selected pairs, out of a total number of:

$$C(19,2) \times C(11,2) \times C(16,2) \times C(14,2) \times C(13,2) = 515$$

possible pairs, are inspected here. The initial value of m_1 is taken as the ratio of the smaller CPR value to the larger CPR value for the two municipios in question.

Results from equation (7) are reported in Table 12.2. The predicted values are the $\hat{Y} = Xb$ values from the transformed data OLS regression whose results appear in Table 1 (n = 73). Of note is that considerable differences exist between many Y and $\hat{Y}$ values; the worst case is Dorado/Carolina. The constraint $1'Y_m = S$ means that additional information is available for, and can be introduced into, the estimation process. Assuming estimates of the regression coefficients α and β, namely a and b, are available, the values $\text{U\hat{S}DA}|a,b$ are produced with equation (7). Of note here is that overall the constrained estimates are considerably better (reducing the aggregate sum-of-squares for this subset of municipios by nearly 85 percent). Nineteen pairs of estimates show improvement, some quite dramatically, such as the conspicuous case of Maricao/Las Marias (a 99.90 percent reduction) and the most improved case of Villalba/Arroyo (a 99.92 percent reduction). Only one pair shows deterioration; the sum-of-squares for Aguas Buenas/San Lorenzo has increased by roughly 35 percent. Many pairs, such as Ciales/Morovis, have a single good prediction from the OLS results, with the imposed missing value constraint impacting mostly on the single markedly deviant estimate. In this case, the binding constraint of a total imposed upon the sum of missing values will prevent both of the estimates from being markedly poor. And, a few pairs of estimates are essentially the same (e.g., Barranquitas/Carolina, and Maunabo/Caguas).

Table 12.2: Missing data estimation (in 1000s) procedure evaluation using 1964 values.

Region	Municipio	Observed	Predicted	$\hat{USDA}\|a,b$	$\hat{USDA}$	a_m	b_m
San Juan	Dorado	8756	12135	10287	10298	0.16306	0.96992
	Carolina	13318	14438	11787	11776		
	Bayamon-Catano	2922	2960	2816	2816	0.16935	0.96747
	Rio Grande	3310	3632	3416	3415		
	Barranquitas	8	95	2	3	0.16944	0.96746
	Ceiba	9	77	14	14		
	Barranquitas	8	95	90	93	0.17614	0.96553
	Carolina	13318	14438	13235	13232		
Arecibo	Arecibo	18708	17961	18879	18881	0.16969	0.96738
	Hatillo	30620	28578	30449	30448		
	Ciales	1648	1634	1673	1673	0.16960	0.96740
	Morovis	2616	2546	2591	2591		
	Lares	12	174	101	102	0.17113	0.96699
	Utuado	196	349	106	106		
	Lares	12	174	180	186	0.17699	0.96529
	Hatillo	30620	28578	30452	30446		
Mayaguez	Lajas	6187	4228	6125	6125	0.16940	0.96747
	Cabo Rojo	6744	4432	6805	6806		
	San Sebastian	443	627	400	400	0.16950	0.96748
	Guanica	525	673	567	567		
	Maricao	1	104	5	5	0.16941	0.96747
	Las Marias	4	133	0	0		
	Maricao	1	104	143	146	0.17897	0.96473
	Cabo Rojo	6744	4432	6601	6599		
Ponce	Salinas	5666	5079	5615	5616	0.16936	0.96748
	Coamo	6416	5764	6467	6466		
	Yauco	870	836	887	887	0.16943	0.96746
	Adjuntas	1242	1162	1226	1226		
	Villalba	2	103	4	4	0.16925	0.96751
	Arroyo	89	111	87	87		
	Villalba	2	103	111	115	0.17710	0.96526
	Coamo	6416	5764	6307	6304		
Caguas	Juncos	10061	9766	9340	9333	0.17185	0.96647
	Caguas	15176	17200	15897	15904		
	Aguas Buenas	2005	1242	1379	1378	0.16636	0.96801
	San Lorenzo	2601	2621	3227	3228		
	Maunabo	13	58	33	33	0.17146	0.96688
	Cayey	805	1165	785	785		
	Maunabo	13	58	56	58	0.17487	0.96592
	Caguas	15176	17200	15133	15131		

But having estimates for α and β beforehand is highly unlikely and, as equation (6) suggests, these estimates cannot be based strictly on the known data, since X_m also is known and contains valuable information that should not be discarded; this complication is why the EM algorithm is iterative (in its first step, it estimates the regression coefficients; in its second step, it substitutes $\hat{Y}_{\tau+1} = \boldsymbol{X}\boldsymbol{b}_{\tau}$, and then returns to the first step), for $(\hat{Y}_{m,\tau+1} - \boldsymbol{X}_m\boldsymbol{b}_{\tau})$ must equal zero at the final iteration in order for the missing values part of the likelihood function to reduce to exp(0) = 1. Moreover, the regression problem here becomes:

$$\left(1 - \sum_{k=1}^{M} \boldsymbol{I}_k\right) \ln\left(\frac{USDA_i}{4000 \times area_i} + 6\right) = \alpha + \beta \ln\left(\frac{CPR_i}{4000 \times area_i} + 6\right)$$
$$- \sum_{k=1}^{M-1} \boldsymbol{I}_k \ln\left(\frac{m_k S}{4000 \times area_k \left(1 + \sum_{j=1}^{M-1} m_j\right)} + 6\right) \tag{8}$$
$$- \boldsymbol{I}_M \ln\left(\frac{S}{4000 \times area_M \left(1 + \sum_{j=1}^{M-1} m_j\right)} + 6\right)$$

where $\boldsymbol{I}_k = 1$ if municipio k has a missing value, and 0 otherwise, $k = (1, 2, ..., M-1)$.

Equation (8) will have a zero entered into its left-hand side for each missing value case. The right-hand side of this equation will be the simple linear regression equation for non-missing value cases, and the difference between the predicted $\boldsymbol{X}\boldsymbol{b}$ value from a simple linear regression and the specified transformed value containing the unknown percentage coefficients m_k, for missing value cases. In the raw data situation, when only two values of Y are missing, the solution is given by:

$$\frac{m_1}{1 + m_1} = \frac{S + (x_1 - x_2)\beta}{2S} \tag{9}$$

and

$$\frac{1}{1 + m_1} = \frac{S - (x_1 - x_2)\beta}{2S} \tag{10}$$

Of note is that when $\beta = 0$, these ratios reduce to 1/2 (a uniform, arithmetic mean allocation of S over the set of missing value areal units). The slope coefficient β

still needs to be estimated, and still would need to be done in an iterative fashion. In addition, the use of a variable transformation increases the complexity of this solution. In general, the set of initial m_j values may be obtained by dividing each CPR value by the largest CPR value for those municipios having missing values, resulting in the one with the largest CPR value being set to unity; this procedure will tend to imposed the non-binding restriction of $0 < m_j < 1$, which will facilitate convergence.

Results from equation (8) also are reported in Table 12.2. Of note is that all of the estimates a_m are within 6 percent of the estimate a, and all of the estimates b_m are within 0.3 percent of the estimate b obtained from the complete data OLS. Two generalizations suggested here are that (1) the absolute deviation $|a_m - a|$ is almost perfectly related to the absolute deviation $|b_m - b|$ ($r = 0.985$), and (2) a negative exponential relation exists between $|a_m - a|$ and the ratio sum-of-squares($\hat{Y}$)/sum-of-squares($\hat{Y}_m$) as well as between $|b_m - b|$ and this same ratio (i.e., $b_m \rightarrow b$ tends to increasingly reduce the sum-of-squares term, for as b_m moves away from b the missing values sum-of-squares term still converges on zero while the relation between Y_0 and X_0 becomes increasingly distorted).

In conclusion, the output summarized in Table 12.2 indicates that missing value estimates given by equation (8) should be sensible. At least in part, how reasonable they are may be indexed by how much the regression coefficient vector converged upon by b_τ deviates from b_0. Of course, this latter perspective assumes that missing values have occurred at random, and are not correlated with the attribute in question, which here has to do with number of farms in a municipio. Applying this methodology, the USDA space-time data series has been completed, on a year-by-year basis, using the corresponding CPR data and an EM-type of algorithm. As mentioned earlier, this algorithm iteratively estimates the regression coefficients (i.e., maximizes the likelihood function), predicts the missing values (here, subject to regional total constraints), substitutes these predicted values into equation (6), and then re-estimates the regression coefficients.

12.4 Implementation of the Spatial Statistical GLM

One key to the traditional GLM mentioned above is that it can be constructed by computing an OLS regression for each Y_j on the matrix of predictors $\boldsymbol{X}$. Paralleling this notion, five spatial AR models will be estimated here, followed by an analysis, jointly, of their sets of residuals.

12.4.1 Preliminary Spatial Analysis of Milk Yields: AR Trend Surface GLMs

As mentioned earlier, the geographic pattern of milk production across the island should be able to be described with a second-order trend surface (or gradient) model. Let the east-west coordinate be denoted by u and the north-south coordinate by v. GLM estimation results for the standard quadratic trend surface specification appear in Table 12.3. For both the USDA and the CPR measures, roughly 30-40 percent of the variation across the island is accounted for by this second-order polynomial model. The noteworthy regression coefficient trends are: (1) east-west differentiation increases, then decreases, becoming less important with the passing of time; (2) north-south differentiation becomes increasingly important with the passing of time; and, (3) second-order polynomial terms essentially are unimportant. The noteworthy geographic trends are: (1) the east-west yield decreases more sharply from the center of the island westward than eastward; and, (2) the north-south trend is concave upward (the gutter-type surface initially anticipated).

Meanwhile, GLM estimation results for a spatial autoregressive trend surface specification also appear in Table 12.3. For both the USDA and the CPR measures, clear trade-offs between spatial autocorrelation and trend surface components materialize over time. The noteworthy regression coefficient trends now are: (1) positive spatial autocorrelation becomes increasingly marked with the passing of time; (2) as the spatial autocorrelation term becomes more pronounced, east-west and second-order polynomial terms become less pronounced; and, (3) north-south differentiation increases, and then decreases in importance. These trends are qualitatively different from those identified with the traditional trend surface specification.

This second set of findings is quite robust, too. Not only are the errors spatially unautocorrelated, but the covariance matrices for the USDA data appear to be constant across the five agricultural regions ($P(\chi^2 > 72.38) \approx 0.13$). Furthermore, the multivariate normality Q-Q plot correlation diagnostics are:

San Juan	Arecibo	Mayaguez	Ponce	Caguas
0.97128	0.91568	0.97917	0.97143	0.94978

Clearly there is little evidence that the individual regional frequency distributions do not conform to *MVN* ones. Finally, because the milk production data are a spatial time series, temporal autocorrelation should be quite strong. Not surprisingly, then, a principal components analysis of the residuals yields two prominent components that account for, respectively, 88.36 percent and 7.91 percent of the milk yield variation across the island; and, a discriminant function analysis (using the five agricultural regions) of the residuals yields two conspicuous components that account for, respectively, 83.94 percent and 13.67 percent of the relative variation of the regional means.

Table 12.3: The GLM trend surface model estimation results.[a]

Model Specification	Year	$\hat{\rho}$	a	b_u	b_v	b_{u2}	b_{uv}	b_{v2}	R^2
USDA data									
OLS	1964	----	2.65***	0.29***	0.43***	-0.08	0.19**	0.32***	41.1%
	1969	----	2.82***	0.35***	0.50***	-0.09	0.14	0.26**	39.7%
	1974	----	2.83***	0.30***	0.53***	-0.09	0.01	0.18	37.4%
	1978	----	3.01***	0.24**	0.55***	-0.10	-0.07	0.09	32.5%
	1982	----	2.82***	0.21*	0.56***	-0.02	-0.14	0.10	34.7%
NLS	1964	0.18	2.61***	0.24**	0.37***	-0.07	0.16*	0.30***	
	1969	0.20	2.80***	0.28**	0.43***	-0.07	0.11	0.23**	
	1974	0.46**	2.77***	0.16	0.34***	-0.04	-0.01	0.14	
	1978	0.51**	2.98***	0.10	0.33**	-0.05	-0.05	0.06	
	1982	0.58***	2.79***	0.08	0.30**	-0.01	-0.09	0.06	
CPR data									
OLS	1964	----	2.60***	0.30***	0.46***	-0.10	0.19**	0.30***	42.4%
	1969	----	2.84***	0.35***	0.50***	-0.11	0.09	0.20*	40.0%
	1974	----	2.82***	0.32***	0.55***	-0.09	-0.002	0.17	39.5%
	1978	----	2.95***	0.26**	0.57***	-0.07	-0.06	0.11	35.5%
	1982	----	2.72***	0.20*	0.60***	0.01	-0.15	0.14	32.4%
	1987	----	2.59***	0.14	0.60***	0.05	-0.18	0.16	34.1%
NLS	1964	0.12	2.58***	0.26**	0.42***	-0.08	0.17**	0.29***	
	1969	0.19	2.82***	0.28**	0.43***	-0.09	0.07	0.18*	
	1974	0.44**	2.77***	0.17*	0.37***	-0.04	-0.02	0.13	
	1978	0.50**	2.91***	0.12	0.36***	-0.03	-0.05	0.08	
	1982	0.53**	2.63***	0.09	0.33***	0.01	-0.10	0.10	
	1987	0.57***	2.49***	0.06	0.33***	0.03	-0.13	0.12	

[a] For OLS estimation, $\hat{\rho}$ is set equal to zero, and is not estimated.

12.4.2 AR GLM Models for the Repeated Measures Case

As was mentioned earlier, the CPR values essentially are repeated measures of the USDA values. Results reported in Table 12.3 suggest that the latent spatial structure in these two data sets is quite similar, as should be the case. And, regressing each USDA variable on its respective CPR variable yields extremely high R^2s (to employ a familiar criterion, one that is useful when the data are well-behaved, and one advocated by Upton and Fingleton (1985, p. 298)) as well as residuals displaying little or no spatial autocorrelation; the estimates obtained are shown in Table 12.4.

Table 12.4: USDA variables regressed on respective CPR variables.

Year	1964	1969	1974	1978	1982
$\hat{\rho}$	0.00768	0.01006	0.00518	-0.00855	0.14640***
R^2	0.9750	0.9701	0.9754	0.9728	0.9573

These percentages of variance accounted for are considerably higher than those obtained with the trend surface models (see Table 12.3). But this improvement in precision is off-set by a serious deterioration in the behavior of the residuals.

Parameter estimates from the GLM are shown in Table 12.5.[1]

Table 12.5: Parameter estimates from the GLM.

			USDA		
	1964	1969	1974	1978	1982
a	0.18503***	0.02376	0.04917	0.06032	-0.07836
CPR b_{1964}	0.92968***	0.23874***	0.12758*	0.04754	-0.07480
CPR b_{1969}	0.02685	0.76997***	0.03249	-0.09808	-0.02759
CPR b_{1974}	0.11733	-0.01283	0.54256***	0.05364	0.09959
CPR b_{1978}	-0.17402*	0.04422	0.28687***	0.94925***	0.36803***
CPR b_{1982}	0.06674	-0.01833	-0.00098	0.03926	0.51069***
R^2	0.97627	0.97451	0.98187	0.97314	0.96786

This general structure mostly is as expected, since each USDA variable is highly significantly related to its corresponding CPR variable. Further, the other markedly significant relations tend to be for adjacent census years (temporal correlation).

Unfortunately, now the covariance matrices appear to be non-constant across the five agricultural regions ($P(\chi^2 > 88.69) \approx 0.009$). Furthermore, the multivariate normality Q-Q plot correlation diagnostics have deteriorated, on average, and now are:

San Juan	Arecibo	Mayaguez	Ponce	Caguas
0.97397	0.88603	0.98049	0.93074	0.92363

[1] For the next stage of the analysis, spatial autocorrelation was filtered out of the 1982 USDA variable, conditional on the presence of its CPR variable counterpart.

A principal components analysis of the residuals renders a much more uniform distribution of variance accounted for across components, respectively yielding 27.81 percent, 21.83 percent, 20.92 percent, 16.73 percent, and 12.71 percent (these variables are reasonably orthogonal); and, a discriminant function analysis (using the five agricultural regions) of the residuals yields three conspicuous components that account for, respectively, 61.71 percent and 26.47 percent, and 9.13 percent of the relative variation of the regional means. Finally, the cross-Moran Coefficients for the residuals are shown in Table 12.6.

Table 12.6: Cross-Moran coefficients for the residuals, USDA.

	1964	1969	1974	1978	1982
1964	0.010814				
1969	0.026305	-0.065040			
1974	0.005806	0.083243	-0.073460		
1978	0.091657	-0.002150	0.091802	-0.133028	
1982	0.029926	-0.054427	0.002593	-0.114803	-0.014855

None of these values are significant, based upon the asymptotic variance (i.e., 0.07625) of independent and identically distributed (i.i.d.) random variables.

Of note is that these cross-MC values can also be calculated using the GLM. The necessary procedure involves the following three steps [after Anselin (1993)].

Step 1: standardize each of the Q sets of regression residuals, e_q, so that e_q* has a mean of zero and a variance of unity;

Step 2: construct the Q corresponding spatially lagged residuals (see lines #6-20 of the *Spatial Statistics Gradient Model* SAS code appearing in Appendix 1), $\mathbf{C}e_q*$, and multiply it by $n/\mathbf{1}'\mathbf{C}\mathbf{1}$ (= 73/344 here), giving $n\mathbf{C}e_q*/\mathbf{1}'\mathbf{C}\mathbf{1}$; and,

Step 3: use PROC GLM Q times to regress the Q variables $n\mathbf{C}e_q*/\mathbf{1}'\mathbf{C}\mathbf{1}$ on each e_q*, resulting in the estimation of Q^2 simple linear regression equations.

As Wartenberg (1985) suggests, if their latent structure is of interest, eigenfunction analysis of the cross-MC matrix can be informative.

12.4.3 A Spatially Adjusted Canonical Correlation Analysis of the Milk Production Data

Arguments summarized and demonstrated by Haining (1991), regarding the need to filter out spatial autocorrelation before undertaking correlation analysis, imply the need to employ a spatially adjusted canonical correlation analysis here.

Table 12.7: Diagnostics for the reduced dimensions spatially adjusted canonical correlation analysis.

GLM parameter estimates for canonical variate $\mathbb{Y}_1$* extracted from matrix **Y***

variable	parameter estimate	VIF
a	-1.76798***	0.00
CPR b_{1964}	0.35850***	9.64
CPR b_{1969}	0.17928***	23.57
CPR b_{1974}	0.33413***	24.80
CPR b_{1978}	0.30262***	24.26
CPR b_{1982}	0.10569**	10.09
R^2	0.99172	

Homogeneity of variance test:
Regional basis: $P(\chi^2 > 4.41) = 0.3528$
Non-regional basis: $P(\chi^2 > 14.36) = 0.8119$

Multivariate normality test:

San Juan	Arecibo	Mayaguez	Ponce	Caguas
0.92229	0.92174	0.96976	0.88387*	0.95922

MC: -0.11662

GLM parameter estimates for canonical variate $\mathbb{X}_1$* extracted from matrix **X***

variable	parameter estimate	VIF
a	-1.75386***	0.00
USDA b_{1964}	0.25688***	7.51
USDA b_{1969}	0.27695***	12.59
USDA b_{1974}	0.42179***	13.04
USDA b_{1978}	0.14879***	15.02
USDA b_{1982}	0.20855***	10.79

Homogeneity of variance test:
Regional basis: $P(\chi^2 > 3.60) = 0.4623$
Non-regional basis: $P(\chi^2 > 19.84) = 0.4680$

Multivariate normality test:

San Juan	Arecibo	Mayaguez	Ponce	Caguas
0.92696	0.93326	0.97145	0.88314	0.95406

MC: -0.11589

This perspective is reinforced by Switzer's (1985) argument concerning spatial principal components analysis, particularly since canonical correlation is equivalent, geometrically, to extracting the principal components from the correlation matrix based upon the *Y*s, followed by an orthogonal rotation based upon the full $(P + Q)$ covariance matrix [see Johnson and Wichern (1992, p. 470)].

The first step in this direction is to estimate the spatial autocorrelation parameters, so that filterings can be computed. According to Table 12.3, only the north-south trend is persistent over time, conditional on the estimation of a spatial autocorrelation parameter. Therefore, this gradient, alone, has been included in the spatial autocorrelation estimation exercise. The resulting $\hat{\rho}$s are shown in Table 12.8.

Table 12.8: Spatial autocorrelation coefficients.

Year	1964	1969	1974	1978	1982
USDA	0.50818	0.48744	0.56942	0.57254	0.62263
CPR	0.47782	0.47072	0.56114	0.56292	0.56727

For the last two years, these estimates are very close to those reported in Table 12.3. For the first three years, these estimates are considerably higher than those reported in Table 12.3, specifically because the spatial autocorrelation parameter estimate is compensating for the trend surface terms that have been dropped from the regression equation. Of note is that these pairs of estimates are very similar.

In keeping with Haining (1991), a spatially adjusted canonical correlation analysis of these data, where $\boldsymbol{Y}^*$ is the 73-by-5 matrix of filtered USDA milk production figures, and $\boldsymbol{X}^*$ is the 73-by-5 filtered matrix of CPR milk production figures, generates the following output, where $\mathbb{X}^*$ is the set of canonical variates extracted from matrix $\boldsymbol{X}^*$ and $\mathbb{Y}^*$ is the set of canonical variates extracted from matrix $\boldsymbol{Y}^*$ (see Table 12.9).

Table 12.9: Analysis of canonical correlations.

$\rho_{\mathbb{X},\mathbb{Y}}$	Approximate *F*-ratio	Probability	Variance Accounted for USDA	CPR
0.995852	75.5226	0.0001	87.66%	88.78%
0.956839	24.7433	0.0001	8.61%	8.03%
0.603470	7.9069	0.0001	1.48%	1.68%
0.524296	8.3365	0.0001	1.00%	1.02%
0.347951	9.2290	0.0034	1.25%	0.49%

All of the canonical correlations are significant, suggesting relationships that may be described as two extremely strong, two moderate, and one weak. Comparing these percentages of variance accounted for reveals that only the first two account for a reasonably sizeable amount of variation; the remaining three canonical variates seem to be capturing noise. Thus, one of the canonical variates (*CV*-I*) is extremely prominent; the second canonical variate (*CV*-II*) is modestly prominent. The canonical structure for the two conspicuous variate pairs is shown in Table 12.10.

Table 12.10: The canonical structure for the two conspicuous variate pairs.

	CV-I*		*CV*-II*	
	r_{USDA}	r_{CPR}	r_{USDA}	r_{CPR}
1964	0.9003	0.9108	-0.3994	-0.3935
1969	0.9577	0.9685	-0.1796	-0.2050
1974	0.9802	0.9789	0.1162	0.0857
1978	0.9339	0.9535	0.2831	0.2742
1982	0.9016	0.8974	0.3482	0.3561

Of the two, only *CV*-I* consistently displays very strong positive correlations with the filtered variables. This pair of variates may be labelled a repeated measures dimension, the one initially expected to run across both of the data sets. Meanwhile, *CV*-II* displays weak correlations, at best, with the filtered variables. This pair of variates may be labelled a time series end points contrast dimension. The respective quadratic and linear patterns displayed by these two dimensions are commonly found uncovered in time series data. Therefore, only a single significant, prominent, meaningful dimension spans both of these filtered data sets.

Diagnostics for this canonical correlation analysis can be obtained by exploiting the following property [Johnson and Wichern (1992, p. 469)]:

> the set of canonical correlations is the set of multiple correlations from a GLM where $\mathbb{Y}$ is regressed on matrix $\boldsymbol{X}$, and/or $\mathbb{X}$ is regressed on matrix $\boldsymbol{Y}$.

Thus, the popular battery of regression diagnostics can aid in model assessment here. Such diagnostic results are reported in Table 12.7. Both regressions suffer from serious multicollinearity (as attested to by most *VIF* > 10), but satisfy the homogeneity of variance, normality, and spatially independent errors assumptions. The spatially adjusted CPR variables contribute somewhat uniformly to the principal canonical variate extracted from the spatially adjusted USDA variables; the converse exhibits considerably more variability.

Consequently, the inferential basis of the reduced dimensions canonical correlation analysis appears to be reasonably sound.

12.5 Conclusions

A number of useful conclusions can be gleaned from this study. With regard to its stated objectives, first, a procedure for implementation of the GLM for spatial statistics has been outlined; it may be summarized as follows:
Step 1: estimate the spatial statistical model for each $\boldsymbol{Y}_q$;
Step 2: use the estimated values of the $\hat{\rho}_q$s in order to specify the GLM; and
Step 3: conduct standard diagnostics on the joint set of Q residuals.

The SAS code for implementing these three steps is presented in the Appendices of this paper, too. Second, the use of repeated measures in this study supports the idea that the 'missing variable' interpretation of spatial autocorrelation is not the only one; spatial autocorrelation still can remain (as with the 1982 data), and errors do not necessarily become well-behaved. Third, the following interesting features of the US Agricultural Census data have been disclosed: (1) statistical distribution properties of the error are bothersome; (2) reasonable missing data values can be imputed; and, (3) the 1987 USDA data display almost no relationship to their CPR counterpart, which is disturbing (and which is why this geo-referenced data series was removed from the study).

Besides enabling diagnostics to be performed for canonical correlation analyses, the GLM also facilitates the computation of cross-MCs, simplifying a sizeable portion of the computation executed with the *Spatial Autocorrelation Indices* SAS code. Furthermore, the marginal filtering approach discussed in Haining (1991) is brought into question here. Such filtering is not necessarily effective. Both Griffith (1988) and Mardia (1988) argue for, as an alternative, a simultaneous approach. Their specifications are far more computationally intensive; however, utilization of marginal filterings appears to convolute and/or perturb spatial structure effects. One question alluded to here, though, asks whether or not the residuals of trivial dimensions merit so much concern. On the one hand, Griffith (1978) treated them as though they do not; on the other hand, valuable insights can be extracted from the trivial dimensions of principal components analyses. Results reported in Table 12.7 are appealing; when expanded to include the trivial dimensions, though, the homogeneity of variance decision is reversed, as are some of the *MVN Q-Q* plot correlation implications.

Finally, several illuminating substantive conclusions have been obtained here, pertaining to milk production in Puerto Rico. Foremost is that positive spatial autocorrelation is increasing, while global trends across the island are disappearing, with the passing of time. Second, a few pathological municipios exist on the island, this set at least including Dorado. Similarly, Arecibo and Mayaguez are problematic regions, in terms of statistical distribution theory. The pathology of Hatillo might account for Arecibo's deviant behavior, whereas San German may well have something to do with Mayaguez's deviant behavior. Considerably more detailed investigations are necessary before a solid understanding of the space-economy of the island's milk industry can be achieved.

References

Anselin, L., *Spatial Econometrics: Methods and Models*, Dordrecht: Kluwer, 1988.

Anselin, L., Moran Scatterplots as a Means to Visualize Instability in Spatial Autocorrelation, in: *Position Papers: NCGIA Exploratory Spatial Data Analysis and GIS Workshop*, Santa Barbara, 1993.

Arbia, G., *Using Spatial Data to Characterize Short-term Dynamic Economic Models*, paper presented to the annual North American Regional Science Association meetings, Santa Barbara, 1989.

Barnett, V. and T. Lewis, *Outliers in Statistical Data*, New York: Wiley, 1984 (2nd ed.).

Belsley, D., E. Kuh and R. Welsch, *Regression Diagnostics*, New York: Wiley, 1980.

Cliff, A. and J. Ord, *Spatial Processes*, London: Pion, 1981.

Cressie, N. and J. Hilterbrand, *Multivariate Spatial Statistics in a GIS*, Iowa State University, Department of Statistics, 1993 (mimeo).

Emerson, J. and M. Stoto, Transforming Data, in: D. Hoaglin, F. Mosteller and J. Turkey (eds.), *Understanding Robust and Exploratory Data Analysis*, New York: Wiley, 1983.

Gastel, M. van and J. Paelinck, Computation of Box-Cox Transform Parameters: A New Method and its Applications to Spatial Econometrics, 1995 (this issue).

Griffith, D., A Spatially Adjusted ANOVA Model, *Geographical Analysis*, 10, 296-301, 1978.

Griffith, D., Urban Dominance, Spatial Structure and Spatial Dynamics: Some Theoretical Conjectures and Empirical Implications, *Economic Geography*, 55, 95-113, 1979.

Griffith, D., *Advanced Spatial Statistics*, Dordrecht: Kluwer, 1988.

Griffith, D., *Spatial Regression Analysis on the PC: Spatial Statistics Using MINITAB*, Ann Arbor, Michigan: Institute of Mathematical Geography, Discussion Paper #1, 1989.

Griffith, D., What is Spatial Autocorrelation? Reflections on the Past 25 Years of Spatial Statistics, *l'Espace Geographique*, 21, 265-280, 1992a.

Griffith, D., A Spatially Adjusted N-way ANOVA Model, *Regional Science and Urban Economics*, 22, 347-369, 1992b.

Griffith, D., Simplifying the Normalizing Factor in Spatial Autoregressions for Irregular Lattices, *Papers in Regional Science*, 71, 71-86, 1992c.

Griffith, D., Estimating Missing Values in Spatial Urban Census Data, *The Operational Geographer*, 10, 23-26, 1992d.

Haining, R., Bivariate Correlation with Spatial Data, *Geographical Analysis*, 23, 210-227, 1991.

Haining, R., D. Griffith and R. Bennett, A Statistical Approach to the Problem of Missing Spatial Data Using a First-Order Markov Model, *The Professional Geographer*, 36, 338-345, 1984.

Johnson, R. and D. Wichern, *Applied Multivariate Statistical Analysis*, Englewood Cliffs: Prentice Hall, 1992 (3rd ed.).

Little, R. and D. Rubin, *Statistical Analysis with Missing Data*, New York: Wiley, 1987.

Mardia, K., Multi-Dimensional Multivariate Gaussian Markov Random Fields with Application to Image Processing, *Journal of Multivariate Analysis*, 24, 265-284, 1988.

Rousseeuw, P. and A. Leroy, *Robust Regression & Outlier Detection*, New York: Wiley, 1987.

Switzer, P., MIN/MAX Autocorrelation Factors for Multivariate Spatial Imagery, in: L. Billard (ed.), *Computer Science and Statistics: The Interface*, Amsterdam: North-Holland, 1985.

Wartenberg, D., Multivariate Spatial Correlation: A Method for Exploratory Geographical Analysis, *Geographical Analysis*, 17, 263-283, 1985.

Appendix 1: SAS Computer Code to Compute the Popular Spatial Autocorrelation Indices.

Suppose the *n*-by-*n* binary geographic weights matrix, say ***C***, has been read into SAS as a set of (*n* =) 73 indicator variables (C1-C73), and the variable of interest in this case is CPR1 (the 1964 CPR variable). Then:

```
***********
* MC & GR *
***********;
PROC TRANSPOSE DATA=STEP3 PREFIX=XT OUT=TRDATA1;
   VAR CPR1;;
DATA PRTEMP (REPLACE=YES);
   SET STEP3;
   X = CPR1;
   XSQ = X*X;
   ARRAY CONN{73} C1-C73;
   ARRAY XCONN{73} XC1-XC73;
   CSUM = 0;
   DO I=1 TO 73;
     CSUM = CSUM + CONN{I};
     XCONN{I} = X*CONN{I};
   END;
DROP C1-C73;
RUN;
DATA PRTEMP (REPLACE=YES);
   SET PRTEMP;
   IF _N_=1 THEN SET TRDATA1;
   ARRAY XTMTX{73} XT1-XT73;
   ARRAY XCONN{73} XC1-XC73;
   SGRX = XSQ*CSUM;
   SMCX = 0;
   DO I=1 TO 73;
     SMCX = SMCX + XTMTX{I}*XCONN{I};
   END;
RUN;
PROC MEANS DATA=PRTEMP NOPRINT;
   VAR CSUM XSQ SGRX SMCX;
   OUTPUT OUT=CH5OUT1 SUM=CSUM XSQ SGRX SMCX;
RUN;
DATA PRTEMP (REPLACE=YES);
   SET CH5OUT1;
N=73;
```

```
37 MCX = N*SMCX/(CSUM*XSQ);
38 ZAPP = (MCX + 1/(N-1))/SQRT(2/CSUM);
39 PROBMC = 1 - PROBNORM(ZAPP);
40 GRX = ((N-1)/N)*(N*SGRX/(CSUM*XSQ) - MCX);
41 PROC PRINT;
42    VAR N CSUM MCX ZAPP PROBMC GRX;
43 RUN;
```

Spatial Autocorrelation Indices SAS Code Description: Lines #4-5 transpose the vector of milk production values, so that subsequently the cross product values $z_i z_j$ can be calculated. Lines #5 and #8 specify the variable in question, namely the 1964 CPR variate. Line #9 squares each attribute value. Lines #13-16 compute $\Sigma_{j=1}^{n} c_{ij}$ and $c_{ij} z_j$. Lines #26-28 compute $z_i \Sigma_{j=1}^{n} c_{ij} z_j$. PROC MEANS sums all of these values over subscript i (i.e., it computes $\Sigma_{i=1}^{n}\Sigma_{j=1}^{n} c_{ij}$, $\Sigma_{i=1}^{n} x_i^2$, $\Sigma_{i=1}^{n} x_i^2 \Sigma_{j=1}^{n} c_{ij}$, and $\Sigma_{j=1}^{n}\Sigma_{j=1}^{n} c_{ij} z_i z_j$). Finally, Line #37 calculates the Moran Coefficient (MC), and Line #40 calculates the Geary Ratio (GR).

Of note is that the variable under study has been standardized (PROC STANDARD MEAN=0 STD=1), or converted to z-scores, before its latent spatial autocorrelation is measured. Further, the asymptotic approximate standard error of MC is used, namely, $\sqrt{(2/\Sigma_{i=1}^{n}\Sigma_{j=1}^{n} c_{ij})}$. And, because MC is a more powerful test statistic, the final statistical decision is based upon it; GR values are reported, too, but without a significance test, solely in order to check for consistency (a discrepancy between these two indices often suggests that the data need a more careful examination).

Appendix 2: SAS Code for Estimating Missing Values in the 1969 Data Set:

This SAS code includes

```
1  DATA STEP2 (REPLACE=YES);
2     SET STEP1;
3  IF USDA2='.' THEN USDA2=0;
4  IF NUMCPR=8 THEN I8=1; ELSE I8=0;
5  IF NUMCPR=59 THEN I59=1; ELSE I59=0;
6  IF NUMCPR=64 THEN I64=1; ELSE I64=0;
7  S = 3694340;
8  Y = (1 - I8 - I59 - I64)*LOG(USDA2+6);
9  RUN;
10 PROC NLIN METHOD=MARQUARDT;
11    PARM A=0.099461 B=0.988177
12      M1=0.00001 M2=0.0886417;
13    BOUNDS 0<B, 0<M1, 0<M2;
14    T = 1 + M1 + M2;
15    YHAT = S*(M1*I8 + I59 + M2*I64)/(T*AREA);
16    MODEL Y = A + B*LOG(CPR2 + 6)
17          - (I8+I59+I64)*LOG(YHAT+6);
18    DER.A=1;
19    DER.B=LOG(CPR2 + 6);
20    DER.M1=(-I8*(1+M2)/(M1*S/T+6*AREA) +
21         I59/(S/T+6*AREA) +
22         I64*M2/(M2*S/T+6*AREA))*S/(T**2);
23    DER.M2=(I8*M1/(M1*S/T+6*AREA) + I59/(S/T+6*AREA) -
24         I64*(1+M1)/(M2*S/T+6*AREA))*S/(T**2);
25 RUN;
```

Missing Data Estimation SAS Code Description: Line #3 replaces each missing value with a zero. Lines #4-6 create an indicator variable for each missing value municipio. Line #7 states the total of the missing values (only one region is involved here). Line #8 replaces the transformed data with zeroes, for missing value municipios. Line #11 initializes the regression coefficients with those computed solely with the known data. Line #12 initializes the missing data proportion factors with ratios obtained from the CPR data. Line #14 is the term $(1 + \Sigma_{j=1}^{M-1} m_j)$ of equation (7). Lines #16-17 are equation (8). Finally, Lines #18-24 are the analytical derivatives.

Convergence was attained relatively quickly and precisely, as is indicated by computing an OLS for $\hat{Y}$ on X to see that the regression coefficients are equivalent to those obtained with PROC NLIN, where $\hat{Y}' = [Y_0': \hat{Y}\ m']$. Slight differences in the fourth digit to the right of the decimal are noticeable in some of the cases, though; presumably these differences arise from rounding errors.

Appendix 3: SAS Code for 1969 USDA Data Analysis

This SAS code includes replacement of the missing values symbol with estimates, utilization of the variance-stabilizing transformation, conversion of the municipio centroid coordinate system to z-scores, and:

```
DATA STEP1 (REPLACE=YES);
   SET STEP1;
U2 = U**2;
UV = U*V;
V2 = V**2;
   ARRAY CON{73} C1-C73;
   ARRAY CONDMLK{73} CDMLK1-CDMLK73;
   CSUM = 0;
   DO I=1 TO 73;
     CSUM = CSUM + CON{I};
     CONDMLK{I} = USDA1*CON{I};
   END;
RUN;
PROC MEANS DATA=STEP1 NOPRINT;
   VAR CDMLK1-CDMLK73;
   OUTPUT OUT=COUT1 SUM=CDMLKS1-CDMLKS73;
RUN;
PROC TRANSPOSE DATA=COUT1 PREFIX=CDMLKT OUT=COUT1;
   VAR CDMLKS1-CDMLKS73;
RUN;
DATA STEP1 (REPLACE=YES);
   SET STEP1;
   SET COUT1;
WDMLK1 = CDMLKT1/CSUM;
DROP CDMLK1-CDMLK73 CDMLKT1;
A1 = 0.16040;
A2 = 0.11805;
D1 = 1.44472;
D2 = 1.04259;
RUN;
**********************************************
* AR MODEL FIT TO REGRESSION RESIDUALS USDA1 *
**********************************************;
PROC NLIN DATA=STEP1 METHOD=MARQUARDT MAXITER=500;
   PARMS RHO=0.5 B0=2.65 B1=.29 B2=.43 B3=-.08 B4=.19 B5=.32;
   BOUNDS -1.273431 < RHO < 1.0;
   JHAT = EXP(A1*LOG(D1) + A2*LOG(D2)
```

```
38        - A1*LOG(D1+RHO) - A2*LOG(D2-RHO));
39     ZX = USDA1*JHAT;
40     MODEL ZX = (RHO*WDMLK1 + (1-RHO)*B0 + B1*U + B2*V +
41           B3*U2 + B4*UV + B5*V2)*JHAT;
42     DER.B0 = (1 - RHO)*JHAT;
43     DER.B1 = U*JHAT;
44     DER.B2 = V*JHAT;
45     DER.B3 = U2*JHAT;
46     DER.B4 = UV*JHAT;
47     DER.B5 = V2*JHAT;
48     DER.RHO = ((RHO*WDMLK1 + (1-RHO)*B0 + B1*U + B2*V +
49          B3*U2 + B4*UV + B5*V2 - USDA1)*
50          (-A1/(D1+RHO)+A2/(D2-RHO)) + WDMLK1 - B0)*JHAT;
51  RUN;
```

Spatial Statistics Gradient Model SAS Code Description: Lines #3-5 calculate the second-order polynomial terms. Lines #6-20 compute the spatially lagged 1969 USDA data. Line #24 converts the spatially lagged values to those based on matrix ***W***. Lines #26-29 are the coefficients for the Jacobian approximation (equation (2)). Line #35 inputs the OLS results as initial estimates for the NLS parameters. Lines #37-38 is the Jacobian approximation. Lines #39-41 are the spatial autoregressive response model estimation equation. Lines #42-50 are the analytical derivatives of the spatial autoregressive response model.

Preceding this portion of the code, `DATA STEP1` has had all of the data, the connectivity matrix, and the variable transformations inputed into it (see lines #4-29 of the SAS code appearing in Appendix 1). Most of this code needs to be repeated for each spatial autoregressive response model component of the GLM.

13 Econometric Models and Spatial Parametric Instability:

Relevant Concepts and an Instability Index

Emilio Casetti[1] and Jessie Poon[2]
[1] Ohio State University, Columbus, OH, U.S.A.
[2] University of Georgia, Athens, GA, U.S.A.

13.1 Introduction

This chapter is concerned with the measurement of parametric instability in general, and spatial parametric instability in particular. Let us begin by placing into focus some pertinent concepts and issues.

The concept of parametric instability needs to be addressed first. Consider an econometric model relating a dependent variable y to independent variables $x_1, \ldots, x_m$, and a space Z spanned by 'contextual' variables $z_1, \ldots, z_k$. The model drifts across (namely, is unstable over) the context defined by the Z variables if it holds with different parameter values at different 'points' in Z space. The notion involved here is that theoretically grounded models should be viewed against backgrounds ('contexts') with respect to which they vary. For example, if a production function varies over time, over space, and by firm size, suitable spatial temporal coordinates and a firm size variable define the production function's context. The model context nexus will be addressed at some length later in this paper.

Assume that a data set with values for y and for the X and Z variables is given; that the model's parametric instability has been specified, tested for, found to exist, and estimated; and that the observations in the data set have been associated with values of the model's parameters that reflect the occurrence of the hypothesized drift. This chapter is concerned with measuring the 'extent' of this instability, and specifically with determining how much instability exists over the data set or over any of its subsets. The approach used is based on isolating the variation of the predicted values of the dependent variable that depends only on the model's drift. The instability coefficient proposed is a function of this variation.

If the model and its drift are estimated from a data set with spatially referenced observations, the instability coefficient measures its spatial instability. For instance, if the data are by country, the instability coefficients proposed will measure the model's instability over the countries included in the data set, or over any substantively meaningful subset of them.

Let us place into perspective the methodological and substantive significance and uses of measuring (and thus, quantifying) the extent to which parametric instability exists over a given set of observations. If some suitable hypothesis testing established that a model drifts, this still does not tell us how much the model drifts over the data set used to estimate it, or over 'interesting' subsets of it. The drift can be negligible, major, or anything in between. An instability measure such as the one discussed in this chapter quantifies the extent of the drift over given data. In the process, it clues us as to how wrong we are if we erroneously consider the model stable over those data. Furthermore, the measurement of parametric instability allows comparisons to be made: across models, for the same model across data sets, and for the same model and data set across data subsets.

In this chapter these and related issues are discussed formally, and then an index of (spatial) parametric instability is defined and demonstrated by an example. In the example it is shown using country level data, that an econometric model relating economic development to exports' growth drifts with respect to development levels. Then the instability of the model is measured for the full data employed in the estimation, and also for four groupings of countries.

The concepts and formalisms of the expansion method will be used. The expansion methodology [Casetti (1972, 1986, 1991)] combines a technique and a research philosophy that are especially well suited to investigate the contextual parametric instability of econometric models. Reviews of a number of applications of the expansion method are contained in Casetti (1986), in Casetti and Jones (1987), and in Jones and Casetti (1992).

As a technique, the expansion method consists of the following steps: (a) an 'initial model' is specified; the model is made of variables and/or random variables and at least some of its parameters are in letter form; (b) at least some of the letter parameters in the initial model are redefined by 'expansion equations' into functions of variables and/or random variables representing a context; (c) the expanded parameters are replaced into the initial model to create a 'terminal model'; (d) the expansions can be iterated, since the terminal model produced by one expansion can become the initial model of the next.

Suppose that we take as the initial model a relationship with strong theoretical grounding, and that the expansion equations model the contextual variation of this relationship. Then the terminal model obtained from the two will encompass into the same entity both the model and its contextual drift. Thus, the identity of the initial model is preserved, but at the same time the initial model can mold itself to complex contextual realities that were not part of it.

As a research philosophy, the expansion methodology suggests that theoretically grounded relationships such as those arising from economic theory should be regarded as the building blocks of more complex theoretical structures encompassing both them and their contexts or environments. These higher structures should reflect both the theory behind the initial model, and the theory about the nexus between the initial model and its contexts.

In the prevalent econometric practice models tend to be investigated and estimated under the implicit presupposition that they possess some form of universal, or quasi universal validity. Certainly, in most cases, they are presumed to be invariant over the data sets from which they are estimated. In contrast, the expansion methodology suggests that presuppositions of invariance are almost always unwarranted, and that the variation of relationships across contexts should be presumed, investigated, tested for, and theorized.

In the sections that follow, first expansion method concepts and formalisms relevant to the measurement of parametric instability are addressed, and an operational index of spatial parametric instability is defined. An example follows. Some concluding thoughts cap the chapter.

13.2 The Expansion Method

The next few paragraphs demonstrate the expansion method formalisms using linear initial models and linear expansion equations. Assume an initial model consisting of a linear relation between a dependent random variable y, m independent variables x_1, x_2, ..., x_m, and the random error u.

$$y = \alpha_0 + \alpha_1 x_1 + \ldots + \alpha_m x_m + u \tag{1}$$

Hypothesize that (1) drifts in a context defined by the variables z_1, ..., z_k, and that this drift is specified by the following linear expansion equations:

$$\begin{aligned} \alpha_0 &= \gamma_{00} + \gamma_{01} z_1 + \ldots + \gamma_{0k} z_k \\ \alpha_1 &= \gamma_{10} + \gamma_{11} z_1 + \ldots + \gamma_{12} z_k \\ &\vdots \\ \alpha_m &= \gamma_{m0} + \gamma_{m1} z_1 + \ldots + \gamma_{mk} z_k \end{aligned} \tag{2}$$

The variables z_1, ..., z_k span a k-dimensional expansion space. The coordinates of any point in this space are a vector of k values of these variables. If at least some of the γ parameters in (2) that are associated with z variables are non-zero, every point in Z space is associated with a distinct realization of the initial model (1), and we can say that (1) 'drifts' in Z space. If all of the γ parameters associated with Z variables have a value of zero, every point in Z space is associated with the same realization of the initial model, in which case the inital model is 'stable' in it. Consequently the occurrence of drift can be ascertained by determining whether appropriate γ parameters are different from zero. By substituting the right hand sides of (2) for the corresponding coefficients in equation (1) the following terminal model is obtained:

$$\begin{aligned} y = \ & \gamma_{00} + \gamma_{01} z_1 + \dots + \gamma_{0k} z_k \ + \\ & \gamma_{10} x_1 + \gamma_{11} x_1 z_1 + \dots + \gamma_{1k} x_1 z_k \ + \\ & \vdots \qquad\qquad\qquad\qquad\qquad \vdots \\ & \gamma_{m0} x_m + \gamma_{m1} x_m z_1 + \dots + \gamma_{mk} x_m z_k + u \end{aligned} \tag{3}$$

The first and second subscripts of the γ's denote respectively which x and z variables are represented in the expression to which the coefficient is attached. For instance, γ_{12}, identifies the presence of variables x_1 and z_2.

Whenever a terminal model has been generated from a linear initial model by linear expansion equations, there is a second linear initial model and associated linear expansion equation(s) that will yield the same terminal model [Casetti (1986)]. If a terminal model is given, as soon as a linear initial model and linear expansion equations capable of producing it are defined, also a second linear initial model and associated linear expansion equations that will produce the same terminal model become defined. Let us call the two initial models and associated expansion equations respectively 'primal' and 'dual'. Which expansion is primal is arbitrary, but when an expansion is defined as primal, the second one becomes the dual of the first one.

The intrinsic duality of the linear expansions is illustrated by the fact that the same terminal model (3) can be arrived from an initial model relating the y to the z variables by expanding its coefficients in terms of the x variables. To show it assume the initial model:

$$y = \beta_0 + \beta_1 z_1 + \dots + \beta_k z_k + u \tag{4}$$

and the expansion equations:

$$\begin{aligned} \beta_0 &= \gamma_{00} + \gamma_{10} x_1 + \dots + \gamma_{m0} x_m \\ \beta_1 &= \gamma_{01} + \gamma_{11} x_1 + \dots + \gamma_{ml} x_m \\ &\vdots \\ \beta_k &= \gamma_{0k} + \gamma_{1k} x_1 + \dots + \gamma_{mk} x_m \end{aligned} \tag{5}$$

By replacing the right hand sides of (5) for the corresponding coefficients in (4) terminal model (3) is indeed arrived at again.

These duality relations imply that when we define a primal initial model and specify its drift in Z space, we also define a dual initial model and specify its drift in X space. When the drift of the primal initial model is tested for, so is simultaneously the drift of the dual initial model.

Denote var(u) = $E(uu')$ by ϕ. For appropriate specifications of the error term u, equations (1) through (5) correspond to a spectrum of situations including some characterized by random coefficients and/or by stochastic expansion equations models. Let's consider a few cases:

(1) in the standard linear model specification $u = \varepsilon$, where $\varepsilon \sim N(0,\sigma^2 I)$;

(2) $u = \varepsilon$, where $\varepsilon \sim N(0,\Sigma)$ is a general formulation that depending upon the structure of Σ can correspond to variance or covariance components specifications, and/or to specifications involving dependencies 'autocorrelation' or heteroscedasticies;

(3) $u = \varepsilon + \eta_0 x_0 + \eta_1 x_1 + ... + \eta_m x_m$, where here and hereafter $\varepsilon \sim N(0,\Sigma)$, corresponds to a random coefficients specification for the x variables, or, equivalently, to stochastic primal expansion equations with error terms $\eta_0, \eta_1, ..., \eta_m$ attached to the expansion equations for α_0 through α_{mi};

(4) $u = \varepsilon + \theta_0 z_1 + \theta_1 z_1 + ... + \theta_k z_k$ corresponds to a random coefficients specification for the z variables, or, equivalently, to stochastic dual expansion equations with error terms $\theta_0, \theta_1, ..., \theta_k$ attached to the dual expansion equations for β_0 through β_k;

(5) $u = \varepsilon + \eta_0 x_0 + \eta_1 \omega_1 + ... + \eta_m x_m + \theta_0 z_1 + \theta_1 z_1 + ... + \theta_k z_k$ corresponds to a random coefficients specification for the x and z variables, or, equivalently, to stochastic primal and dual expansion equations.

In cases (3) through (5) the η's and θ's are n by 1 normally distributed random vectors, the x's and z's are n by 1 vectors, and the expression $\eta_i x_i$ or $\theta_i z_i$ signify 'element by element' products. When all the elements of the η's and θ's vectors are distinct, for a wide range of assumptions cases (3) through (5) will be characterized by diagonal Φ's, that correspond to heteroscedastic u's. When instead the n by 1 η's and θ's vectors are made of J distinct elements (namely, one each for J blocks of consecutive observations), for a wide range of assumptions cases (3) through (5) will be characterized by block diagonal Φ's. This latter situation arises in 'multilevel' analyses [Mason et al. (1983), Jones (1991)] and in 'hierarchical linear models' [Bryk and Raudenbush (1987, 1992)]. Spatial, temporal, or spatial-temporal autocorrelation will result into non-diagonal Σ's and Φ's.

The cases under (1) fall within the standard linear model tradition. The cases under (2) correspond to extensions of the linear model reviewed for instance in Judge et al. (1985) or, for their geographical applications, in Anselin (1988) and Anselin and Griffith (1988). Some phases of estimation and inference for cases under (3) through (5) are addressed in the literatures on random coefficients [Swamy (1971), Cooley and Prescott (1976), Fomby et al. (1984 p. 316)]; on varying parameters [Maddala (1977), Raj and Ullah (1981), Judge et al. (1985, chapter 19)]; on Bayesian linear models [Zellner (1971), Lindley and Smith (1972), Smith (1973)]; and on empirical Bayes [Maritz (1970), Rubin (1981), Morris (1983), Raudenbush and Bryk (1985), Bryk and Raudenbush (1987)].

In the presence of heteroscedasticity, estimation can be carried out by Maximum Likelihood (ML) or by Estimated Generalized Least Square (EGLS) [Fomby et al.

(1984 p. 170), Amemiya (1985 p. 181)]; if heteroscedasticity is assumed but its specification is uncertain parameter estimates may be based on OLS and tests of significance on heteroscedasticity-robust, consistent estimates of the variance-covariance of the regression coefficients [White (1980, 1984)].

In the presence of a block diagonal Φ of the type arising in multilevel or hierarchical Bayes formulations, estimation and inference has been recently based on a two phase procedure involving the use of the EM algorithm [Dempster et al. (1977, 1981)] to obtain REstricted Maximum Likelihood (REML) estimates of Φ, that are then used to produce EGLS estimates of the model's parameters [Mason et al. (1983), Entwisle and Mason (1985), Wong and Mason (1985), Bryk and Raudenbush (1987)]. The estimation of spatial models with heteroscedastic and/or autocorrelated error terms has been reviewed and discussed in Anselin (1988).

The tests for contextual parametric drift involve estimating the population parameter vector, γ, and testing null hypotheses to the effect that elements and sets of elements of γ are identically zero. Specifically:

(1) If none of the coefficients associated with terms in which the z variables appear is different from zero, but all or some of the coefficients of the x variables are different from zero, the primal model (or a subset of it) holds, and is stable.

(2) If none of the coefficents of terms in which the x variables appear is different from zero, but all or some of the coefficients of the z variables are different from zero, the dual initial model (or a subset of it) holds and is stable.

(3) If all or some of the coefficients of the terms in which the x variables appear and all or some of the coefficients of the terms in which the z variables appear are different from zero both the primal and the dual initial models (or subsets of them) hold and drift.

13.3 Parametric Instability

Assume that we have a data set consisting of a vector with the values of the dependent variable, and of a design matrix made of a column of 1's, the values of the primal variables, of the dual variables, and of the products of the two. Suppose also that the population parameters have been estimated by by whatever estimation procedure is appropriate in the case in point. Let equations (6) and (7) denote the estimated primal initial model and the estimated primal expansion equations:

$$\hat{y} = a_0 + a_1 x_1 + \ldots + a_m x_m \tag{6}$$

$$\begin{aligned} a_0 &= c_{00} + c_{01} z_1 + \ldots + c_{0k} z_k \\ a_1 &= c_{10} + c_{11} z_1 + \ldots + c_{12} z_k \\ &\vdots \\ a_m &= c_{m0} + c_{m1} z_1 + \ldots + c_{mk} z_k \end{aligned} \tag{7}$$

where $\hat{y}$ denotes the predicted values of y, and the c's identify the estimates of the corresponding population parameters. Let us focus on what causes $\hat{y}$ to 'vary'. Assume first that the initial model is 'stable' and that all the c's associated with z variables are equal to zero. In this case the estimated initial model is the same for all observations, and $\hat{y}$ varies in response to the x's variables. Instead, if we have parametric drift, the a's will vary in response to the values of the z variables and different observations will be associated with different realizations of the initial model. Hence, in an 'instability' situation the variation of $\hat{y}$ results from two (not necessarily uncorrelated) components, one depending upon the initial model, and the second associated with the variation of the initial model.

Suppose we suppress the variation due to the initial model by locking each of the x variables at its mean value. In this case, if the initial model is stable, the predicted values of y produced by the mean values of the x's will be identical for all observations. Instead, if the initial model is unstable, $\hat{y}$ will vary across the observations but only in response to the variation of the initial model.

Denote the mean values of the x's variables by bar superscripts. Then, the predicted values of y that reflect the parametric instability of the primal initial model, written $\hat{y}_p$, can be calculated from equations (7) and (8):

$$\hat{y}_p = a_0 + a_1\bar{x}_1 + \ldots + a_m\bar{x}_m \tag{8}$$

It is useful to reiterate that $\hat{y}_p$ denotes an n by 1 vector of predicted y values generated from a modified data set in which the values of each x variable have been replaced by the variable's mean. If the initial model is stable these predicted values are identical for all the observations, if the initial model is unstable, they will vary across the observations in response to the variation of the initial model across the observations. The variance of $\hat{y}_p$, written var($\hat{y}_p$), constitutes a measure of the initial model's instability because it will have a value of zero if the model is stable, it will have small values if the initial models changes only slightly across the observations, and will grow large if the initial model drifts a great deal across the observations.

However, as a measure of parametric instability, var($\hat{y}_p$) has the major shortcoming that is not 'dimensionless'. It has the dimension of y (squared), but so does the variance of the predicted y, var($\hat{y}$). Hence the index:

$$I_p = \text{var}(\hat{y}_p) \,/\, \text{var}(\hat{y}) \tag{9}$$

is a dimensionless index of the parametric instability of the primal initial model. The 'p' subscripts denote 'primal' initial model's instability and distinguish the I_p index from its dual counterpart, I_d, that will be defined in the paragraphs that follow.

Let us start from the estimated dual initial model (10) and expansion equations (11):

$$\hat{y} = b_0 + b_1z_1 + \ldots + b_kz_k \tag{10}$$

$$
\begin{aligned}
b_0 &= c_{00} + c_{10}x_1 + \dots + c_{m0}x_m \\
b_1 &= c_{01} + c_{11}x_1 + \dots + c_{m1}x_m \\
&\vdots \\
b_k &= c_{0k} + c_{1k}x_1 + \dots + c_{mk}x_m
\end{aligned} \tag{11}
$$

where, again, c's denote the estimates of the corresponding population parameters. The estimated y's produced by setting all the z variables to their mean values can be easily computed from equations (11) and (12):

$$\hat{y}_d = b_0 + b_1\bar{z}_1 + \dots + b_k\bar{z}_k \tag{12}$$

By a reasoning analogous to the one articulated previously, we posit that:

$$I_d = \text{var}(\hat{y}_d) \;/\; \text{var}(\hat{y}) \tag{13}$$

constitutes a dimensionless index of the parametric instability of the dual initial model.

If the observations in the data set employed in the estimation are spatially referenced, I_p and I_d are indices of the spatial instability of the primal and dual initial models. The concept of spatial referencing however needs to be clarified. Observations associated with sites, geographical areas, or regions, or with points the location of which is specified by coordinates such as latitudes and longitudes, are spatially referenced. But so are observations associated with links in networks, or with points in three dimensional coordinate systems (namely, including altitude or depth) such as those employed in climatology and geology. Indices of spatial instability can be calculated with any kinds of spatially referenced observations, although the usual cross-sectional data in which the observations are associated with countries, counties, and the like suggest themselves as the most obvious candidates for the application of these indices.

Instability and spatial instability indices can be calculated on the entire data set employed in an analysis, but also on any subset of it. If we take a group of observations and calculate the values of $\hat{y}$, $\hat{y}_p$, and $\hat{y}_d$ for them, obviously we can also calculate the I_p and I_d indices for these observations. The application of the instability indices technique to subsets of observations is potentially quite useful to determine how much the instability of relationships varies, for instance, over different regions. A demonstration of this point is given in connection with the example that follows.

13.3.1 Example

In the section that follows the I_p and I_d indices are demonstrated by applying them to measure the spatial and temporal instability of the relationship between economic growth and exports' growth. This relationship has policy relevance, because it has provided the justification of the export-promotion development strategies on which a number of World Bank activities have been predicated [World Bank (1987)]. In fact, these export-promotion strategies have been advocated as a generally valid prescription for economic growth in the presupposition that the relationship between economic growth and growth of exports is stable, namely, is generally valid. This discussion builds upon Poon's (1994) investigation of the relationship's drift with respect to development levels. Let us address briefly the theoretical bases of the relationship, as well as of the notion that the relationship should be expected to drift with development levels.

A number of theoretical propositions suggest that the growth of exports induces economic growth. Exports help overcome the limitations of a country's domestic market. They provide an outlet for commodities requiring locally abundant resources and that can be produced in excess to internal demand. When idle or surplus resources are absorbed into 'exportables' this has the effect of stimulating economic growth [Myint (1957)]. Exports' growth ensures greater capacity utilization as well as increased specialization through economies of scale. Balassa (1985) and Bhagwati (1988) argued that export-oriented economies are more efficient, so that income growth can be achieved at a much lower costs especially in terms of investments. Feder (1982) concluded that the export sector affects economic growth by enhancing the quantity and quality of labor and capital stocks.

Countries producing for export are exposed to international consumption patterns, which leads to an acceleration in their learning processes. In turn this results into higher quality products [Krueger (1985)] and encourages the rise of entrepreneurship. However, as regard the developing countries, the most compelling explanation of the importance of exports to growth lies in the alleviation of the foreign exchange bottlenecks that they produce. Developing countries need to earn foreign currencies to purchase the capital goods necessary for industrial development. Exports provide the financial means to attain these material prerequisites which are themselves essential for the overall growth of the national income.

A number of empirical studies sought to verify the contribution of exports to growth [Emery (1967), Maizels (1968), Michalopoulos and Jay (1973), Michaely (1977), Balassa (1978, 1985), Tyler (1981), Ram (1985), Moschos (1989), Fosu (1990)]. However, these studies revealed wide differences in the effect of exports on growth, as evidenced by the wide variation in the size of the regression parameters associated with export factors.

Poon (1994) argued that these differences are partly due to the fact that the economic growth export growth relationship drifts with development levels. A number of theoretical reasons suggest why this is the case. Economies at different development levels have very different structural characteristics and productive capabilities [Syrquin (1988)]: it does not make sense to assume that a one per cent growth in exports produces the same growth response irrespective of these differences. Bruton (1989, p. 1616) noted that high exports growth alone does not assure growth. Kavoussi (1984) stated that it is only when export expansion is accompanied by a rapid growth of resources with major gains in factor productivity that the effects of exports on growth are strongest. This occurs especially during the transition from low to intermediate development levels which is characterized by an acceleration in the growth of labor and of capital stocks [Syrquin (1988)]. Thus, middle income countries tend to grow faster than lower or higher income countries [Syrquin (1986)]. On the other hand, some minimum level of development is necessary before exports are likely to benefit a country's growth [Michaely (1977), Tyler (1981)].

The staple theory of export-led growth also suggests that the effect of exports growth on economic growth depends on development levels. In terms of this theory [North (1961), Lewis (1969)] the exports from underdeveloped regions or countries influence economic growth by processes in which we can recognise distinct stages: early on, the exports are resource rich and promote the expansion of infrastructures and the demand for domestically produced inputs. Later, the exports encompass intermediate products, and encourage the processing of raw materials. Eventually the exports become comparatively less important as import substitution leads to the domestic production of consumer goods [Balassa (1989, p. 1665)]. The developmental effects of exports growth increase during the first of these stages and declines throughout the second and third.

The econometric studies in the literature lend occasional support to the notion that the effect of exports growth are development dependent. For instance it has been pointed out that the export variables tend to have a smaller effect on growth at lower development levels [Syron and Walsh (1968), Tyler (1981), Kavoussi (1984)], and Kavoussi (1984) noted that with development the correlation between manufacturing exports and economic growth changes from negative to positive.

Starting from these premises Poon (1994) expanded the parameters of a linear initial model relating the percentage growth rate of GDP per capita, y, to percentage growth rate of exports, x, into quadratics in the logarithm of GDP per capita, z. Which led to the primal initial model:

$$y = \alpha_0 + \alpha_1 x_1 + \varepsilon \tag{14}$$

with $\varepsilon \sim N(0,\sigma^2 \boldsymbol{I})$, and to the expansion equations:

$$\begin{aligned} \alpha_0 &= \gamma_{00} + \gamma_{01} z + \gamma_{02} z^2 \\ \alpha_1 &= \gamma_{10} + \gamma_{11} z + \gamma_{12} z^2 \end{aligned} \tag{15}$$

By replacing the right hand sides of expansion equations (15) in equation (14) the following terminal model is obtained:

$$y = \gamma_{00} + \gamma_{01} z + \gamma_{02} z^2 + \gamma_{10} x + \gamma_{11} xz + \gamma_{12} xz^2 + \varepsilon \tag{16}$$

This formulation implies the following dual initial model:

$$y = \beta_0 + \beta_1 z + \beta_2 z^2 + \varepsilon \tag{17}$$

and expansion equations:

$$\begin{aligned} \beta_0 &= \gamma_{00} + \gamma_{10} x \\ \beta_1 &= \gamma_{01} + \gamma_{11} x \\ \beta_2 &= \gamma_{02} + \gamma_{12} x \end{aligned} \tag{18}$$

which also produce the terminal model (16). This dual formulation has a strong theoretical support in its own right. The tendency for countries at intermediate development levels to develop faster, first noted by Russett et al. (1964), is congruent with the suggestions emanating from several theoretical strands that were reviewed in Casetti (1986, pp. 34-35). The most compelling of these is the notion that the developing countries at the low end of the development spectrum grow slowly because of the shortage of capital and trained labor force irrespective of the large stock of available well established technologies available to them. When the capital and trained manpower bottlenecks are alleviated, these countries experience the growth surge that appears to characterize intermediate development levels. However, as the developing countries approach maturity, the tertiarization of their economies, and an increasing dependence on current rather than past technological progress pulls their rates of development down. For appropriate parameter values the dual initial model (17) is congruent with these propositions.

On the other hand, the dual expansion equations (18) reflect the general suggestion that the theoretical relationship (17) may hold with different parameter values in response to factors promoting or hindering economic development. Here the development-promoting exports' growth is considered. In Casetti (1986) a growth hindering factor (population growth) was investigated.

The data set employed in the estimation of equation (16) includes 91 countries, extracted from a larger sample of 109 by removing the countries for which the 1988 population was less than two million people. The source data are average annual growth rates of exports and of GDP per capita over the time intervals 1960-70, 1970-80, and 1980-88, and GDP per capita for 1980 in 1980 US dollars. GDP per capita at the midpoints of the three time intervals were obtained by projecting the 1980 GDP's per capita backward and forward. The data for the three time intervals were then stacked to produce a total of 273 observations. From these, the observations with negative growth of product per capita were removed, under the presupposition that they would not reflect the conditions of social economic and political 'normalcy' in which the effects of growth of exports on the economy can be expected to materialize. This left 201 usable observations. The source data are from UNCTAD (1983, Table 6.1) and UNCTAD (1990, Tables 1.5, 1.6, and 6.2).

Equation (16) implements the expansion of both intercept and slopes of equation (14). Following a 'testing down' approach, the estimation of (16) was carried out first on the full equation, then by restricting to zero subsets of its variables so as to implement an intercept expansion and a slopes' expansion of (14), and finally on the unexpanded equation (14). The best of the four equations was selected on the basis of criteria involving the significance of the variables included, the lack of significance of the variables excluded, and the R-squares. The OLS estimate of the equation selected is:

$$\begin{array}{lllll} \hat{y} = 2.226 & - 1.318\,x & + 0.385\,xz & - 0.026\,xz^2 \\ \quad (11.5) & (-5.1) & (5.4) & (-5.5) \end{array} \tag{19}$$

where $R^2 = 0.246$. The t-values in parentheses under the corresponding regression coefficients are based on White's robust variance-covariance matrix since the null hypothesis of no heteroscedasticity was rejected at the 5 percent level on the basis of the appropriate LM test.

It is useful to note that equation (19) can be interpreted in terms of the primal estimated initial model and primal estimated expansion equations in (20):

$$\begin{aligned} \hat{y} &= a_0 + a_1 x \\ a_0 &= 2.226 \\ a_1 &= -1.318 + 0.385z - 0.026z^2 \end{aligned} \tag{20}$$

or in terms of the dual estimated initial model and dual estimated expansion equations in (21):

$$\begin{aligned} \hat{y} &= b_0 + b_1 z + b_2 z^2 \\ b_0 &= 2.226 - 1.318x \\ b_1 &= 0.385x \\ b_2 &= -0.026x \end{aligned} \tag{21}$$

The instability coefficients discussed later in this chapter are calculated using (20) and (21).

The results above are substantively meaningful. To show it, let us focus on the $a_1(z)$ relationship in (20), and on the $\hat{y}(z)$ relationship in (21), in the order. Let us recall that a_1 is the estimate of the percentage increase in product per capita associated with a 1 percent growth of exports, and z is the logarithm of GDP per capita. A plot of a_1 versus GDP per capita based on the $a_1(z)$ relation in (20) is shown in Figure 13.1. The plot shows that the effect of export growth on economic growth increases at first and then decreases as higher levels of product per capita are considered, which is indeed consistent with theory and expectations.

As noted earlier on, the growth of product per capita can be expected to be largest at intermediate development levels. Figure 13.2 shows the plots of the estimated relationships between growth and levels of per capita product based on (21) for low and high growth rates of exports, operationalized as $x = 6.1$ and $x = 19.2$. These are the mean values of the export growth data respectively below and above the overall average. Both curves in Figure 13.2 display the expected maximum at an intermediate level of product per capita.

$$y = \beta_0 + \beta_1 z + \beta_2 z^2 + \varepsilon \tag{17}$$

and expansion equations:

$$\begin{aligned} \beta_0 &= \gamma_{00} + \gamma_{10} x \\ \beta_1 &= \gamma_{01} + \gamma_{11} x \\ \beta_2 &= \gamma_{02} + \gamma_{12} x \end{aligned} \tag{18}$$

which also produce the terminal model (16). This dual formulation has a strong theoretical support in its own right. The tendency for countries at intermediate development levels to develop faster, first noted by Russett et al. (1964), is congruent with the suggestions emanating from several theoretical strands that were reviewed in Casetti (1986, pp. 34-35). The most compelling of these is the notion that the developing countries at the low end of the development spectrum grow slowly because of the shortage of capital and trained labor force irrespective of the large stock of available well established technologies available to them. When the capital and trained manpower bottlenecks are alleviated, these countries experience the growth surge that appears to characterize intermediate development levels. However, as the developing countries approach maturity, the tertiarization of their economies, and an increasing dependence on current rather than past technological progress pulls their rates of development down. For appropriate parameter values the dual initial model (17) is congruent with these propositions.

On the other hand, the dual expansion equations (18) reflect the general suggestion that the theoretical relationship (17) may hold with different parameter values in response to factors promoting or hindering economic development. Here the development-promoting exports' growth is considered. In Casetti (1986) a growth hindering factor (population growth) was investigated.

The data set employed in the estimation of equation (16) includes 91 countries, extracted from a larger sample of 109 by removing the countries for which the 1988 population was less than two million people. The source data are average annual growth rates of exports and of GDP per capita over the time intervals 1960-70, 1970-80, and 1980-88, and GDP per capita for 1980 in 1980 US dollars. GDP per capita at the midpoints of the three time intervals were obtained by projecting the 1980 GDP's per capita backward and forward. The data for the three time intervals were then stacked to produce a total of 273 observations. From these, the observations with negative growth of product per capita were removed, under the presupposition that they would not reflect the conditions of social economic and political 'normalcy' in which the effects of growth of exports on the economy can be expected to materialize. This left 201 usable observations. The source data are from UNCTAD (1983, Table 6.1) and UNCTAD (1990, Tables 1.5, 1.6, and 6.2).

Equation (16) implements the expansion of both intercept and slopes of equation (14). Following a 'testing down' approach, the estimation of (16) was carried out first on the full equation, then by restricting to zero subsets of its variables so as to implement an intercept expansion and a slopes' expansion of (14), and finally on the unexpanded equation (14). The best of the four equations was selected on the basis of criteria involving the significance of the variables included, the lack of significance of the variables excluded, and the R-squares. The OLS estimate of the equation selected is:

$$\hat{y} = \underset{(11.5)}{2.226} - \underset{(-5.1)}{1.318\,x} + \underset{(5.4)}{0.385\,x\,z} - \underset{(-5.5)}{0.026\,x\,z^2} \tag{19}$$

where $R^2 = 0.246$. The t-values in parentheses under the corresponding regression coefficients are based on White's robust variance-covariance matrix since the null hypothesis of no heteroscedasticity was rejected at the 5 percent level on the basis of the appropriate LM test.

It is useful to note that equation (19) can be interpreted in terms of the primal estimated initial model and primal estimated expansion equations in (20):

$$\begin{aligned} \hat{y} &= a_0 + a_1 x \\ a_0 &= 2.226 \\ a_1 &= -1.318 + 0.385z - 0.026z^2 \end{aligned} \tag{20}$$

or in terms of the dual estimated initial model and dual estimated expansion equations in (21):

$$\begin{aligned} \hat{y} &= b_0 + b_1 z + b_2 z^2 \\ b_0 &= 2.226 - 1.318x \\ b_1 &= 0.385x \\ b_2 &= -0.026x \end{aligned} \tag{21}$$

The instability coefficients discussed later in this chapter are calculated using (20) and (21).

The results above are substantively meaningful. To show it, let us focus on the $a_1(z)$ relationship in (20), and on the $\hat{y}(z)$ relationship in (21), in the order. Let us recall that a_1 is the estimate of the percentage increase in product per capita associated with a 1 percent growth of exports, and z is the logarithm of GDP per capita. A plot of a_1 versus GDP per capita based on the $a_1(z)$ relation in (20) is shown in Figure 13.1. The plot shows that the effect of export growth on economic growth increases at first and then decreases as higher levels of product per capita are considered, which is indeed consistent with theory and expectations.

As noted earlier on, the growth of product per capita can be expected to be largest at intermediate development levels. Figure 13.2 shows the plots of the estimated relationships between growth and levels of per capita product based on (21) for low and high growth rates of exports, operationalized as $x = 6.1$ and $x = 19.2$. These are the mean values of the export growth data respectively below and above the overall average. Both curves in Figure 13.2 display the expected maximum at an intermediate level of product per capita.

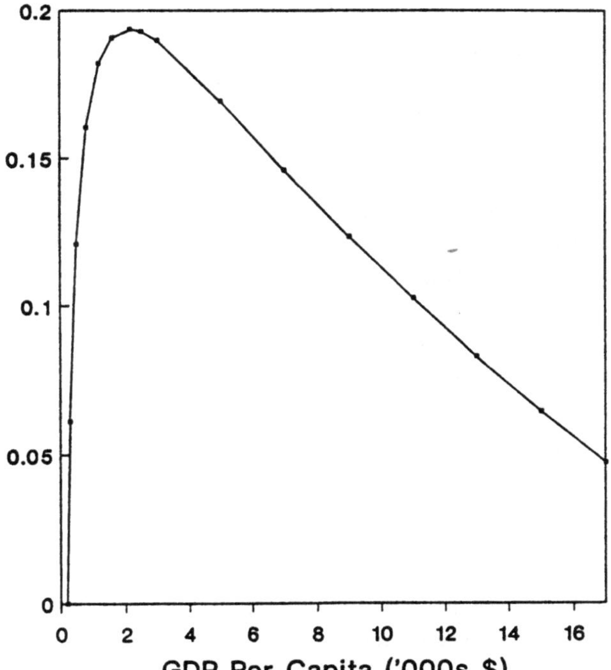

Figure 13.1: Export growth parameter (a_1).

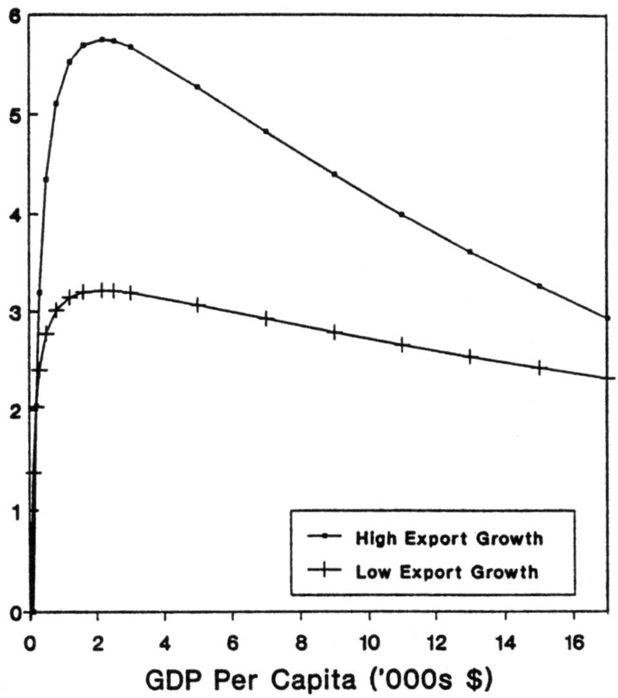

Figure 13.2: GPD per capita growth percentage.

However, the curve corresponding to high exports growth shows a maximum almost three percentage points larger than that of the low export growth curve. Again, this is consistent with theory and expectations, and demonstrates that the expansion of the exports enhances very considerably the spurt of economic growth that the developing countries have tended to experience after breaking out of the bottlenecks that dominate the lowest development levels.

Let us now turn to the instability coefficients I_p and I_d, and refocus on their meaning and significance. Both I_p and I_d are standardized measures of the 'variation', or 'drift', or 'instability' of the primal and dual initial models. An estimated relationship is unstable over a data set if it holds with different parameter values over the observations constituting it. The instability considered here reflects the variation of relationships functionally related to expansion variables. I_p and I_d index respectively the variation of the primal initial model as function of the primal expansion variables, and of the dual initial model as function of the dual expansion variables.

In the example, the primal initial model is a relationship between growth of product per capita and growth of exports. After establishing via tests of statistical hypotheses that this relationship drifts with development levels, it is substantively relevant to ask how important such drift is. An 'instability' measure refers to a specific data set. In data sets made of spatially referenced observations the measure concerns spatial instability. Here I_p tells us how much the relationship between growth of product per capita and of exports varies across countries in response to development levels. The meaning of the dual instability coefficient I_d can be easily arrived at by a parallel reasoning. I_d indexes the spatial variation of the relationship between rates and levels of product per capita, in response to exports' growth.

The values of these coefficients computed using (20) and (21) are as follows: for the entire data set $I_p = 0.41$ and $I_d = 0.29$; for the developed countries (including Japan) $I_p = 0.62$ and $I_d = 0.21$; for Latin America $I_p = 0.10$ and $I_d = 0.82$; for Africa $I_p = 0.37$ and $I_d = 0.21$; and for Asia (minus Japan) $I_p = 0.38$ and $I_d = 0.30$. These figures show that for the full data sets and for all its subsets except Latin America the I_p's are larger than their I_d's counterparts. Namely, the relationship between growth of product per capita and growth of exports is generally more unstable (that is to say, varies across countries more) than the relationship between rates and levels of product per capita. However, for the Latin American countries the opposite is true.

As regard the exports growth relationship, it is for the developed countries that its variation is greatest. This is indeed interesting: it suggests that the developed countries are the least suited to provide a basis for making general statements in support of exports based development strategies. The anomalous results for the Latin American countries should be perhaps linked to their comparatively low growth performances especially after the 1973 oil crisis.

The instability coefficients can make us aware of the extent to which we are unjustified to estimate an econometric relationship under assumptions of invariance. An I_p of 0.41 roughly tells us that the variation of the relationship between growth of exports and product per capita is 41 percent the size of the variance of $\hat{y}$. Clearly, this relationship should not be estimated under the presupposition of invariance that has characterized most of the empirical econometric work in this area. But then, invariance presuppositions do permeate the prevalent econometric practice.

The instability coefficients presented in this chapter are dimensionless, and as such allow comparisons involving different relationships and data sets. If a body of instability estimates were to become available from a variety of analyses involving different econometric relationships, these relationships could be then compared in terms of their comparative instability in different situations. Such comparisons would indeed make a useful addition to empirical econometrics.

13.4 Conclusions

In this conclusion two points will be addressed, concerning (1) the range of applicability of parametric instability measures, and (2) the relation of these measures to the model-context expansion method perspective. Let us begin by articulating some thoughts about the scope of the approach to the measurement of parametric instability discussed in this paper.

13.4.1 Instability Measures: Scope

Within the expansion methodology frame of reference, the minimal ingredients required to calculate the I_p and I_d coefficients are three: (1) a terminal model the parameters of which are in numerical form, (2) the specification of an initial model and of expansion equations capable of producing the terminal model, and (3) a suitable data set. The terminal model should involve drift in terms of the expansion structure attached to it, else the measurement of instability is unfeasible. On the other hand many models existing in the literature can be interpreted as the result of expanding an initial model via contextual variables.

In particular, any multiple regression model can be regarded as the result of expanding the intercept of an initial model into the remaining variables. Such interpretation is tantamount to superimposing on these models a logical structure differentiating model and context, and seems to be especially well suited to conceptualize the addition of non-economic variables to functional relationships originated from economic theory. At any rate, given a published estimated equation from the literature, if we interpret it as an estimated terminal model and specify an initial model within it, the initial model's instability becomes measurable, provided that we can link the equation to a suitable data set.

A data set, to be suitable for the measurement of instability needs to have 'enough' observations. However, it needs not be the one employed in the estimation of the model, or a subset of it. Even a data set consisting of artificial data spanning substantively interesting regions in the X and Z spaces can be used to measure the parametric instability over these regions. However, the X and Z variables should have non-zero variances over the data sets/subsets on which instabilities are calculated. These considerations concerning the minimal ingredients required to measure parametric instability suggest that the range of applicability of the I_p and I_d indices is very wide indeed, and extends over already estimated models found in the literature.

Especially well suited for the measurement of primal and dual instabilities is the class of models that can be conceptualized as involving the expansion of any arbitrary relationship in terms of dummy variables linking spatially referenced observations to geographical regions. With respect to these models I_p provides an intuitively appealing measure of the geographical instability of the relationship, and I_d gives an equally appealing measure of the extent to which the dual relationship between the dependent variable and the regional dummies drifts in response to the independent variable(s) appearing in the primal initial model.

Throughout this chapter the instability measures proposed have been discussed within an expansions frame of reference whereby the observations in a data set are associated with observation specific realizations of an initial model's coefficients via estimated expansion equations. From these observation-specific coefficients the instability measures is easily arrived at. However, observation specific estimated coefficients can be also produced by other approaches, such as Foster and Gorr's (1984, 1986) 'spatial adaptive filtering'. Instability measures can be just as easily calculated using the techniques outlined in this paper from observation specific regression coefficients generated using these approaches.

The I_p and I_d coefficients have been dealt with so far as instability 'measures' rather than statistics. In terms of the reasoning employed here, first the parametric instability of an initial model is determined by testing the hypothesis that the coefficients of terms in which expansion variables appear are significantly different from zero. If instability is thus proven, the extent to which it occurs is measured by the instability coefficients. An interesting line of enquiry suggested by a reviewer would involve treating I_p and I_d as statistics and investigating their probability distributions by Monte Carlo simulations, or by resampling techniques. Such investigations could provide the basis to test for spatial instability in cases such as the 'spatial adaptive filtering' in which standard regression based tests are not applicable.

13.4.2 Instability Measures: Significance

The significance of spatial parametric instability measures is best understood within the model-context expansion method perspective, that holds the investigation of parametric instability as one of its central tenets. The expansion method perspective challenges the presuppositions of parametric stability that dominate the construction and estimation of econometric models. It is useful to conclude this chapter with some thoughts on the historical and disciplinary roots of these presuppositions.

In the prevalent econometric practice models are estimated under implicit or explicit assumptions of 'parametric stability' or 'invariance'. The estimation terminology itself betrays implicit stability assumptions. Consider the implication of presupposing that our data reflect a 'true' model plus white noise. A 'true model' has to be valid, namely, invariant, over the data to be used for estimating it. The expression 'estimation of population parameters' also presupposes that that the model referred to 'holds' over the data employed to estimate it.

These presuppositions of parametric invariance have deep historical roots. Generally, the econometric models are made of relationships arising from economic theory plus an error term designed to accomodate the less than perfect congruence to be expected between any theory, no matter how valid, and the empirical world. Economic theory reflects an aspiration to uncover economic truth. The presumption that the relationships arising from theory are true (and invariant) embodies this aspiration. The stability assumptions have an even deeper foundation and reach whenever a mathematical model is viewed as the expression of a 'law', or a candidate for law status. Invariance and universal validity are the distinguishing traits of laws.

To geographers (or regional scientists) working at the interface with economics, the models that originate from economic theory are cast into an altogether different frame of reference. Economic theory and the relationships it produces result from a process of abstraction that isolated economic constructs from aspects and complexities of reality, many of which are close to the core of the geographers' own disciplinary interests. Geographers are interested in spatial differentiation, in the spatial distribution of phenomena, and in what causes these. For instance, they might focus upon variables indexing the characteristics of geographic environments, and employ such indices, or variables such as distances and functions of spatial coordinates to model the spatial distributions of these characteristics. Yet, economic geographers are also driven to incorporate in their work economic theory, and the many relationships and models spawned by economic theory.

The expansion methodology originated from the tension between these conflicting drives as an approach capable of resolving it. It suggested with statements general in scope and applicability how any simple, theoretically grounded, a-spatial model can be expanded to encompass complexities and spatial dimensions that were not part of it. It also suggested why such expansions should be carried out, why the estimation of models under routine presuppostions of invariance is suboptimal, why the potential parametric drift of theoretically grounded relationships should be investigated, tested for, and possibly modeled and theorized. The point being emphasized here, is that the economist and the economic geographer are likely to be driven by different conscious and subsonscious disciplinary aspirations. Possibly, the economist aspires to uncover into economic empirical realities a mix of theoretically grounded relationships plus noise, while instead the economic geographer's aspirations center on a synthesis between these relationships and spatial geographical contexts plus, also, noise.

However, conflicting disciplinary aspirations are not all that is involved here. It can be argued that while, in general there is no question that the abstraction of simple important relations from complex contexts can provide very significant additions to knowledge, many limitations and shortcomings of the social sciences and social science models can be traced to the same processes of abstraction that are responsible for these advances. Simple and elegant models can yield important insights, but they are in all likelyhood inadequate for understanding complex realities. There is an objective need to reintroduce the complexities of the real world into simple theoretically grounded mathematical models without destroying these models in the process. In fact, the simple models and the simplified important relationships prevalent in the contemporary social sciences, should be regarded as early steps in the growth of knowledge, rather than the end point and culmination of it.

References

Amemiya, T., *Advanced Econometrics*, Cambridge: Harvard University Press, 1985.

Anselin, L., *Spatial Econometrics: Methods and Models*, Dordrecht: Kluwer Academic Publishers, 1988.

Anselin, L. and D.A. Griffith, Do Spatial Effects Really Matter in Regression Analysis?, *Papers of the Regional Science Association*, 65, 11-34, 1988.

Balassa, B., Exports, Policy Choices, and Economic Growth in Developing Countries after the 1973 Oil Shock, *Journal of Development economics*, 18, 23-35, 1985.

Balassa, B., Exports and Economic Growth: Further Evidence, *Journal of Development Economics*, 5, 181-189, 1978.

Balassa, B., Outward Orientation, in: H.B. Chenery and T.N. Srinivasan (eds.), *Handbook of Development Economics*, Vol. II, Amsterdam: North Holland, 1989.

Bhagwati, J., Export-Promoting Trade Strategy: Issues and Evidence, *The World Bank Research Observer*, 3, 27-57, 1988.

Bruton, H., Import-Substitution, in: H.B. Chenery and T.N. Srinivasan (eds.), *Handbook of Development Economics*, Vol. II, Amsterdam: North Holland, 1989.

Bryk, A.S. and S.W. Raudenbush, Application of Hierarchical Linear Models to Assessing Change, *Psychological Bulletin*, 101, 147-158, 1987.

Bryk, A.S. and S.W. Raudenbush, *Hierarchical Linear Models*, London: Sage, 1992.

Casetti, E., Generating Models by the Expansion Method: Applications to Geographic Research, *Geographical Analysis*, 4, 81-91, 1972.

Casetti, E., The Dual Expansion Method: An Application for Evaluating the Effects of Population Growth on Development, *IEEE Transactions on Systems, Man, and Cybernetics*, vol. SMC-16, 29-39, 1986.

Casetti, E., The Investigation of Parameter Drift by Expanded Regressions: Generalities and a Family Planning Example, *Environment and Planning A*, 23, 1045-1051, 1991.

Casetti, E. and J.P. Jones III, Spatial Applications of the Expansion Method Paradigm, in: C. Dufournaud and D. Dudycha (eds.), *Quantitative Analysis in Geography*, University of Waterloo, Waterloo, Ontario, Canada, 1987.

Cooley, T. and E. Prescott, Estimation in the Presence of Stochastic Parameter Variation, *Econometrica*, 44, 167-184, 1976.

Dempster, A.P., N.M. Laird and D.B. Rubin, Maximum Likelihood from Incomplete Data via the EM Algorithm (with discussion), *Journal of the Royal Statistical Society B*, 39, 1-38, 1977.

Dempster, A.P., D.B. Rubin and R.K. Tsutakawa, Estimation in Covariance Components Models, *Journal of the American Statistical Association*, 76, 341-353, 1981.

Emery, R.F., The Relation of Exports and Economic Growth, *Kyklos*, 20, 470-485, 1967.

Entwisle, B. and W.M. Mason, Multilevel Effects of Socio-economic Development and Family Planning Programs on Children Ever Born, *American Journal of Sociology*, 91, 616-649, 1985.

Feder, G., On Exports and Economic Growth, *Journal of Development Economics*, 12, 59-73, 1982.

Fomby, T.B., R.C. Hill and S.R. Johnson, *Advanced Econometric Methods*, New York: Springer-Verlag, 1984.

Foster, S. and W. Gorr, Spatial Adaptive Filtering, *Modelling and Simulation* 15, 29-34, 1984.

Foster, S. and W. Gorr, An Adaptive Filtering for Estimating Spatially Varying Parameters: Application to Modeling Police Hours Spent in Response to Calls for Service, *Management Science*, 32, 878-889, 1986.

Fosu, A.K., Exports and Economic Growth, The African Case, *World Development*, 18, 831-835, 1990.

Jones, K., Specifying and Estimating Multi-Level Models for Geographical Research, *Transactions of the Institute of British Geographers* 16, 148-160, 1991.

Jones III, J.P. and E. Casetti, *Applications of the Expansion Method*, London: Routledge, 1992.

Judge, G.G., W.E. Griffiths, R. Carter Hill, H. Lütkepohl and T.C. Lee, *The Theory and Practice of Econometrics*, New York: Wiley, 1985.

Kavoussi, R., Export Espansion and Economic Growth, Further Empirical Evidence, *Journal of Development Economics*, 14, 241-250, 1984.

Krueger O', A., The Experiences and Lessons of Asia's Superexporters, in: V. Corbo, A. O'Krueger and F. Ossa (eds.), *Export-Oriented Development Strategies*,

The Success of Five Newly Industrializing Countries, Boulder: Westview Press, 1985.

Lewis, W.A., *Aspects of Tropical Trade, 1883-1967*, Uppsala: Wiksell, 1969.

Lindley, D.V. and A.F.M. Smith, Bayes Estimates for the Linear Model, *Journal of the Royal Statistical Society B*, 34, 1-41, 1972.

Maddala, G.S., *Econometrics*, New York: McGraw-Hill, 1977.

Maizels, A., *Exports and Economic Growth of Developing Countries*, London: Cambridge University Press, 1968.

Maritz, J.S., *Empirical Bayes Methods*, London: Methuen, 1970.

Mason, W.M., G.Y. Wong and B. Entwisle, Contextual Analysis Through the Multilevel Linear Model, in: S. Leinhart (ed.), *Sociological Methodology 1983-1984,* San Francisco: Jossey-Bass, 1983.

Michalopoulos, C. and K. Jay, *Growth of Exports and Income in the Developing World, A Neoclassical View*, AID Discussion Paper No. 28, Washington, D.C.: Agency for International Development, 1973.

Michaely, M., Exports and Growth: An Empirical Investigation, *Journal of Development Economics*, 4, 49-53, 1977.

Morris, C., Parametric Empirical Bayes Inference: Theory and Applications (with discussion), *Journal of the American Statistical Association*, 78, 47-65, 1983.

Moschos, D., Export Expansion, Growth and Level of Economic Development, *Journal of Development Economics*, 30, 93-102, 1989.

Myint, H., The Classical Theory of Internation Trade and the Underdeveloped Countries, *The Economic Journal*, 67, 317-337, 1957.

North, D.C., *The Economic Growth of the United States, 1790-1860*, Englewood Cliffs: Prentice Hall, 1961.

Poon, J., Export Growth, Economic Growth and Development Levels: An Empirical Analysis, *Geógraphical Analysis*, 26, 37-53, 1994.

Raj, B. and A. Ullah, *Econometrics, A Varying Coefficients Approach*, New York: St. Martins Press, 1981.

Ram, R., Exports and Economic Growth: Some Additional Evidence, *Economic Development and Cultural Change*, 33, 415-425, 1985.

Raudenbush, S.W. and A.S. Bryk, Empirical Bayes Meta-analysis, *Journal of Educational Statistics*, 10, 75-98, 1985.

Rubin, D., Using Empirical Bayes Techniques in the Law School Validity Studies, *Journal of the American Statistical Association*, 75, 801-827, 1981.

Russett Jr., B.A., K.W. Deutsch, and H.D. Lasswell, *World Handbook of Political and Social Indicators*, New Haven: Yale University Press, 1964.

Smith, A.F.M., A General Bayesian Linear Model, *Journal of the Royal Statistical Society B*, 35, 67-75, 1973.

Swamy, P.A.V.B., *Statistical Inference in Random Coefficients Models*, New York: Springer-Verlag, 1971.

Syron, R.F. and B.M. Walsh, The Relation of Exports and Economic Growth, A Note, *Kyklos*, 21, 541-545, 1968.

Syrquin, M., Patterns of Structural Change, in: H.B.Chenery and T.N. Srinivasan (eds.), *Handbook of Development Economics*, Vol. I, Amsterdam: North-Holland, 1988.

Syrquin, M., Productivity Growth and Factor Reallocation, in: H.B. Chenery, S. Robinson, and M. Syrquin (eds.), *Industrialization and Growth*, New York: Oxford University Press, 1986.

Tyler, W.G., Growth and Export Expansion in Developing Countries, Some Empirical Evidence, *Journal of Development Economics*, 9, 121-130, 1981.

UNCTAD (United Nations Conference on Trade and Development), *Handbook of International Trade and Development Statistics 1983*, New York: United Nations, 1983.

UNCTAD, *Handbook of International Trade and Development Statistics 1990*, New York: United Nations, 1990.

White, H., A Heteroskedastic-Consistent Covariance Matrix Estimator and Direct test for Heteroskedasticity, *Econometrica*, 48, 817-838, 1980.

White, H., *Asymptotic Theory for Econometricians*, New York: Academic Press, 1984.

World Bank, *World Development Report 1987*, Washington D.C.: World Bank, 1987.

Wong, G.Y., and W.M. Mason, The Hierarchical Logistic Regression Model for Multilevel Analysis, *Journal of the American Statistical Association*, 80, 513-524, 1985.

Zellner, A., *An Introduction to Bayuesian Analysis in Econometrics*, New York: Wiley, 1971.

14 Bayesian Hierarchical Forecasts for Dynamic Systems:

Case Study on Backcasting School District Income Tax Revenues

George T. Duncan, Wilpen Gorr and Janusz Szczypula
Carnegie Mellon University, Pittsburgh, PA, U.S.A.

14.1 Introduction

Time series forecasting for related units is common practice. Examples include sales of a chain of fast-food restaurants in a metropolitan area, precipitation in neighboring sections of farm land, and, as explored in this chapter, tax revenues for the school districts of a county. Our premise is that there is information in the cross-sectional data to be exploited for forecasting. The geographic scale of our research is much smaller and therefore our data are more dynamic than those of related studies. Garcia-Ferrer et al. (1987) and Zellner and Hong (1989) used Bayesian shrinkage methods for pooling time series forecasts – similar to the one we develop in this chapter – to forecast growth in the economies of several countries. Lesage and Magura (1990) applied the same and additional methods to multi-regional data, seven metropolitan areas in Ohio. Our interest is in small local government economies at the sub-regional level. We forecast (and backcast) income tax revenues for forty school districts in Allegheny County, Pennsylvania.

At such a small scale, discrete exogenous events can produce exceptionally noisy data and even large and sharply defined changes in the pattern of time series. Moreover, spatial heterogeneity and underlying spatial processes [see Anselin (1988, pp. 13-14), Haining (1990, pp. 24-25)] can also lead to sharply-defined and widely-ranging spatial structural variations at small scales. For example, three out of five integrated steel mills closed in Allegheny County during the severe 1981-1982 recession (which is included in our case study). The impacts on school district income tax revenues ranged from permanent downturns, to pronounced cyclical downturns, to hardly any impact at all (see Figures 14.4-14.6).

As the geographic scale decreases, it becomes increasingly imperative to use time-varying-parameter forecasting models to accommodate structural change. Variation in regional leading indicators is unlikely to accurately adjust for sub-regional structural changes (given spatial heterogeneity). Furthermore, local government officials (e.g., school district business managers) lack both the needed *sub-regional* leading indicator data and sophistication for multivariate forecasting models that can handle cyclical behavior.

If there are known future discrete events (e.g., a shopping mall opening), then local government officials typically make a judgmental adjustment to a 'business as usual' extrapolative forecast [Lee (1990), Bunn and Wright (1991)]. In this case, officials need a reliable extrapolative forecast as the basis for adjustment. If there are limited data for a structural change at the end of the historical time series, then it is desirable to have a method that quickly and reliably adjusts estimated time trends. Lastly, if there are structural changes earlier in the historical data, it is desirable to have an automatic 'data cleaning' method that ignores irrelevant data.

'Borrowing strength' through cross-sectional pooling of time series data and resulting shrinkage estimation (combining individual location parameter estimates with cross-sectional averages) is highly promising for this context. Shrinkage toward cross-sectional averages can screen out noise and transients, providing more reliable trend lines for extrapolation. Shrinkage can also speed up adaptation to area-wide time series pattern changes by providing more reliable adjustments, selectively, to level and slope components of time trend models.

Such shrinkage can handle moderate spatial heterogeneity, but not the large variations encountered at the sub-regional level. We therefore introduce clustering (similarity or exchangeability determinations) of sub-regional aerial units into reference groups –in our application, low, medium and high sensitivity to economic cycles. Each reference group is modeled independently with pooling within each group.

In Duncan, Gorr and Szczypula (1993), hereafter referred to as DGS, we introduced a model called C-MSKF. We combined the Multi-State Kalman Filter (MSKF) [Harrison and Stevens (1971)] and Conditionally Independent Hierarchical Model (CIHM) [e.g., Kass and Steffey (1989)] to produce the Cross-Sectional Multi-State Kalman Filter (C-MSKF). The MSKF is a univariate time-varying-parameter forecasting method that mixes a no-state-change time series model and several state change time series models using Bayesian posterior probabilities for state changes. The CIHM provides a Bayesian hierarchical, random effects model for cross-sectional pooling and resulting estimates.

Relative to DGS, this chapter makes additional ties to various literatures and presents new empirical results on the performance of the C-MSKF. DGS's forecasts of the school district data provided results for the case of a structural change at the end of the historical time series. This chapter's backcasts of the same data provide results on data cleaning.[1]

---

[1] Note that we use backcasting to extend the value of limited data in obtaining experience with alternative forecasting methods. Many practitioners may wish to do the same, gaining twice the experience as available in forecasting alone, on their limited data sets. The wide-

The outline of the rest of this chapter is as follows. Section two provides a further overview of the C-MSKF and related methods. Section three presents the C-MSKF in detail. Sections four and five cover the case study on backcasting school district revenues. Finally, Section six concludes the chapter and covers future work.

14.2 Literature Review

Earlier attempts to pool time series data using multivariate Kalman Filters were hampered by unrestricted variance covariance matrices for the Kalman Filter observation equation [e.g., Jones (1966)]. In such models, the number of parameters for estimation increases with the square of the number of cross-sectional units. This has the perverse effect of a net loss of information as more cross-sectional units are added; e.g., 100 series of length 15 data points each has 1,500 observations and approximately 5,000 parameters to estimate. In contrast, Harvey (1986, 1989) showed that highly specialized forms of 'homogeneous systems' [e.g., Enns et al. (1982)] draw very little information from cross-sectional data.

More promising is an approach described by Laird and Ware (1982) that has a history tracing back to random-effects models [e.g., Lindley and Smith (1972), Harville (1977)]. The basic assumption is that the probability distribution for the cross-sectional measurements has the same form for each observational unit, but the parameters of that distribution vary over observational units. The distribution of these parameters, or 'random effects,' can be thought of as giving a second stage to the model, and leads to 'borrowing strength'. The mean parameter estimates for the entire cross-section contain information for an individual location's estimates. Thus, as a result of the model, variation in individual locations' estimates is shrunk to the mean by combining cross-sectional mean and individual location estimates. This approach, with Bayesian estimation, has substantial model structure drawing on cross-sectional data, even with strong restrictions on the observation equation variance covariance matrix [see Highfield (1992)].

Garcia-Ferrer et al. (1987), Zellner and Hong (1989), Lesage and Magura (1990) and Highfield (1992) used constant-parameter multivariate time series models with leading indicators and pooling to borrow strength for the constant parameter estimates; thereby obtaining better point estimates and narrower predictive confidence intervals. In contrast, we use time-varying-parameter univariate models with pooling not only for the usual purpose, but also to more quickly and reliably adjust

---

spread availability of forecasting software makes ex ante forecast experiments feasible and valuable. Additionally, backcasting may be useful in estimating missing data in the beginning and other portions of a time series.

univariate time series model parameters for structural changes. While the constant parameter models are estimated using batch processing methods, we use a Kalman Filter and therefore recursive calculations to adjust to time series pattern changes.

We note that Zellner and Hong (1989) concluded with plans to extend their multivariate model to time-varying-parameter forecasts using a simple Kalman Filter, and Zellner (1992) reported on initial successes with such forecasts. Also, Gamerman and Migon (1993) have provided a general framework incorporating the multivariate Kalman Filter and Bayesian pooling. They, however, do not revise variance estimates. Updating variance estimates is important in order to respond to structural change: the variances are used directly in shrinkage estimates.

In regard to multivariate versus univariate modeling, note further that there is no cumulative evidence that constant parameter multivariate time series models outperform constant or time-varying parameter univariate models for short or medium-term forecasting horizons [Armstrong (1978), Fildes (1985), Makridakis et al. (1993)]. Nevertheless, Zellner's 'sophisticatedly simple' multivariate models [e.g., Zellner and Hong (1989)] are attractive, especially in time-varying form and for the multi-regional scale and larger.

The C-MSKF is best described as a sequential and recursive estimation *system* with four modular components: 1) transformation of time series data to remove cross-sectional variation due to scale differences; 2) clustering to yield independent reference groups, based on similarity and exchangeability; 3) Kalman Filter time series technique; and 4) cross-sectional pooling to borrow strength and increase reliability of estimate.

For transformation, we have standardized each time series by subtracting the historical sample mean, dividing by the sample standard deviation of each series, and adding a constant to make the standardized values positive for convenience. Zellner [e.g., Zellner and Hong (1989)] has used logs of growth rates for the same purpose.

We have clustered data using a school district expert's judgment. Since the underlying spatial structure of regions and subregions, transportation networks, locations of work versus residences, etcetera, is likely to be relatively slow changing, it seems promising that cross-sectional multivariate clustering methods using available census and other data could provide appropriate reference groups. Alternatively and more directly, one could base grouping on similarity of time series movement.

We use the MSKF because it was designed for estimating time series pattern changes. We could just as easily, however, use a single-state Kalman Filter in univariate or multivariate form. The MSKF should adjust more quickly and accurately than a single state Kalman Filter. Fildes (1983) has made the most comprehensive comparative study of the MSKF and simpler time series methods (deseasonalized simple exponential smoothing), and found little difference in their performance. However, he did not segment forecast error data by change versus no change in the historical period. We suspect that the MSKF is superior for series with structural changes (the rarer case in the 111 data series used), but somewhat poorer for no change cases, being overreactive to transients.

Our estimation system is sequential and recursive. We start at the beginning of the time series. Given priors for each time series' distribution, we apply the MSKF independently to each time series for the first time period, thereby obtaining Bayesian posterior distributions. The CIHM takes these distributions as priors, and makes a cross-sectional adjustment to produce posteriors for each time series. These posteriors become priors for the next time period and time series adjustment. This process continues, sequentially and recursively switching between time series and cross-sectional perspectives, through to the last time period.

14.3 The C-MSKF Model: Time Series Prediction with Spatial Adjustments

The objective of this section is to present the models and estimation methods comprising the C-MSKF. This objective is approached in three steps. First, subsection one, briefly describes the MSKF. Second, subsection two describes the C-MSKF that allows prediction to 'borrow strength from its neighbors'. Third, subsection three ties the two models together to produce the C-MSKF model.

14.3.1 Multi-State Kalman Filter

Harrison and Stevens (1971) proposed the MSKF as a univariate time series forecasting method using Bayesian inference. The method can quickly recognize and respond appropriately to sudden step and slope changes in time series as was shown by Harrison and Stevens (1971) using artificial data, and by Bomhoff and Kool (1983) using quarterly Swiss money supply data. While the mechanisms of the MSKF methodology are Bayesian, the assessment of prior parameter estimates is through heuristic default values rather than through subjective judgment [Ameen and Harrison (1984)].

The approach taken by Harrison and Stevens to estimating a system with structural changes was to construct a small number, $J = 4$, of univariate Dynamic Linear Models (DLMs) for a given time series i, $i = 1, ..., I$ [West and Harrison (1989)]. In particular, it is assumed that the linear time series model, at any given time, is in one of four states: no change, step change, slope change, or transient (see Figure 14.1). The DLMs differ from each other in the values of the measurement error and process error variances.

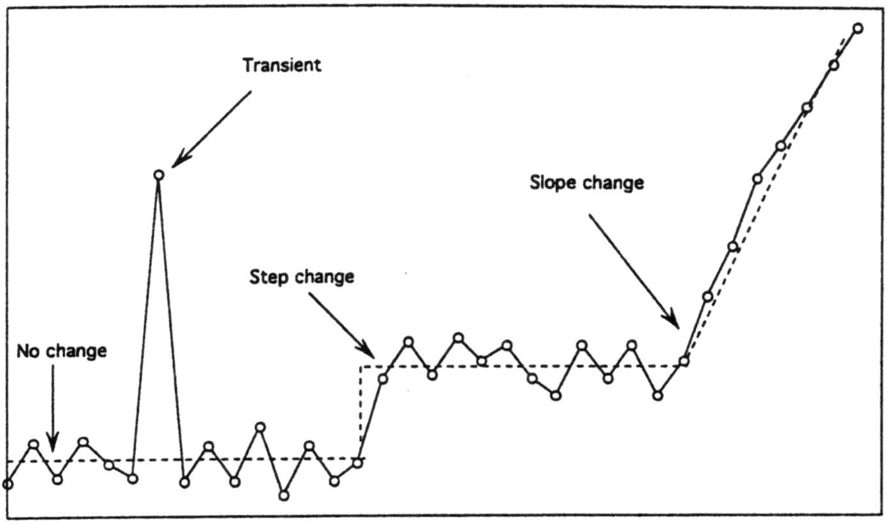

Figure 14.1: Illustration of the four states used in the Multi-State Kalman Filter.

The resulting model for each state $j = 1, ..., 4$, time t ($t = 1, ..., T$), and unit i can be represented by the following linear equations with normally distributed error terms. The observation equation:

$$Y_{it} = T_{it} + \varepsilon_{it} \tag{1}$$

with $\varepsilon_{it} \mid j \sim N(0, V_{\varepsilon^{ji}})$; the system equations:

$$T_{it} = T_{i,t-1} + S_{it} + \gamma_{it} \tag{2}$$

where $\gamma_{it} \mid j \sim N(0, V_{\gamma^{ji}})$, and

$$S_{it} = S_{i,t-1} + \rho_{it} \tag{3}$$

where $\rho_{it} \mid j \sim N(0, V_{\rho^{ji}})$, and the prior:

$$(T_{io}, S_{io} \mid Y_{io}) \sim \sum_{j=1}^{J} q_{jio} N((m_{jio}, b_{jio}), C_{jio}) \tag{4}$$

where Y_{it} = observation for unit i at time t, T_{it} = expected value of Y_{it} ('current' level of Y_{it}), S_{it} = time trend slope (growth rate of Y_{it}), $\varepsilon_{it} \mid j$, $\rho_{it} \mid j$ = serially uncorrelated and mutually independent error terms, and q_{jio}, m_{jio}, C_{jio} are the initial values.

In general m_{jit}, b_{jit} are the means of T_{it} and S_{it} in state j, and:

$$C_{jit} = \begin{bmatrix} V_{jit}[1,1] & V_{jit}[1,2] \\ V_{jit}[1,2] & V_{jit}[2,2] \end{bmatrix}$$

is the covariance matrix of (T_{it}, S_{it}) in state j for unit i at time t, and q_{jit} is the posterior probability of unit i being in state j in period t. The notation used here is based on that of Harrison and Stevens (1971) and Bomhoff and Kool (1983).

The observation equation (1) describes the generation of the observations Y_{it} from a given time-dependent parameter vector (T_{it}, S_{it}). The system equations (2) and (3) describe the evolution of the parameter vector. The vectors (T_{it}, S_{it}) give parameters of the system needed for predicting future observations. Since they are not known or observable, however, it is necessary to make inferences about the (T_{it}, S_{it}).

Given the observations $\{Y_{ik}\}$, $k \leq t$-1, the joint distribution of ($T_{i,t-1}$, $S_{i,t-1}$) for unit i at time t-1 is a convex mixture of bivariate normal distributions defined for each of the J states and is given by:

$$\left(T_{i,\,t-1},\, S_{i,\,t-1} \mid D_{i,\,t-1}\right) \sim \sum_{j=1}^{J} q_{ji,\,t-1}\, N\left((m_{ji,\,t-1},\, b_{ji,\,t-1}),\, C_{ji,\,t-1}\right) \tag{5}$$

where $D_{i,\,t-1} = \left(Y_{i,\,t-1},\, D_{i,\,t-2}\right)$ denotes information at time t-1.

After the observation of Y_{it} , we apply the Kalman Filter algorithm to each of the J current components of the prior distribution J times for J states in time t (since each of the components from time t-1 can be in any state at time t). This operation creates J^2 = 16 components of posterior normal distributions:

$$\left(T_{it},\, S_{it} \mid D_{it}\right) \sim \sum_{k=1}^{J} \sum_{j=1}^{J} p_{(jk)it}\, N\left((m_{(jk)it},\, b_{(jk)it}),\, C_{(jk)it}\right) \tag{6}$$

where $p_{(jk)it}$ = posterior probability with respect to observation Y_{it} that the process was in state j in the period t-1 and is currently in state k.

The Kalman Filter recursive equations for the terms in the above formulae are:

$$m_{(jk)it} = m_{ji,\,t-1} + b_{ji,\,t-1} + A_{(jk)it}[1]\, e_{jit} \tag{7}$$

$$b_{(jk)it} = b_{ji,\,t-1} + A_{(jk)it}[2]\, e_{jit} \tag{8}$$

$$V_{(jk)it}[1,1] = r_{(jk)it}[1,1] - \left(A_{(jk)it}[1]\right)^2 V_{e_{(jk)it}} \tag{9}$$

$$V_{(jk)it}[1,2] = r_{(jk)it}[1,2] - A_{(jk)it}[1] A_{(jk)it}[2] V_{e_{(jk)it}} \tag{10}$$

$$V_{(jk)it}[2,2] = r_{(jk)it}[2,2] - \left(A_{(jk)it}[2]\right)^2 V_{e_{(jk)it}} \tag{11}$$

$$p_{(jk)it} = s\left(2\pi V_{e_{(jk)it}}\right)^{-1/2} \exp\left[\frac{-\left(Y_{it} - m_{ji,\,t-1} - b_{ij,\,t-1}\right)^2}{2 V_{e_{(jk)it}}}\right] \pi_j q_{ji,\,t-1} \tag{12}$$

where π_j is the probability of occurrence of state j (constant over time for each state j), and s = probability normalization factor,

$$e_{jit} = Y_{it} - \left(m_{ji,\,t-1} + b_{ji,\,t-1}\right)$$

$$A_{(jk)it}[1] = \frac{r_{(jk)it}[1,1]}{V_{e_{(jk)it}}}$$

$$A_{(jk)it}[2] = \frac{r_{(jk)it}[1,2]}{V_{e_{(jk)it}}}$$

and where:

$$V_{e_{(jk)it}} = r_{(jk)it}[1,1] + V_{e_{ki}}$$

$$r_{(jk)it}[1,1] = V_{ji,\,t-1}[1,1] + 2V_{ji,\,t-1}[1,2] + V_{ji,\,t-1}[2,2] + V_{\gamma^{ki}} + V_{\rho^{ki}}$$

$$r_{(jk)it}[1,2] = V_{ji,\,t-1}[1,2] + V_{ji,\,t-1}[2,2] + V_{\rho^{ki}}$$

$$r_{(jk)it}[2,2] = V_{ji,\,t-1}[2,2] + V_{\rho^{ki}}$$

The e_{jit} is the one-step-ahead forecast error. An element of the vector $A_{(jk)it}$ is analogous to the 'smoothing factor' of exponential smoothing systems [Brown (1962)] and regulates the proportion of the forecast signal e_{jit} used in updating the mean $m_{(jk)it}$ (see equations (7) and (8)). More precisely, since $A_{(jk)it}$ is time varying and adaptive, its elements are analogous to the Trigg and Leach (1967) tracking signal. The values of the variances adjust the learning rate (smoothing) vector $A_{(jk)it}$ for the particular state at hand. For example, in the presence of a time series step change, the elements of $A_{(jk)it}$ adjusting the level term of the observation equation increase relative to those elements of $A_{(jk)it}$ adjusting the time trend slope.

In order to achieve the form required by MSKF at each time period, we must collapse J^2 components into a J - component normal distribution given by:

$$(T_{it}, S_{it} \mid D_{it}) \sim \sum_{k=1}^{J} q_{kit} N\left((m_{kit}, b_{kit}), C_{kit}\right) \tag{13}$$

Equations for collapsing densities are the following [for details and rationale see Bomhoff and Kool (1983)]:

$$q_{kit} = \sum_{j=1}^{J} p_{(jk)it}$$

$$m_{kit} = \sum_{j=1}^{J} \frac{p_{(jk)it} m_{(jk)it}}{q_{kit}}$$

$$b_{kit} = \sum_{j=1}^{J} \frac{p_{(jk)it} b_{(jk)it}}{q_{kit}}$$

$$V_{kit}[1,1] = \sum_{j=1}^{J} p_{(jk)it} \frac{V_{(jk)it}[1,1] + \left(m_{(jk)it} - m_{kit}\right)^2}{q_{kit}}$$

$$V_{kit}[1,2] = \sum_{j=1}^{J} p_{(jk)it} \frac{V_{(jk)it}[1,2] + \left(m_{(jk)it} - m_{kit}\right)\left(b_{(jk)it} - b_{kit}\right)}{q_{kit}}$$

$$V_{kit}[2,2] = \sum_{j=1}^{J} p_{(jk)it} \frac{V_{(jk)it}[2,2] + \left(b_{(jk)it} - b_{kit}\right)^2}{q_{kit}}$$

These updating and collapsing equations are used independently for each cross-sectional unit i and repeated at every time period $t = 1, 2, \ldots, T$.

14.3.2 Spatial Adjustment via Hierarchical Random Effects Model

Many forecasting problems have multiple time series generated in the context of social and environmental systems with spatial structure in the forecasted variable. Forecasting for a particular observational unit should achieve better accuracy if effective use of spatial context is made. This would obviously be the case if the units could be put together into homogeneous groups and considered identical and subject to independent noise. Then simple statistical aggregation could be employed. Even with substantial variation among observational units, however, we can still gain some useful information (and improve forecasts). In this section we construct a way to 'borrow strength' from neighbors in forecasting for several parallel observational units. We first illustrate the approach in the simplest case. We then develop this approach for the parameters of the MSKF model at a specific time t.

The simple hierarchical model: An effective way of both incorporating heterogeneity into models for observational units and enabling the 'borrowing strength' from the relatedness of these units, is through a hierarchical or 'random effects' model [see, e.g., Laird and Ware (1982)]. The central idea of a hierarchical statistical model is that the data generating process is governed by a parameter vector that in turn is drawn from a probability distribution whose parameters (i.e., hyperparameters of the process) are not completely known. This structure is related to parametric empirical Bayes concepts [Efron and Morris (1973), Morris (1983)]. The Bayesian version specifies a probability distribution for the hyperparameters [Deely and Lindley (1981)]. The hierarchical model simplifies when the observable quantities are assumed to be independent conditionally on the parameter vector. The unit-specific parameter vectors are taken to be independent and identically distributed, conditionally on the values of the hyperparameters, yielding the conditionally independent hierarchical model (CIHM).

A time series hierarchical model: For the example developed in this chapter we have time series data collected on each of several similar observational units. Specifically, there are I cross-sectionally similar observational units and corresponding univariate time series $\{Y_{it}: t = 1, ..., T\}$, $i = 1, ..., I$. Notice that time t is fixed for the whole hierarchical analysis. We develop a three-stage hierarchical structure to relate the I time series with a random effects model. At the lowest stage, for each individual observational unit i, a time series $\{Y_{it}\}$, is generated according to the DLM's given by Equations (1) – (3). At this stage and time, all parameters –the (T_{it}, S_{it})– are considered fixed but generally different for each time series i. Each time series i is taken to be conditionally independent given these parameters.

At the middle stage, the (T_{it}, S_{it}) are generated, according to Equation (13). The parameters (m_{jit}, b_{jit}), and C_{jit} of the distributions are considered fixed but generally different for each unit i and each state j. Over i (with fixed t and fixed j), the parameters (T_{it}, S_{it}) are taken to be conditionally independent given the parameters (m_{jit}, b_{jit}), and C_{jit}.

At the highest stage, the parameters (m_{jit}, b_{jit}), and C_{jit} are generated according to hyperparameters λ_{jt1} and λ_{jt2}. The hyperparameters are considered fixed for all units i and different for each j and every t. This stage is the key to 'borrowing strength from neighbors' in forecasting. The CIHM assigns probability distributions to the hyperparameters λ_{jt1} and λ_{jt2} and to the parameters (m_{jit}, b_{jit}), and C_{jit}. The means of the respective posterior distributions of (m_{jit}, b_{jit}) can then be used as posterior values of the parameters given the values of the hyperparameters λ_{jt1} and λ_{jt2} and the vector of observations.

14.3.3 CIHM Method

The following discussion develops the spatial adjustment for the distribution of the current level T_{it} of the observations Y_{it} at given time t for every $i = 1, 2, ..., I$ and each state $j = 1, ..., J$. An analogous adjustment is performed for the parameters of the distribution of the slope S_{it}. Consider the cross-section of parameters T_{it}, T_{2t}, ..., T_{It}. We assume that, in the middle stage, each T_{it} is normally distributed with mean m_{jit} and variance $V_{jit}[1,1]$ for every $i = 1, 2, ..., I$. The highest stage is divided into two parts. In the first part, we treat the variance $V_{jit}[1,1]$ as known and assume that each mean m_{jit} is normally distributed conditionally on the values of hyperparameter $\lambda_{jt1} = (\mu_{jt}, \tau_{jt}^2)$ for every $i = 1, 2, ..., I$. In the second part, we treat the mean m_{jit} as known and assume that the variance $V_{jit}[1,1]$ is inverse chi-squared distributed conditionally on hyperparameter $\lambda_{jt2} = (\vartheta_{jt}, \upsilon_{jt})$ for every $i = 1, 2, ..., I$.

Consistent with Kass and Steffey (1989), we implement the CIHM method of incorporating spatial adjustments of the means m_{jit} of T_{it} by using maximum likelihood to estimate all the hyperparameters of the last stage. Having the estimates, the expected values $E(m_{jit} | \boldsymbol{T}_t, \mu_{jt}, \tau_{jt}^2)$ can be approximated with $E(m_{jit} | T_{it}, \mu_{jto}, \tau_{jto}^2)$ where μ_{jto} and τ_{jto}^2 are the maximum likelihood estimates of the hyperparameters μ_{jt} and ϑ_{jt}^2 and $\boldsymbol{T}_t$ denotes the vector $(T_{1t}, T_{2t}, ..., T_{It})$ of cross-sectional units at time t. $E(m_{jit} | T_{it}, \mu_{jto}, \tau_{jto}^2)$ is the spatially adjusted value of the mean m_{jit} for each $i = 1, ..., I$ and for each state $j = 1, ..., J$ at time t. Similarly, the expected value $E(V_{jit}[1,1] | \boldsymbol{T}_t, \vartheta_{jt}, \upsilon_{jt})$ can be estimated with $E(V_{jit}[1,1] | T_{it}, \vartheta_{jto}, \upsilon_{jto})$ where ϑ_{jto} and υ_{jto} denote the MLEs of the hyperparameters ϑ_{jt} and υ_{jt} for each state j at time t.

The calculations of the posterior means follow. In the first part of the highest stage, the random means $m_{j1t}, m_{j2t}, ..., m_{jIt}$ are assumed to be i.i.d. $N(\mu_{jt}, \vartheta_{jt}^2)$. Given the estimated values of m_{jit} $(i = 1, ..., I)$ from the MSKF, the MLEs $(\mu_{jto}, \tau_{jto}^2)$ of the hyperparameters $(\mu_{jt}, \vartheta_{jt}^2)$ are simply the sample mean and the sample variance of $m_{j1t}, m_{j2t}, ..., m_{jIt}$. Furthermore, the spatially adjusted posterior distribution of m_{jit} given T_{it} will also be a normal distribution. It is known [DeGroot (1986, p. 410)] that the means of the posterior distributions are:

$$E\left(m_{jit} \mid T_{it}, \mu_{jto}, \tau^2_{jto}\right) = \frac{\mu_{jto} * V_{jit}[1,1] + T_{it} * \tau^2_{tjo}}{\left(V_{jit}[1,1] + \tau^2_{jto}\right)} \tag{14}$$

Bayesian estimation of the hyperparameters and time series means in the CIHM part of the model results in the intuitively appealing update equation (14). The revised means m_{jit} are composed of weighted averages of two terms: 1) the overall average, μ_{jto}, of the means of the cross-sectional distributions of the sample for state j and 2) the mean current level, T_{it}, of observation unit i from the time series. The weights are inversely proportional to the estimated variance of each component. Thus if the variance of the overall cross-sectional mean is less than that of the individual time series mean, the cross section has the greater influence. This is effective for the case where most or all cross-sectional units experience a common time series pattern change. The opposite case places emphasis on the time series trend of each observational unit. Thus, the C-MSKF automatically emphasizes the correct perspective: time series or cross-sectional.

Estimates of the spatially adjusted variances $V_{jit}[1,1]$ (the second part of the highest stage) are achieved in an analogous way by finding MLEs of their hyperparameters and using the resulting estimated values to calculate the posterior distributions of the variances. The variances are taken to be inverse chi-squared distributed with density function as follows [Lee (1986, p. 236)]:

$$p\left(V_{jit}[1,1] \mid \vartheta_{jt}, \upsilon_{jt}\right) = \frac{\vartheta_{jt}^{\upsilon_j/2}}{2^{\upsilon_j/2}\,\Gamma\left(\frac{\upsilon_{jt}}{2}\right)} \left(V_{jit}[1,1]\right)^{-\upsilon_j/2} \exp\left(\frac{-\vartheta_{jt}}{2\,V_{jit}[1,1]}\right) \tag{15}$$

where ϑ_{jt} and υ_{jt} are the hyperparameters and $\Gamma(u)$ is the gamma function. Finding MLEs ϑ_{jto} and υ_{jto} for ϑ_{jt} and υ_{jt} is achieved by solving the likelihood equations:

$$\vartheta_{jt} = \frac{I\,\upsilon_{jt}}{\sum_{i=1}^{I} \frac{1}{V_{jit}[1,1]}} \tag{16}$$

and

$$\frac{\Gamma'\left(\frac{\upsilon_{jt}}{2}\right)}{\Gamma\left(\frac{\upsilon_{jt}}{2}\right)} = \frac{1}{2}\left(\log\vartheta_{jt} - \log 2 - \left(\frac{1}{I}\right)\sum_{i=1}^{I} \log V_{jit}[1,1]\right) \tag{17}$$

Finding the solution to the two simultaneous equations requires an evaluation of the di-gamma function $\Psi(u) = \Gamma'(u) / \Gamma(u)$ [DeGroot (1986, p. 350)]. We employ an approximation of Ψ given by Kotz and Johnson (1985, p. 373) and used by Bernardo (1976). Since $\Psi(u)$ is a concave increasing function, it is easy to find the solution to the simultaneous equations numerically. Having the estimates ϑ_{jto} and υ_{jto}, we can estimate the adjusted value of the variance $V_{jit}[1,1]$ from its posterior distribution for unit i. The posterior distribution is also the inverse chi-squared distribution with parameters $\vartheta_{jt}' = (\vartheta_{jto} + (T_{it} - m_{jit})^2\,)$ and $\upsilon_{jt}' = (\upsilon_{jto} + 1)$. Finally the mean of the posterior distribution for $i = 1, ..., I$ and at time t and for state j is:

$$E\left(V_{jit}[1,1] \mid T_{it}, \vartheta_{tjo}, \upsilon_{jto}\right) = \frac{\vartheta_{jto} + \left(T_{it} - m_{jit}\right)^2}{V_{jto} - 1} \tag{18}$$

Thus, the new estimate of the variance $V_{jit}[1,1]$ is approximately equal to its prior mean increased by the variation of the T_{it} from its mean represented by the term $(T_{it} - m_{jit})^2$.

14.3.4 C-MSKF

We put forth C-MSKF as a forecasting method that integrates the MSKF model described in sections 14.3.1 and 14.3.2 with the hierarchical random effects CIHM model of section 14.3.3. C-MSKF takes advantage of both the time series structure through MSKF and the 'borrowing strength' from the similar cross-sectional units through the CIHM. In doing this, it works sequentially through the data for $t = 1, ..., T$.

Figure 14.2 provides a schematic representation of an iteration at time t of the C-MSKF method. It consists of an MSKF step followed by a CIHM step. Beginning with time t-1, we apply the MSKF method separately to each time series i ($i = 1, ..., I$) to update the state vector $(T_{i,t-1}, S_{i,t-1})$ given an observation $Y_{i,t-1}$ on that unit. Then, the CIHM method is used to make spatial adjustments to the updated parameters of each time series model based solely on the cross-sectional data for the given time period. The cross-sectional data represents a group of units having certain common characteristics. The output from the CIHM step is then used as a starting point for the MSKF revision of each time series model at time t, etcetera. The recursive parameter updates ends when the observations Y_{iT} at the last historical point are processed.

We can calculate the forecasted value based on the mixture of the distributions of (T_{it}, S_{it}) for each j with their probabilities q_{jit} available at the forecast origin T. The k-step-ahead forecast (i.e., expected value of the forecast distribution) of the observations Y_{T+k} can be calculated using formulas (7) and (8), with $e_{ji,T+k} = 0$, as a weighted sum $\Sigma_j\, q_{jiT}\; m_{ji,T=k}$ where $m_{ji,T=k} = m_{jiT} + k\, b_{jiT}$.

Before the complete method can be used it is necessary to set initial values of some parameters. The starting values for the means of the distributions for each unit i and state j are the first data point for the m_{jio} and difference of the second and the first data point for the b_{jio}. The initial value of C_{jio} for unit i is derived from the individual time series data based on the algorithm provided by Bomhoff and Kool (1983, p. 238) and on certain parameters (variance law) given in Harrison and Stevens (1971). This specification of starting values is somewhat arbitrary. Using Monte Carlo methods, DGS, however, showed that the C-MSKF is insensitive to the starting values. One alternative approach would make use of the modified Kalman filter of Ansley and Kohn (1985) which puts a diffuse prior on the starting values.

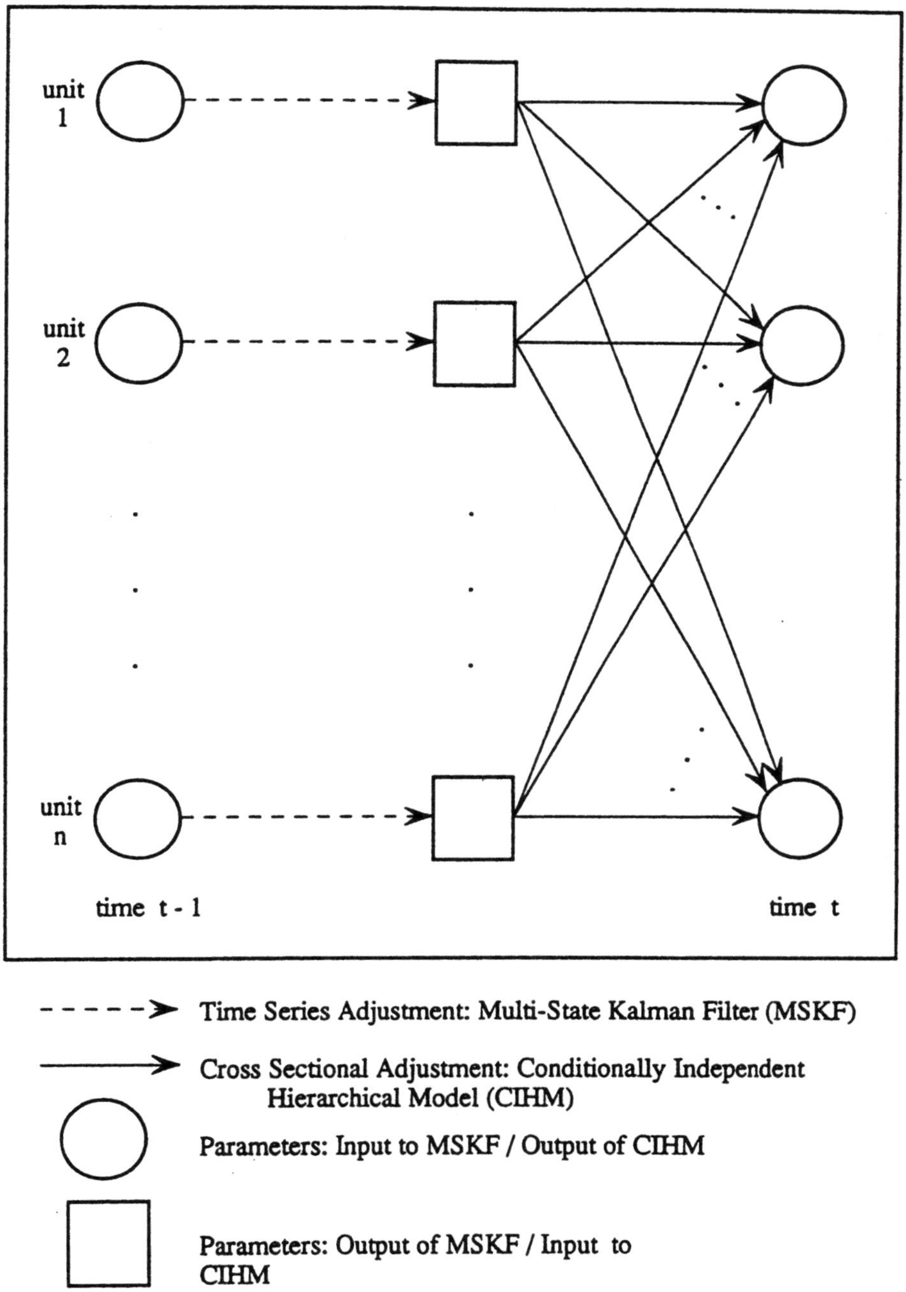

Figure 14.2: Representation of the cross-sectionally Adjusted Multi-State Kalman Filter from time t - 1 to time t.

14.4 Case Study and Observational Setting

14.4.1 Data

The data used are the same as used in DGS annual income tax collections from the forty school districts over a fifteen-year period, from 1972 to 1986, obtained from the Pennsylvania Department of Education. Since the income tax collection rate is constant for each school district over the study period, we used total income tax collections instead of the income tax base that is usually used for forecasting. Income tax collections have the characteristic of being strongly influenced by economic cycles.

Figure 14.3 is a map of Allegheny County with school-district boundaries and a school expert's judgmental classification of school districts as having low, medium, or high sensitivity to economic recessions. Figures 14.4 through 14.6 show overlaid time series plots, from all 15 years of data available, by class. The data in each class are standardized, as described above, to remove scale differences between school districts' revenue and thereby facilitate CIHM–based estimates. The primary forecast simulations carried out by DGS used 1983 as the last historical data point; forecasted one, two, and three years ahead; and successively dropped the oldest data points until there were only four historical observations.

The time series data for each school district are similarly divided into two periods for backcasting in this chapter: 1) the historical period (1986 - 1975), available for estimation of the method parameters and 2) the backcast period (1974 - 1972), for which backcasts of one, two, and three year horizons are performed. As in DGS we varied the estimation sample size by systematically dropping the earliest historical data points, one at a time. These historical data sets had a maximum of twelve data points down to a minimum of only four data points.

The historical period of the backcast data, from 1986 through 1975, has a change in time trend slope in 1981 for the low sensitivity group (Figure 14.4) and turning points in the same year for the other two groups (Figures 14.5 and 14.6). The data for 1986 through 1981 are thus irrelevant for determining the appropriate time trend slope for backcasting, so that the challenge for forecasting methods is to forget the 1986-1981 trend as soon as possible, while quickly learning the new 1981-1975 trend. The case is even more challenging for the high sensitivity group (Figure 14.6). This group was the only of the three visibly impacted by the 1974-1975 economic recession, seen as a temporary flattening of growth from 1974-1976. The situation this presents is one where the last historical data point (1975) signals a slowing of trend which, however, only persists one period into the backcast period (1974), after which the overall 1981-1976 trend resumes.

Relative to MSKF, we expect the C-MSKF to more quickly forget the 1986-1981 trend for simulated backcasts including historical data from this period. Furthermore, we expect the C-MSKF to be better on the first step ahead backcast, for 1974. It is difficult to speculate on relative performance for the two and three year ahead forecasts.

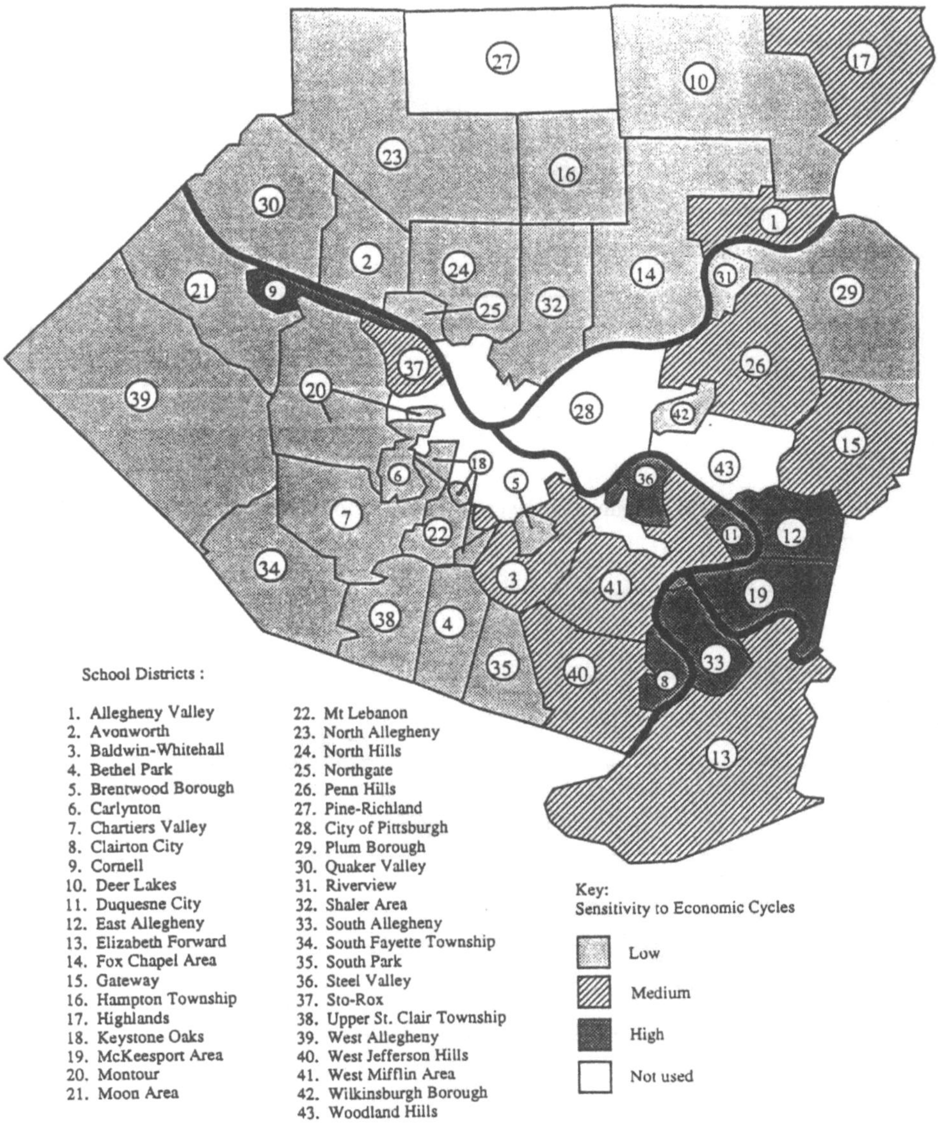

Figure 14.3: Map of school districts in Allegheny county, Pennsylvania showing cross-sectional reference groups from expert judgment regarding sensitivity to economic cycles.

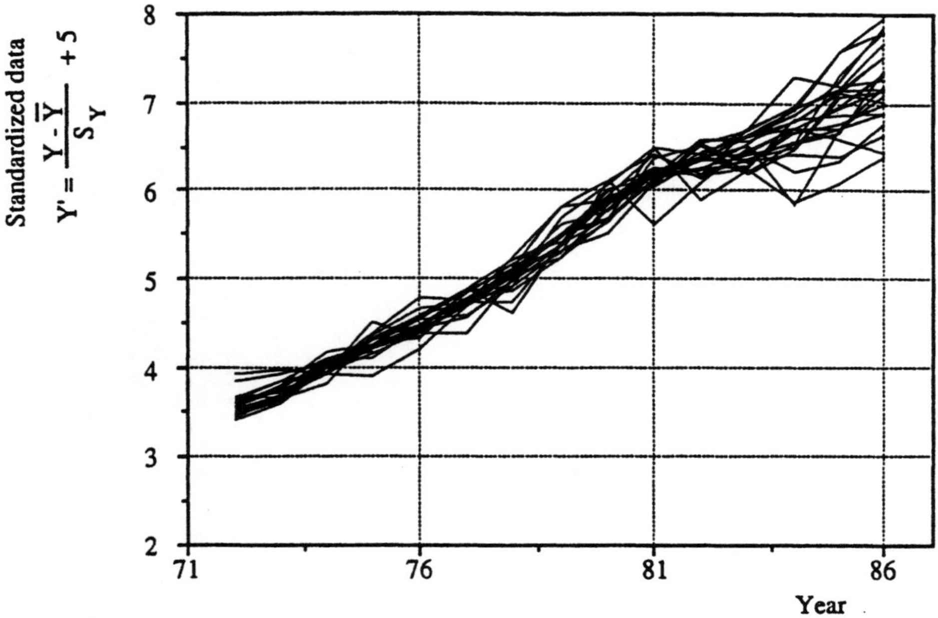

Figure 14.4: Standardized income tax revenue for the 'low' sensitivity cross-sectional reference group.

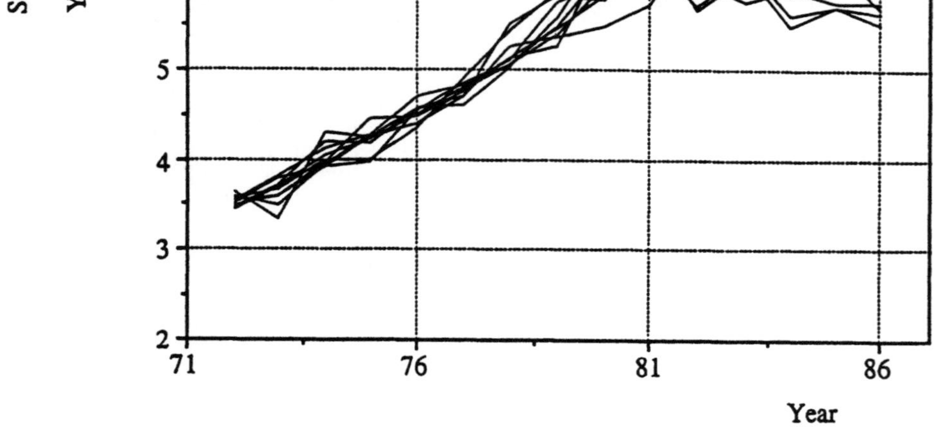

Figure 14.5: Standardized income tax revenue for the 'medium' sensitivity cross-sectional reference group.

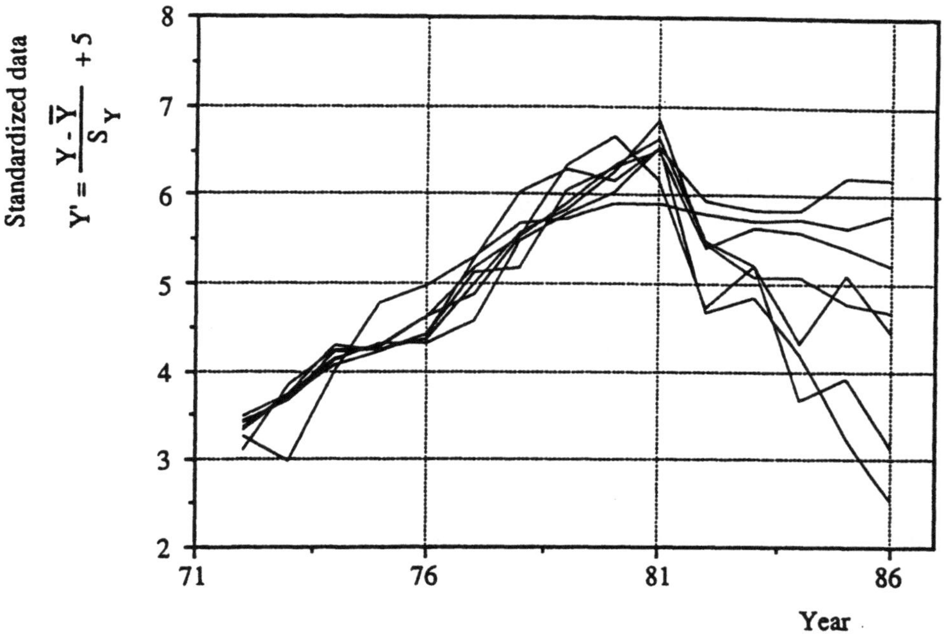

Figure 14.6: Standardized income tax revenue for the 'high' sensitivity cross-sectional reference group.

14.4.2. Treatments

First, we applied the MSKF separately to each individual school district. Second, we used the C-MSKF independently for each of the three economic-cycle based cross-sectional reference groups.

The overall goal was to investigate the advantage of the C-MSKF over the univariate MSKF method. It was anticipated that the C-MSKF method would perform better than the simple univariate MSKF and that the C-MSKF would bring increasing improvements in backcasting accuracy over the univariate MSKF as the sensitivity of school districts to economic cycles increases, backcast horizon increases, and as time series length decreases.

The two Bayesian methods were compared in terms of their accuracy as represented by the mean absolute percentage errors (MAPE) and bias represented by the mean percentage errors (MPE) computed by backcast horizon (one, two, and three years ahead):

$$MAPE_{T+k} = \left(\frac{100}{I}\right) \sum_{i=1}^{I} \frac{| Y_{i,T+k} - Y_{i,T+k}^{f} |}{Y_{i,T+k}} \tag{19}$$

$$MPE_{T+k} = \left(\frac{100}{I}\right) \sum_{i=1}^{I} \frac{Y_{i,T+k} - Y^f_{i,T+k}}{Y_{i,T+k}} \tag{20}$$

where T = backcast origin, k = backcast horizon (1, 2, or 3), $Y_{i,T+k}$ = data value for origin T, horizon k, and unit i, $Y^f_{i,T+k}$ = backcast for origin T, horizon k, and unit i, and I = number of observational units in the cross-sectional reference group.

14.5 Results

First, note that the results of this chapter stem from a single case study, and so must be interpreted with some caution. Like all empirical research, an accumulation of evidence from many trials is needed. In favor of the empirical work that we have done, however, is that our case study has substantial and reasonably well-known spatial structure and time series structural changes. Furthermore, we have designed computational experiments that clearly present pertinent conditions for testing time series methods that draw on cross-sectional data that include structural changes.

Figure 14.7 presents the comparison of backcasting accuracy of the MSKF and C-MSKF by time series length, horizon, and cross-sectional reference group.[2] As was expected, the MAPE generally increases for both methods as the horizon increases from one to three years.

The C-MSKF method has lower MAPE than the univariate MSKF in every case. Also, in support of the utility of the C-MSKF method, the gap between the methods increases as the horizon increases, sensitivity to economic cycle increases, and the length of the time series data decreases (except for the case of high sensitivity districts). The MAPEs for the C-MSKF results are statistically better (lower) in 52 percent of the cases (42 out of 81 cases for the three groups combined), according to the 5 percent significance level of the Wilcoxon matched pairs, signed rank, two-tailed test.

The two methods' MAPEs are closer to each other and lower in the 'low' cross-sectional reference group than they are in other groups. Nevertheless, for short historical time series there is a significant improvement in the accuracy. This suggests that spatial information can compensate for shortness of time series in achieving higher backcast accuracy.

---

[2] Notice that part of Figure 7 representing backcasting accuracy for the case of high sensitivity to economic cycle and backcasting horizon of three years is truncated on the vertical scale. We used the same scale for each graph of Figure 7 in order to facilitate direct comparison between the parts (across backcasting horizons and sensitivity to economic cycle) and to clearly recognize the differences. For the same reason we also truncated Figure 8.

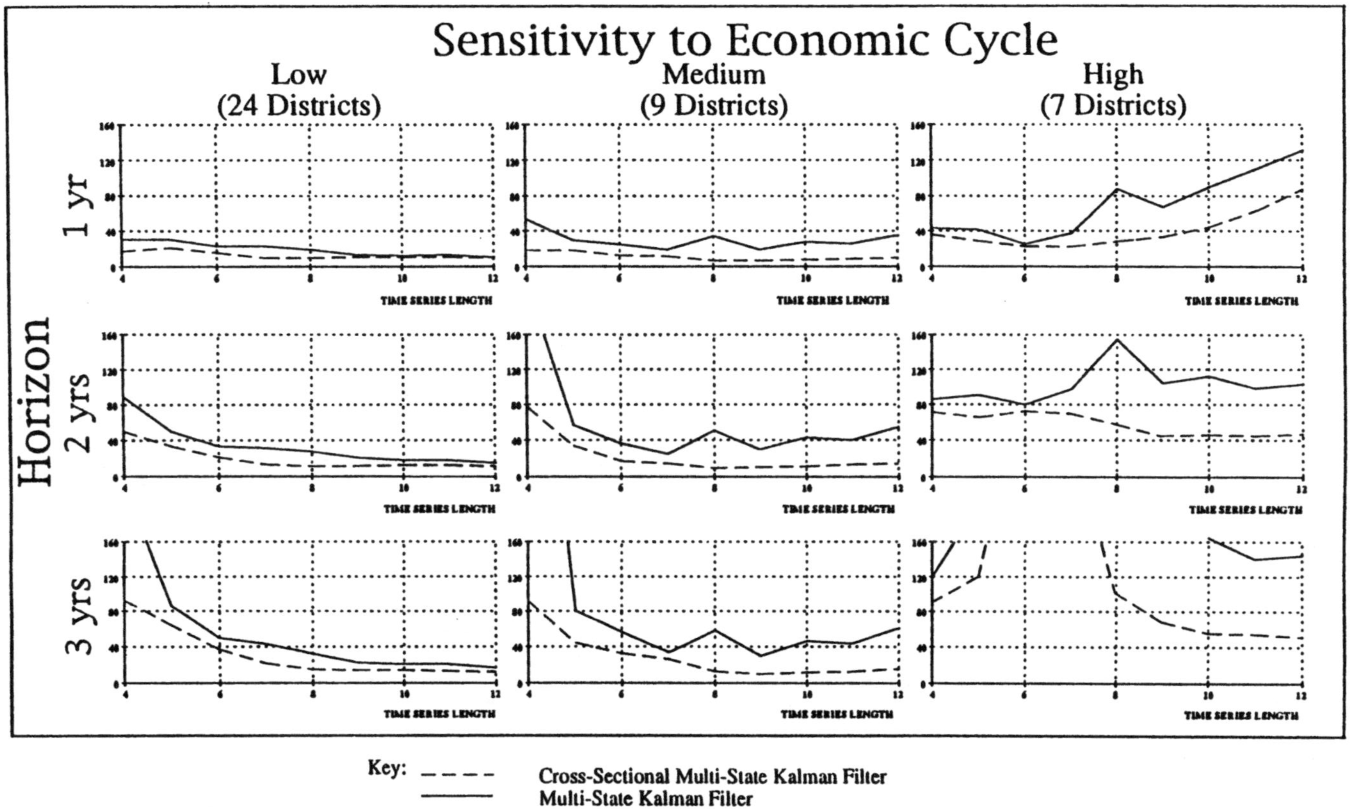

Figure 14.7: MAPE backcasting accuracy (1975 origin) of the Multi-State Kalman Filter and cross-sectional Multi-State Kalman Filter by time series length, horizon and cross sectional reference group.

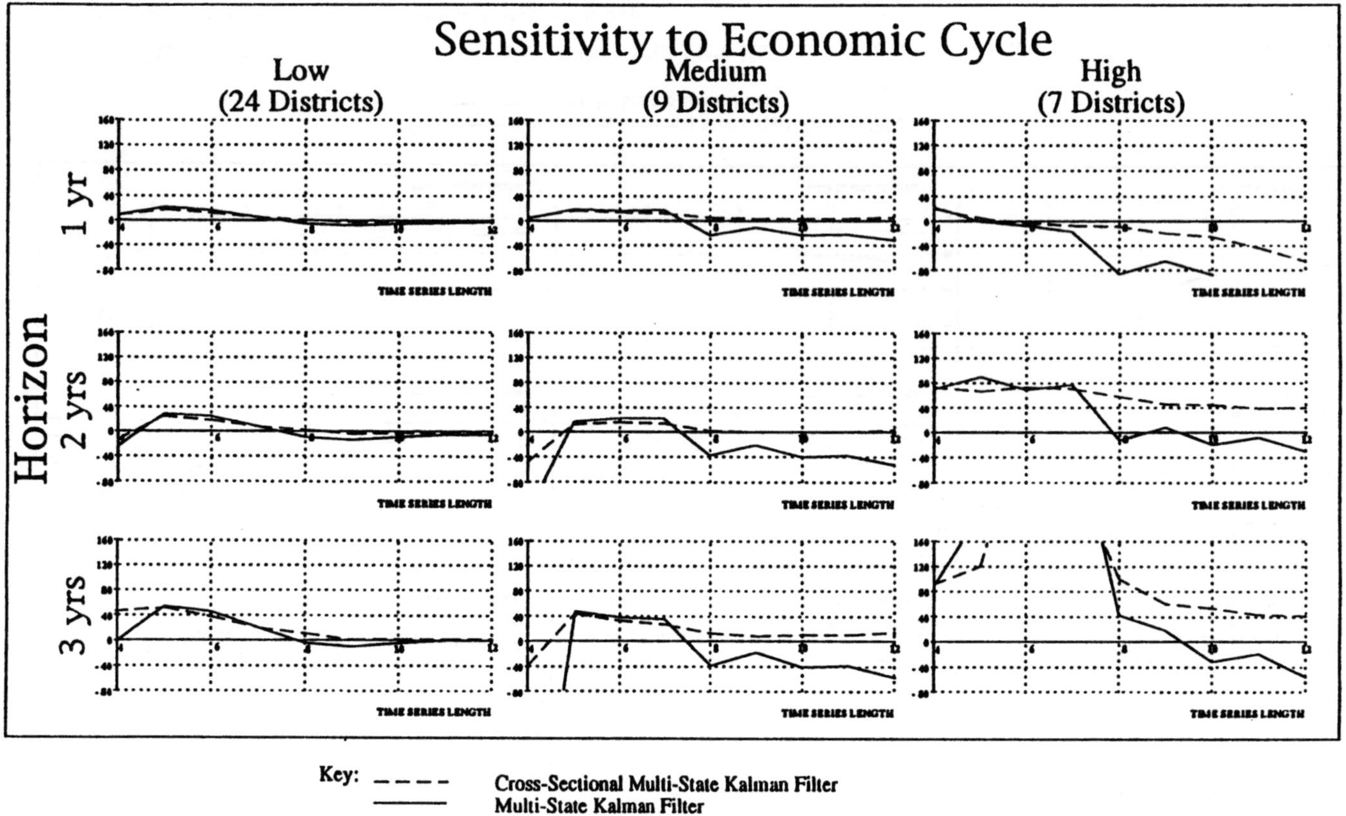

Figure 14.8: MPE backcasting accuracy (1975 origin) of the Multi-State Kalman Filter and cross-sectional Multi-State Kalman Filter by time series length, horizon and cross sectional reference group.

The analysis of the 'medium' cross-sectional reference group shows that the gap between the two MAPE measures of accuracy is larger (as was expected) than the respective difference in the 'low' group. For time series length increasing to more than seven years we notice that the accuracy of the MSKF decreases. This is because the data at the beginning are messy due to the structural change. This situation is challenging for the MSKF which must forget the first few data points with a different trend before it can recognize the relevant pattern in the data. The C-MSKF handles this case much better. Its accuracy does not deteriorate for long time series.

The graphs for the high sensitivity group present a different pattern. For simulated forecasts from models estimated using data including portions of the 'irrelevant' 1986-1981 trend (time series length greater than or equal to 7 years in the graphs), it is evident that the C-MSKF is much more successful than the MSKF in forgetting the irrelevant data. The MAPEs for C-MSKF are less than half of the MAPEs for MSKF. This is a strong result in favor of using space-time data for time series forecasting. It suggests that the C-MSKF can automatically perform 'data cleaning', in effect removing irrelevant data.

For the same group, the C-MSKF is slightly better than the MSKF for historical data sets with less than 7 data points, for the one and two year horizon forecasts. In this case, we helped the MSKF forget the past by directly dropping the irrelevant data. The graph for the third year horizon shows very high errors for both methods for historical data sets excluding the irrelevant data. This is caused by the 1974-1975 recession which, as discussed above, temporally changed the time trend slope at the end of the historical period and part way through the forecast horizon. Historical data sets long enough to establish the 1981-1976 trend in estimates widely underforecast 1972, even though making some adjustment based on the last historical data point, 1975. Nevertheless, the C-MSKF is still the more accurate method during this time.

Graphs similar to those in Figure 14.7 are presented in Figure 14.8 for the MPE measure of bias. Figure 14.8 shows MPE backcast accuracy for the two methods by time series length, horizon, and cross-sectional reference group. The C-MSKF appears to give less biased forecasts since its MPE is closer to zero and more often is positive as well as negative. In only a few cases the MSKF is less biased, but the C-MSKF is overall more consistent. Paired comparison statistical tests for this measure are not meaningful (they indicate consensus, but not relative backcast accuracy), so we did not perform them. There are similarities in the behavior of the MPE in Figure 14.8 as compared to the MAPE in Figure 14.7. The bias increases with decreasing time series length, increasing horizon, and increasing sensitivity of the cross-sectional reference group.

14.6 Conclusions

This chapter continues the work of DGS by addressing the problem of backcasting 'short' and volatile time series data (i.e., series with relatively few historical data points that have turning points and other structural changes) where cross-sectional data are available. A case study on income tax collection data for school districts in Allegheny County, Pennsylvania, illustrates the application of the C-MSKF method and is used to compare the accuracy of the univariate MSKF method with the C-MSKF method. We have reduced the heterogeneity of school districts by grouping them accordingly to sensitivity to economic cycles and by standardizing values to eliminate scale differences.

Results showed that the C-MSKF is more accurate than the univariate MSKF for every factor used in the comparison (backcasting origin, time series length, backcasting horizon, and the reference group). The relative MAPE accuracy of the C-MSKF method proved to be statistically better than the MSKF method in 52 percent of the cases (42 out of 81 cases for the three groups) according to the 5 percent significance level of the Wilcoxon matched pairs, signed rank, two-tail test. In general the gap between the two accuracy measures increases, favoring the C-MSKF method, when the time series length decreases, the sensitivity to economic cycles increases, and the backcasting horizon increases. Finally, the results also showed that the C-MSKF method is robust and gives the same or better accuracy than the MSKF method when the time series are stable. We conclude that the above analyses provide evidence that the C-MSKF method, by incorporating spatial information of time series data, improves forecast accuracy relative to the MSKF univariate time series method.

Future work includes: 1) more extensive empirical comparisons, 2) substituting various Kalman Filters and other time series methods for the MSKF in our system, 3) using a moving window method for accelerating adjustment to structural changes, and 4) experimenting with objective clustering methods for very large cross-sections.

References

Ameen, J. and P. Harrison, Discounted Weighted Estimation, *Journal of Forecasting*, 3, 285-296, 1984.

Anselin, L., *Spatial Econometrics, Methods and Models*, Dordrecht: Kluwer Academic Publishers, 1988.

Ansley, C.F. and R. Kohn, Estimation, Filtering, and Smoothing in State Space Models with Incompletely Specified Initial Conditions, *The Annals of Statistics*, 43, 1286-1317, 1985.

Armstrong, J.S., Forecasting with Econometric Methods: Folklore versus Fact, *Journal of Business*, 51, 549-564, 1978.

Bernardo, J.M., Algorithm AS 103: Psi (Digamma) Function, *Applied Statistics*, 25, 315-317, 1976.

Bomhoff, E.J. and C.J.M. Kool, Forecasts with Multi-State Kalman Filters, Appendix 1, in: E.J. Bomhoff, *Monetary Uncertainty*, Amsterdam: North-Holland, 1983.

Brown, R.G., *Smoothing, Forecasting, and Prediction of Discrete Time Series*, Englewood Cliffs: Prentice Hall, 1962.

Bunn, D. and G. Wright, Interaction of Judgmental and Statistical Forecasting Methods: Issues and Analysis, *Management Science*, 37, 501-518, 1991.

Deely, G.G. and D.V. Lindley, Bayes Empirical Bayes, *Journal of the American Statistical Association*, 76, 833-841, 1981.

DeGroot, M.H., *Probability and Statistics*, Reading: Addison-Wesley, 1986.

Duncan, G., W. Gorr and J. Szczypula, Bayesian Forecasting for Seemingly Unrelated Time Series: Application to Local Government Revenue Forecasting, *Management Science*, 39, 275-293, 1993.

Efron B. and C. Morris, Combining Possibly Related Estimation Problems, *Journal of the Royal Statistical Society Series B*, 35, 379-398, 1973.

Enns, P.G., J.A. Machak, W.A. Spivey and W.J. Wroblewski, Forecasting Applications of an Adaptive Multiple Exponential Smoothing Model, *Management Science*, 28, 1035-1044, 1982.

Fildes, R., An Evaluation of Bayesian Forecasting, *Journal of Forecasting*, 2, 137-150, 1983.

Fildes, R., Quantitative Forecasting – the State of the Art: Econometric Models, *Journal of Operational Research Society*, 36, 549-580, 1985.

Gamerman, D. and H.S. Migon, Dynamic Hierarchical Models, *Journal of the Royal Statisticl Society Series B*, 55, 629-642, 1993.

Garcia-Ferrer, A., R.A. Highfield, F. Pam and A. Zellner, Macroeconomic Forecasting Using Pooled International Data, *Journal of Business and Economic Statistics*, 5, 53-67, 1987.

Haining, R., *Spatial Data Analysis in the Social and Environmental Science*, Cambridge: Cambridge University Press, 1990.

Harrison, P.J. and C.F. Stevens, A Bayesian Approach to Short-term Forecasting, *Operations Research Quarterly*, 22, 341-362, 1971.

Harvey, A.C., Analysis and Generalization of a Multivariate Exponential Smoothing Model, *Management Science*, 32, 374-380, 1986.

Harvey, A.C., *Forecasting, Structural Time Series Models and the Kalman Filter*, Cambridge: Cambridge University Press, 1989.

Harville, D.A., Maximum Likelihood Approaches to Variance Component Estimation and to Related Problems, *Journal of the American Statistical Association*, 72, 320-340, 1977.

Highfield, R.A., Forecasting Similar Time Series with Bayesian Pooling Methods: Application to Forecasting European Output Growth, in: P.K. Goel and N.S.

Iyengar (eds.), *Bayesian Analysis in Statistics and Econometrics*, New York: Springer-Verlag, 1992.

Jones, R.H., Exponential Smoothing for Multivariate Time Series, *Journal of the Royal Statistical Society Series B*, 28, 241-251, 1966.

Kass, R.E. and D. Steffey, Approximate Bayesian Inference in Conditionally Independent Hierarchical Models (Parametric Empirical Bayes Models), *Journal of the American Statistical Association*, 84, 717-726, 1989.

Kotz, S. and N.L. Johnson (eds.), *Encyclopedia of Statistical Sciences*, New York: John Wiley & Sons, 1985.

Laird, N.M. and J.H. Ware, Random Effects Models for Longitudinal Data, *Biometrics*, 38, 963-974, 1982.

Lee, P.M., *Bayesian Statistics: an Introduction*, New York: Oxford University Press, 1986.

Lee, Y., *Conceptual Design of a Decision Support System for Local Government Revenue Forecasting*, Pittsburgh: Carnegie Mellon University, 1990 (unpublished PhD dissertation).

Lesage, J.P. and M. Magura, Using Bayesian Techniques for Data Pooling in Regional Payroll Forecasting, *Journal of Business & Economic Statistics*, 8, 127-135, 1990.

Lindley, D.V. and A.F.M. Smith, Bayes Estimates for the Linear Model, *Journal of the Royal Statistical Society Series B*, 34, 1-41, 1972.

Makridakis, S., C. Chatfield, M. Hibon, M. Lawrence, T. Mills, K. Ord, and L.F. Simmons, The M2-Competition: A Real-Time Judgmentally Based Forecasting Study, *International Journal of Forecasting*, 9, 5-22, 1993.

Morris C.N., Parametric Empirical Bayes Inference: Theory and Applications, *Journal of the American Statistical Association*, 78, 47-55, 1983.

Trigg, D.W. and D.H. Leach, Experimental Smoothing with an Adaptive Response Rate, *Operations Research Quarterly*, 18, 53-59, 1967.

West, M. and P.J. Harrison, *Bayesian Forecasting and Dynamic Models*, New York: Springer-Verlag, 1989.

Zellner, A., Comments, in: P.K. Goel and N.S. Iyengar (eds.), *Bayesian Analysis in Statistics and Econometrics*, New York: Springer-Verlag, 1992.

Zellner, A. and C. Hong, Forecasting International Growth Rates Using Bayesian Shrinkage and other Procedures, *Journal of Econometrics*, 40, 183-202, 1989.

Appendix 1: Poolbayes Program

```
c ***********************************************************************
c
c                                    POOLBAYES PROGRAM
c
c
c     PoolX(t,j)  - table with data for every
c                   cross sectional unit 'j' and time 't'
c     Dist(k,i,j) - parameters for the bivariate distributions
c                   for current time 't' and for
c                   every c.s. unit 'j' and state 'i'
c          k=1       Mi  -- mean of trend             E(T)
c          k=2       Bi  -- mean of growth rate       E(S)
c          k=3       Vmm -- variance of trend       Var(T)
c          k=4       Vmb -- covariance              Cov(T,S)
c          k=5       Vbb -- variance of growth rate Var(S)
c          k=6       Qi  -- probability that process was
c                           in state 'i' at time 't-1'
c     Vars(k,i,j) - variances for disturbance terms
c                   for every c.s. unit 'j' and state 'i'
c          k=1      Vet
c          k=2      Vg
c          k=3      Vr
c     Ncs         - number of cross sectional units
c     Top         - number of elements of a time series
c     Top1        - forecasting horizon
c     Pie(j)      - probability of occurance of state 'j'
c     Xhat(t,j)   - estimated or forecasted value of PoolX(t,j)
c
c
      program poolbayes

      implicit real*8 (a-h),(o-z)

      integer Indik(1000),year(1000),ST(1000)
     1        ,choice,SchDis(50),StateN(50),full,nstate
     2        ,yearcode(15),sc,t,top,top1,Ncs,jo,k,nine,ten,datagr

      real*8 income(1000),PoolX(50,50),Vars(3,4,50),Vlaw(50)
     1       ,Dist(6,4,50),Pie(4),Rg(4),Rr(4),Re(4),X1,X2,avrX
     2       ,Buf1(10),Buf2(10),NiuT,NiuS,SoS,SoT,const1,const2
     3       ,Mi(4,50),Bi(4,50),Vmm(4,50),Vmb(4,50),Vbb(4,50),Ei(4,50)
```

```
4       .XF(51,50),r11,r12,r22,A1,A2,invK,Vg(4),Vr(4),Vet(4)
5       .Ve(4,4),Mf(4,4,50),Bf(4,4,50),Vmmf(4,4,50),Vmbf(4,4,50)
6       .Vbbf(4,4,50),arg,arg1,arg2,Q(4,50),P(4,4,50),Xhat(51,50)
7       .bayeser1,bayeser2,BigT,BigS,err,Value,Xh,methoder
8       .BayMAPE,BayMSPE,BayMSE,med1,med2
9       .NiuT1,NiuS1,SoS1,SoT1

  common/all/Indik,year,ST,choice,SchDis,StateN
1           .yearcode,sc,t,top,top1,Ncs,jo,k,nine,ten
2           .income,PoolX,Vars,Vlaw,Dist,Pie,Rg,Rr,Re,X1,X2,avrX
3           .Buf1,Buf2,NiuT,NiuS,SoS,SoT,const1,const2
4           .Mi,Bi,Vmm,Vmb,Vbb,Ei,XF,r11,r12,r22,A1,A2,invK
5           .Vg,Vr,Vet,Ve,Mf,Bf,Vmmf,Vmbf,Vbbf,arg,arg1,arg2,Q,P
7           .bayeser1,bayeser2,BigT,BigS,err,Value,Xh,methoder
8           .BayMAPE,BayMSPE,BayMSE,med1,med2
9           .NiuT1,NiuS1,SoS1,SoT1,Xhat,datagr

  common/sizeofX/nstate

  data YearCode /86,85,84,83,82,81,80,79,78,77,76,75,74,73,72/
  data Pie,Rg,Rr,Re /0.9, 0.003, 0.003, 0.094, 0.0, 100.0, 0.0, 0.0,
1 0.0, 0.0, 1.0, 0.0, 1.0, 1.0, 1.0, 101.0/

  Character Filein*80,Fileout*80

  Print *,'Type Name of Data File?'
  Accept 10,Filein
  Print *,'Type Name of Output File?'
  Accept 10,Fileout
  Print *,'Number of Steps of Forecasting?'
  Accept *,Nf
  Print *,'Indicator of Group: (0-All Data;1-Low;2-Midium;3-High)?'
  Accept *,Datagr
  Print *,'Mode: (1 - Cross Section; 0 - Bayesian Only)?'
  Accept *,Full
  Print *,'Forecating Origin ?'
  Accept *,Top1
  If ( Top1.Lt.3) Top = 3

  Print *,'What is the Size of Time Series for a School District?'
  Accept *,Choice
  Top = Choice

  open(unit=1,status='old',file=filein)
```

```
      open(unit=2,status='new',file=fileout)
      do 1 k=1,1000
         read(1,*,end=2)Indik(k),year(k),income(k),ST(k)
1     continue

2     close(unit=1)
      n=k-1

c     ==========================================================
c     Preparation of data.  Input transformation into matrices
c     needed in the algorithm.

      Ncs = int(n/choice)
      k=0

      do 11 j=1,Ncs
         do 101 ii=1,choice
            k = k + 1
 101        PoolX(ii,j) = income(k)
         SchDis(j) = Indik(k)
         StateN(j) = ST(k)
 11   continue

c     ==========================================================
c     Normalization of data  { z = (x-miu) / sd + Const}

      do 50 j=1,Ncs
         sum = 0.0
         do 51 i=jj,top1
 51         sum = sum + PoolX(i,j)
         sum = sum / ( top1 - jj + 1)
         sum1 = 0.0
         do 52 i=jj,top1
 52         sum1 = sum1 + ( PoolX(i,j) - Sum )**2
         sum1 = sum1 / ( top1 - jj + 1)

         do 50 i=jj,top
            PoolX(i,j) = ( PoolX(i,j) - sum ) / sqrt(sum1)
            PoolX(i,j) =   PoolX(i,j) + 5.0
  50   continue

c     ==========================================================
c     Variance law (Vlaw) for each cross sectional unit:
```

```
      do 5 j=1,Ncs
         ten  = min(10,top1-jj+1)
         nine = ten - 1
         sum = 0.0
         do 24 i=1,nine
 24         sum = sum + PoolX(jj+i,j)
         sum = sum / nine

         do 303 i=1,10
              Buf1(i) = 0
 303          Buf2(i) = 0

         do 3 i=1,nine
              k = jj + i
              Buf1(i) = abs(PoolX(k,j) - PoolX(k-1,j))
 3            Buf2(i) = abs(PoolX(k,j) - sum)

         med1 = getmedian(Buf1,nine)
         med2 = getmedian(Buf2,nine)

          Vlaw(j) = med1
         if ( med1.gt.med2 ) Vlaw(j) = med2
         Vlaw(j) = (Vlaw(j)/0.6745)**2
  5   continue

      sum = 0.0
      do 53 j=1,Ncs
 53      sum = sum + Vlaw(j)
      sum = sum / Ncs
      do 54 j=1,Ncs
 54      Vlaw(j) = sum * 10

c     Calculation of variance law (end).
c
c     ====================================================
c     Initial values:
c

      do 81 j=1,Ncs
      do 81 i=1,4
         Vars(1,i,j) = Re(i) * Vlaw(j)
         Vars(2,i,j) = Rg(i) * Vlaw(j)
         Vars(3,i,j) = Rr(i) * Vlaw(j)
         Dist(1,i,j) = PoolX(1,j)
```

```
         Dist(2,i,j) = PoolX(2,j) - PoolX(1,j)
         Dist(3,i,j) = Vars(3,i,j) + Vars(2,i,j)
         Dist(4,i,j) = Vars(3,i,j)
         Dist(5,i,j) = Vars(3,i,j)
 81      Dist(6,i,j) = 0.25

c     ======================================================
c     Main Loop:

      do 300 t=jj+1,top1

      call bayesian
      if ( full.eq.1 ) call cross

300   continue

      call results

c     ===========================================

      stop
 10   format(a)
      end

c ***************************************************************
c
       function getmedian(a,n)
      real*8 a(10),s(10),maxim

      do 2 i=1,n
         ind = 1
         maxim = a(1)
         do 3 ii=1,n
            if ( a(ii).gt.maxim ) then
            ind = ii
            maxim = a(ii)
         endif
  3      continue
         s(i) = maxim
         a(ind) = -1
  2   continue

      ind = int(n/2)
      getmedian = s(ind+1)
```

```
      if ( (n/2)*2 .eq. n ) getmedian = (s(ind)+s(ind+1))/2

      Return
      end

c ******************************************************************
c
      subroutine bayesian

c     Declarations and commons as in main program should be included here!
c     ======================================================================

      do 400 jo=1,Ncs
c
c     Bayesian Forecasting -- Multi State Kalman Filter (MSKF)
c

      do 80 i=1,4
         Mi(i,t-1)  = Dist(1,i,jo)
         Bi(i,t-1)  = Dist(2,i,jo)
         Vmm(i,t-1) = Dist(3,i,jo)
         Vmb(i,t-1) = Dist(4,i,jo)
         Vbb(i,t-1) = Dist(5,i,jo)
         Q(i,t-1)   = Dist(6,i,jo)
         Vet(i)     = Vars(1,i,jo)
         Vg(i)      = Vars(2,i,jo)
         Vr(i)      = Vars(3,i,jo)
 80     continue

c     ========================================================
c     Recursive formulas:

      do 100 i=1,4
      do 100 j=1,4
         Ei(i,t) = PoolX(t,jo) - Mi(i,t-1) - Bi(i,t-1)
         r11 = Vmm(i,t-1) + 2*Vmb(i,t-1) + Vbb(i,t-1) + Vg(j) + Vr(j)
         r12 = Vmb(i,t-1) + Vbb(i,t-1) + Vr(j)
         r22 = Vbb(i,t-1) + Vr(j)
         Ve(i,j) = r11 + Vet(j)

         A1 = r11 / Ve(i,j)
         A2 = r12 / Ve(i,j)

         Mf(i,j,t) = Mi(i,t-1) + Bi(i,t-1) + A1*Ei(i,t)
```

```
        Bf(i,j,t) = Bi(i,t-1) + A2*Ei(i,t)
        Vmmf(i,j,t) = r11 - A1*A1*Ve(i,j)
        if ( Vmmf(i,j,t).lt.0.0 ) Vmmf(i,j,t) = 0.0
        Vmbf(i,j,t) = r12 - A1*A2*Ve(i,j)
        if ( Vmbf(i,j,t).lt.0.0 ) Vmbf(i,j,t) = 0.0
        Vbbf(i,j,t) = r22 - A2*A2*Ve(i,j)
        if ( Vbbf(i,j,t).lt.0.0 ) Vbbf(i,j,t) = 0.0
100   continue

       invK=0.0                ! Inverse of Normalization Constant
       do 34 ii=1,4
       do 34 jj=1,4
          arg = -(PoolX(t,jo)-Mi(ii,t-1)-Bi(ii,t-1))**2/(2*Ve(ii,jj))
          arg2 = 2 * 3.1415926536 * Ve(ii,jj)
          invK = invK + Q(ii,t-1)*Pie(jj)*exp(arg)/sqrt(arg2)
 34    continue

      do 121 i = 1,4
      do 121 j = 1,4
          arg = -(PoolX(t,jo) - Mi(i,t-1) - Bi(i,t-1))**2 / (2*Ve(i,j))
          arg2 = 2 * 3.14 * Ve(i,j)
121       P(i,j,t) =  Q(i,t-1)*Pie(j)*exp(arg) / (sqrt(arg2) * invK)

c     =========================================================
c     Compressing posterior densities from N*N into N

      DO 6 j=1,4
         Q(j,t) = 0.0
         Bi(j,t) = 0.0
         Mi(j,t) = 0.0
         Vmm(j,t) = 0.0
         Vmb(j,t) = 0.0
         Vbb(j,t) = 0.0
         Do 7 i=1,4
 7           Q(j,t) = Q(j,t) + P(i,j,t)
         Do 8 i=1,4
              Bi(j,t) = Bi(j,t) + (P(i,j,t) * Bf(i,j,t))/Q(j,t)
 8           Mi(j,t) = Mi(j,t) + (P(i,j,t) * Mf(i,j,t))/Q(j,t)
         Do 6 i=1,4
           arg1 = Mf(i,j,t)-Mi(j,t)
           arg2 = Bf(i,j,t)-Bi(j,t)
           Vmm(j,t) = Vmm(j,t) + P(i,j,t)*(Vmmf(i,j,t) + arg1*arg1)/Q(j,t)
           Vmb(j,t) = Vmb(j,t) + P(i,j,t)*(Vmbf(i,j,t) + arg1*arg2)/Q(j,t)
 6         Vbb(j,t) = Vbb(j,t) + P(i,j,t)*(Vbbf(i,j,t) + arg2*arg2)/Q(j,t)
```

```
c     Storing values for C.S. adjustment and for the next step of MSKF
c
       do 400 i=1,4
          Dist(1,i,jo) = Mi(i,t)
          Dist(2,i,jo) = Bi(i,t)
          Dist(3,i,jo) = Vmm(i,t)
          Dist(4,i,jo) = Vmb(i,t)
          Dist(5,i,jo) = Vbb(i,t)
 400      Dist(6,i,jo) = Q(i,t)

c     MSKF (end - all c.s. units have the distributions prepared)

      return
      end

c *******************************************************************
c
      subroutine cross

c     Declarations and commons as in main program should be included here!
c     ====================================================================

c     Estimation of hyperparameters (MLE) for Vmm and Vbb
.
      do 91  i=1,4

      sum1 = 0.0
      sum2 = 0.0
      sum3 = 0.0
      sum4 = 0.0
         do 55 jj=1,Ncs
         sum1 = sum1 + 1.0 /Dist(3,i,jj)
         sum2 = sum2 + DLOG(Dist(3,i,jj))
         sum3 = sum3 + 1.0 /Dist(5,i,jj)
 55      sum4 = sum4 + Dlog(Dist(5,i,jj))
      Const1 = 0.5 * (log(real(Ncs)) - dlog(sum1) - sum2/Ncs)
       Const2 = 0.5 * (log(real(Ncs)) - dlog(sum3) - sum4/Ncs)

c     ======================================================
      ii = 0
      x1 = 0.001
      x2 = 50.0
 56     avrX  = (x1 + x2) / 2
        Value = digamma(avrX,ifault) - 0.5*dlog(avrX) - Const1
```

```
        ii = ii + 1
        if ( Value.gt.0.0 ) then
           x2 = avrX
        else
           x1 = avrX
        endif
      if ((Value.gt.0.01.or.Value.lt.-0.01).and.ii.lt.20) go to 56

c     The hyperparameters of Vmm:

      NiuT = avrX * 2
      if ( Niut.lt.2.0 ) Niut = 2.0
      SoT = ( NiuT * Ncs ) / sum1

c     ==========================================================
      ii = 0
      x1 = 0.001
      x2 = 50.0
 57        avrX  = (x1 + x2) / 2
           Value = digamma(avrX,ifault) - 0.5*dlog(avrX) - Const2
           ii = ii + 1
           if ( Value.gt.0.0 ) then
               x2 = avrX
           else
               x1 = avrX
           endif
      if ((Value.gt.0.01.or.Value.lt.-0.01).and.ii.lt.20) go to 57

c     The hyperparameters of Vbb:

      NiuS = avrX * 2
      if ( NiuS.lt.2.0 ) NiuS = 2.0
      SoS = ( NiuS * Ncs ) / sum3

c     ==========================================================
c     modification of variances (Vmm,Vbb)
c
      do 91  j=1,Ncs
         Dist(3,i,j) = (SoT + (PoolX(t-1,j) - Dist(1,i,j))**2 )/(NiuT-1)
         Dist(5,i,j) = (SoS + ((PoolX(t,j) - PoolX(t-1,j))
     1                       -Dist(2,i,j))**2 ) / (NiuS - 1)
 91   continue

c     ==========================================================
```

```
c     Estimation of Hyperparameters (MLE) for  MEANS (Mi,Bi)
c
c     Sample mean and sample variance (same for every  C.S. unit)
c
      do 90 i=1,4
      SMofT = 0.0
      SVofT = 0.0
      SMofS = 0.0
      SVofS = 0.0
          do 45 jj=1,Ncs
          SMofT = SMofT + Dist(1,i,jj)
 45       SMofS = SMofS + Dist(2,i,jj)
      SMofT = SMofT / Ncs
      SMofS = SMofS / Ncs
         do 46 jj=1,Ncs
            SVofT = SVofT + (Dist(1,i,jj) - SMofT)**2
 46         SVofS = SVofS + (Dist(2,i,jj) - SMofS)**2
      SVofT = SVofT / Ncs
      SVofS = SVofS / Ncs

c     =======================================================
c     Modification of means for each state (i) and each c.s. unit (j)
c
      do 90 j=1,Ncs
         Dist(1,i,j) = ( Dist(3,i,j) * SMofT + PoolX(t-1,j) * SVofT ) /
     1               ( Dist(3,i,j) + SVofT )
         Dist(2,i,j) = ( Dist(5,i,j) * SMofS + (PoolX(t,j)-PoolX(t-1,j))*
     1                  SVofS ) / ( Dist(5,i,j) + SVofS )
 90   Continue

      return
      end

c ***********************************************************************
c
      function digamma(x,ifault)
c
c     Algorithm as 103 Appl. Statist. (1976) Vol.25, no.3
c     Calculates Digamma(x) = d(log(gamma(x)))/dx
c     Set constants.  sn = Nth Stirling coefficient, D1 = digamma(1.0)
c
      data s,c,s3,s4,s5,d1 /1.0e-5, 8.5, 8.333333333e-2,
     * 8.333333333e-3, 3.968253968e-3, -0.5772156649/
c
```

```
      digamma = 0.0
      y = x
      ifault = 1
       if ( y .le. 0.0 ) return
      ifault = 0.0
c
c             use approximation if argument .LE. S
c
      if ( y .gt. S ) go to 1
      digamma = d1 - 1.0 / y
      return
c
c             reduce to digamma(x+n),  (x+n) .GE. C
c
   1  if ( y .ge. C ) go to 2
      digamma = digamma - 1.0 / y
      y = y + 1.0
      go to 1
c
c             use Stirling if argument .GE. C
c
   2  r = 1.0 / y
      digamma = digamma + alog(y) - 0.5 * r
      r = r * r
      digamma = digamma - r * ( S3 - r * (S4 - r * S5) )

      Return
      end

c *****************************************************************
c
      subroutine results

c     Declarations and commons as in main program should be included here!
c     ====================================================================

      do 85 j=1,Ncs

      BigT = 0.0
      BigS = 0.0
      Do 84 ii=1,4
         BigT = BigT + Dist(6,ii,j) * Dist(1,ii,j)
  84     BigS = BigS + Dist(6,ii,j) * Dist(2,ii,j)
         i = 1
```

```
       Do 94 t=top1+1,top
          i = i + 1
 94       Xhat(t+1,j) = BigT + i*BigS          !  Forecasting
 85    Continue

c      ============================================================

       do 102 t=top1+1,top
          pererr = (PoolX(t,j) - Xhat(t,j))/PoolX(t,j)*100.0
           write(2,200)SchDis(j),Yearcode(t),t,nstate
      1          ,PoolX(t,j),Xhat(t,j),pererr,datagr
 200      format(1x,4i4,2f9.3,f10.2,1i4)
 102   Continue

       return
       end
```

15 A Multiprocess Mixture Model to Estimate Space-Time Dimensions of Weekly Pricing of Certificates of Deposit

James P. LeSage
University of Toledo, Toledo, OH, U.S.A.

15.1 Introduction

The spatial dimension of banking markets is often discussed in the banking and finance literature, but neglected in empirical studies of behavior by depository institutions. As Anselin and Griffith (1988) point out, neglect of spatial aspects of economic behavior is not unusual, even by those working in regional science. Traditional depository institution regulatory policy, based on a structural analysis of local markets, generally operates under the assumption of the U.S. banking system as a collection of segmented markets versus an integrated national banking system. From this perspective, banks and thrifts in local regions have little or no effect on the pricing decisions of firms operating in other localities. In the deregulated banking environment of the 1980s, some bank retail deposit markets may be better characterized as operating in such a way as to generate spatial effects. As of October 1983, all depository institutions were permitted to offer competitive market rates on interest-sensitive deposits, including retail certificates of deposit (CDs). In turn, many institutions, particularly large banks, have used national advertising and brokers to attract large retail (insured) deposits from other regions. With greater deregulation and competition for deposit funds and increased reliance of banks on interest-sensitive deposits, significant spatial effects in the deposit-rate setting decisions of banks across regions might be expected. The presence or lack of spatial effects in regional bank markets is important to bank managers, analysts, and regulators in terms of defining relevant markets and measuring the competitive effects of greater deregulation.

Kenen (1976) suggests that one way to conceptualize the extent of spatial segmentation is to examine whether market participants are 'enabled and obliged to take notice of events occurring in other markets'. This study develops a model of deposit-seeking behavior by banks and thrifts and attempts to empirically quantify the extent to which market participants take notice of events occurring in spatially related markets.

The alternative to the 'spatially segmented markets hypothesis' has come to be known in the banking literature as, the 'integrated markets hypothesis'. Cooperman et al. (1990) provide evidence of increasing market integration using weekly data on six-month retail CD rates for a sample of large banks and thrifts in six metropolitan areas ranging from the east to the west coasts. Their evidence consists of Granger-causal patterns of relationships indicative of co-movements over time in CD rate deviations from the Treasury Bill rate across the cities they examine. In this study, a more powerful test of the segmented/integrated market hypotheses is undertaken. Using a spatially distributed sample of weekly retail six-month CD rate offers by forty banks in Ohio, the parameters of an optimal CD rate-setting rule for banks and thrifts are estimated. The time period of the data sample covers the Ohio Deposit Insurance crisis in March, 1985. In the absence of an unusual event such as this, it may well be that, empirical evidence in favor of the segmented/integrated markets hypothesis is quite difficult to detect using statistical methods. Consequently, the crisis event provides a unique opportunity to employ a newly proposed spatial econometric method to shed light on the important issue of market segmentation/integration.

The Ohio deposit insurance crisis was the first failure of a deposit insurance fund since the 1930's. It began on March 4, 1985 when Florida-based ESM Securities defaulted on its reverse repurchase (repo) agreements with Home State Savings in Cincinnati, threatening the thrift's solvency. On March 6-8, large deposit runs occurred at Home State, resulting in its closure on March 9. Home State's losses undermined the solvency of the private, state-chartered Ohio Deposit Guaranty Fund (ODGF), and runs occurred on March 11-14 at other ODGF-insured thrifts. On March 15, a bank holiday for the 70 ODGF thrifts was declared. On March 18, thrifts were allowed to reopen when they obtained federal insurance, and regulators agreed to ease regulations for conversion.[1]

This chapter is unique in two regards. First, it provides a theoretical motivation for the importance of spatial influences on CD rate-setting behavior. The theoretical model suggests that the parameters in an optimal decision-rule for CD rate-offers will depend on three factors; adjustment costs, the desired level of deposits, and substitution effects arising from competitive institutions. Second, an innovative Bayesian parameter estimation method is used that allows time-variation in the parameters to take on a pattern which models competitive influences as spatially dependent. The parameter variation scheme can accommodate diffusion of competitive influences across the spatial dimension of the sample.

In section two, an optimal CD rate-setting decision rule for banks and thrifts is derived using a dynamic stochastic optimal control model. The econometric proce-

---

[1] See Kane (1989) for a more detailed description and analysis of the events of the Ohio insurance crisis.

dure used to estimate the parameters of this decision rule is described in Section three of the chapter. Section four presents the results of empirical tests for a significant spatial element in the rate-setting behavior of institutions. In addition, a test for spatial patterns of influence during the crisis event is carried out using a sample of forty Ohio depository institutions.

15.2 A Dynamic Targeting Model of CD Rate-Setting Behavior

One of the problems with past studies of market integration versus segmentation is that they are primarily statistical studies of correlations or more sophisticated statistical measures of interrelationships which are not based on an underlying economic model of rate-setting behavior. Here, a simple optimal control view of the rate-setting decisions made by depository institutions is set forth.

15.2.1 The Model

Consider a single depository institution which has past data relating the deposits in each time period t, to its own CD rate (R_{1t}) offer in time t, and a competing institution's rate offer (R_{2t}) in time t. We might plausibly assume that the historical relation between deposits and rate offers takes a linear stochastic form, such as:

$$D_{1t} = \beta_0 + R_{1t}\beta_1 + R_{2t}\beta_2 + U_t \tag{1}$$

where D_{1t} represents deposits at time of institution number 1, R_{1t} reflects the CD rate offer at time by institution 1, R_{2t} represents the CD rate offer of a competing institution (number 2) at time t, U_t denotes a zero mean, constant variance, σ^2 disturbance, and β_0, β_1, β_2 are scalar parameters.

The deposit variable D_1, is assumed to be controllable, and institution 1 uses R_1, its own CD rate offer, as the control variable. Further, R_2, the CD rate offer of institution 2 is assumed not under the control of institution 1, representing an uncontrollable influence on institution 1's level of deposits. (We restrict β_2 to be non-zero in order that the uncontrollable influence be non-trivial.) There is also a stochastic influence on institution 1's attempt to control it's level of deposits reflected by U_t, perhaps arising from national money market forces that are out of the institution's sphere of control.

In the first future period $T + 1$ we define $z_1 \equiv D_{1t+1}$ and $w_{11} \equiv R_{1t+1}, w_{12} \equiv R_{2t+1}$, and for the future observation we have:

$$z_1 = \beta_0 + \beta_1 w_{11} + \beta_2 w_{12} + U_{T+1} \tag{2}$$

where again U_{T+1} is normally distributed with zero mean and constant variance, σ^2. The motivation for defining z_1, w_{11} and w_{12} is that, we assume z_1 has not yet been observed, nor has a value been determined for the control variable w_{11} in the z_1 expression.

The criterion function involves a desire to set a rate R_{1T+1}, such that future deposits, z_1, will be close to a target value, ϕ, which is assumed known. The loss associated with being off target is assumed to be quadratic and of the form shown in (3). The assumption of quadratic loss may be realistic since excess deposits may be just as undesirable as a deficiency of deposits for the institution's management of its portfolio.[2]

$$L = (z_1 - \phi)^2 + (w_{11} - R_{1T})'G(w - R_{1T}) \tag{3}$$

Where R_{1T} denotes the CD rate offer in time T and G is a positive scalar. In order to provide a very general presentation where w_{11} and w_{12} might denote vectors of controllable and non-controllable variables the notation used in (3) is that of a quadratic form, in which case $\boldsymbol{G}$ would denote a positive definite symmetric matrix of conformable dimension. The loss function in (3) takes into account both the cost associated with being off target, $(z_1 - \phi)^2$, and the costs associated with changing the CD rate offer over time, $(w_{11} - R_{IT})'\boldsymbol{G}(w_{11} - R_{IT})$.[3] The institution will use its knowledge of the predictive probability density function for $z_1 \equiv D_{IT+1}$, based on the historical sample information up to time T in order to minimize the mathematical expectation of the random function in (3). The expression for the expected loss takes the form shown in (4).[4]

---

[2] Additional reserve requirements, interest expenses, and portfolio adjustment costs, for instance, make unexpected excess deposits undesirable in time period $T + 1$. It should be noted that changing this assumption to an asymmetric loss function would not change the variables that enter into the resulting optimal decision rule, only the form of the parameters in the resulting rule would change.

[3] This cost could be associated with alienating customers by changing CD rates abruptly from week to week. There may also be administrative costs associated with rapid changes in rate offers.

[4] We delete the intercept term from (2) for clarity of exposition on (4), (5) and (6). Zellner (1971) provides a detailed analysis of this type of optimization problem and its solution.

$$EL = \bar{s}^2\left[1 + (w_{11}\ w_{12})'\begin{pmatrix} M^{11} & M^{12} \\ M^{21} & M^{22}\end{pmatrix}\begin{pmatrix} w_{11} \\ w_{22}\end{pmatrix}\right] \tag{4}$$

$$+ (\phi - w_{11}\bar{\beta}_1 - w_{12}\bar{\beta}_2)^2 + (w_{11} - R_{1T})'G(w_{11} - R_{1T})$$

where $\bar{\beta}_1$ and $\bar{\beta}_2$ are the least-squares estimates of the parameters β_1 and β_2 from the regression shown in (1) and the term $\bar{s}^2 = e'e/(t-2)$, represents the sum of squared residuals. From equation (1), we form a partitioned matrix $R = [R_1 \vdots R_2]$ so that the terms M^{ij} are from the matrix inverse shown in (5).

$$(R'R)^{-1} = \begin{pmatrix} M^{11} & M^{12} \\ M^{21} & M^{22}\end{pmatrix} \tag{5}$$

Differentiating *EL* with respect to the CD rate control variable, w_{11}, we find the minimizing value w^*_{11} to be:

$$w^*_{11} = \theta_1\bar{\beta}_1\phi - (\bar{\beta}_1\bar{\beta}_2 + \bar{s}^2M^{12})w_{12} + \theta_1 GR_{1T} \tag{6}$$

where:

$$\theta_1 = \frac{1}{\left(\bar{s}^2M^{11} + G\right)} - \left[\frac{\bar{\beta}_1^2}{\left(\bar{s}^2M^{11}\right)} + G + \bar{\beta}_1^2\right]\left(\bar{s}^2M^{11} + G\right)$$

15.2.2 The Decision Rule

An intuitive explanation of the optimal CD rate-setting relationship shown in (6) is as follows. First, we can without loss of generality treat $\bar{\beta}_1$ and $\bar{\beta}_2$ as if they were elasticities reflecting the historical responsiveness of the deposits to the CD rate offers. This elasticity interpretation follows from considering that the regression relationship in (1) could be expressed in terms of percentage changes in the deposits and CD rate offers. It then would seem reasonable to assume that, $\bar{\beta}_1 > 0$ and $\bar{\beta}_2 < 0$, since the CD rate offer R_2 of institution 2, would on average over the historical time period have a negative elasticity relationship with the deposits of institution 1. (Institution 2 provides a substitute product for that of institution 1).

Second, we see that a higher target rate ϕ for the level of deposits in time T + 1, should intuitively result in a higher optimal CD rate offer in time, *ceteris paribus*, implying that $\theta_1 > 0$. Third, the impact of institution 2's CD rate offer w_{12} in period $T + 1$, on the optimal rate offer of institution 1, in period $T + 1$, will depend on the cross-price elasticity magnitude reflected in $\bar{\beta}_2$. To see this, consider the simple case where $M^{12} = 0$. From the matrix expression in (5), we see that this represents no covariation between the historical rate offers of institutions 1 and 2; they are perfectly orthogonal. In this case, the *ceteris paribus* impact of the rate offer w_{12} by institution 2 in period $T+1$, on the optimal rate offer w_{11} of institution 1, will equal $-\theta_1\bar{\beta}_1\bar{\beta}_2$ which is positive if, as we have assumed, $\theta_1 > 0$, $\beta_1 > 0$, $\beta_2 < 0$. We conclude that, the more sensitive institution 1's deposits have been historically to the rate offers of institution 2, (the larger in absolute value terms $\bar{\beta}_2$) the greater will be the impact of institution 2's CD rate offer on the optimal rate offer of institution 1. This seems intuitively correct.

Fourth, the importance of the previous period rate offer R_{1T} of institution 1, will be determined by the cost of adjustment parameter G and the term θ_1 . The larger the cost G, the more important the previous period R_{1T} will be in determining the optimal rate offer for period $T+1$, which seems intuitively appealing.

For estimation purposes we write the optimal CD rate-setting rule for the representative depository institution from (6) as shown in (7), suppressing most of the underlying structural parameters. Because multiple underlying structural parameters were collapsed into a single intercept and two regression coefficients, it seems desirable to use a time-varying estimation technique that allows the parameters to change over time. This is indicated in expression (7) by the time subscripts on the parameters a_{1t}, b_{1t} and c_{1t}. To be explicit here consider that, $\boldsymbol{M}^{11}$, $\boldsymbol{M}^{12}$, $\boldsymbol{G}$, $\bar{\beta}_1$, $\bar{\beta}_2$, and $\bar{s}^2$ are all subject to change as further historical information is processed by the institution solving the optimization problem.[5]

$$R_{1t} = a_{1t} + b_{1t}R_{2t} + c_{1t}R_{1t-1} + \varepsilon_t \tag{7}$$

where:

$$\begin{aligned}
a_{1t} &= \theta_1\beta_1\phi \\
b_{1t} &= -\theta(\beta_1\beta_2 + \bar{s}^2\boldsymbol{M}^{12}) \\
c_{1t} &= \theta_1\boldsymbol{G} \\
\theta_1 &= \frac{1}{(\bar{s}^2\boldsymbol{M}^{11} + \boldsymbol{G})} - \left[\frac{\beta_1^2}{(\bar{s}^2\boldsymbol{M}^{11})} + \boldsymbol{G} + \beta_1^2\right](\bar{s}^2\boldsymbol{M}^{11} + \boldsymbol{G})
\end{aligned}$$

By way of conclusion, the simple dynamic optimization model of rate-setting behavior suggests a result that can be used to test which depository institutions exert a competitive influence. Any depository institution that has had a historical influen-

[5] The deregulation of retail certificate of deposits (CDs) in October 1983, was followed by a rapid development of agencies providing national deposit rate information, deposit brokers, 800-numbers and national advertising by banks and thrifts. It seems reasonable that institutions were going through a period of learning during the 1984 through 1987 time covered by the data sample used here.

ce on the deposit levels of institution 1 will influence the parameters of expression (7), as illustrated by the term β_2 in the definition of b_{1t}. Those institutions which have not historically exerted an influence on deposit levels of institution 1, i.e., where $\beta_2 = 0$, $M^{12} = 0$, will not affect the rate-setting decision of institution 1.

15.3 The Spatial Econometric Model

The basis for the estimation procedure is a dynamic linear model (DLM) first suggested by Harrison and Stevens (1976) and later refined by West and Harrison (1989). This section develops a spatial DLM model based on the optimal CD rate-setting rule shown in expression (7).

15.3.1 Spatial Time-Varying Parameters

For ease of exposition, we consider the representative institution i in setting forth the DLM model and let n denote the number of institutions in our data sample. The notation used to present the model requires that we partition the explanatory variables matrix for institution i as $X_{it} = [1 \ \vdots \ r_{it\text{-}1} \ \vdots \ W_i R_t]$, where $r_{it\text{-}1}$ represents last period's CD rate offer by institution i. Let W_i denote the ith row of an $n \times n$ spatial weighting matrix, say a first order contiguity matrix, making W_i a row vector of dimension $1 \times n$. Finally, let R_t represent an n vector of period t CD-rates for all institutions in the sample. With this notation, the observation equation for institution i in our model can be written as in (8).

$$r_{it} = X_{it}\beta_{it} + \varepsilon_{it} \tag{8}$$

where $\beta_{it} = (a_{it}\ b_{it}\ c_{it})'$ and ε_{it} is a scalar stochastic disturbance term.

Equation (8) models institution i's CD rate offer at time t as a linear time-varying parameter model, with the rate offer dependent on last week's rate offer and a linear combination of competing institutions' rate offers. If the spatial weight matrix is normalized by dividing each row element by the number of non-zero elements in the row, the linear combination of spatially contiguous institutions' rate offers will take the form of an arithmetic average. Equation (8) represents a somewhat specialized version of the CD rate-setting rule from the previous section, since the spatial weight matrix W plays a role in selecting which of the sample CD rate offers represent 'competing institutions'. By competing institutions, we mean those that should have had an historical influence on the deposit levels of institution i. We can test alternative hypotheses concerning the spatial extent of competition by altering the spatial weighting matrix, W, from which we derive the row W_i in equation (8). This topic is covered in section 3.3.

To complete the DLM model specification we require a parameter transition equation describing how the parameters a_{it}, b_{it}, c_{it} varying over time, which is shown in (9).

$$\begin{pmatrix} a_{it} \\ b_{it} \\ c_{it} \end{pmatrix} = \begin{pmatrix} 1 & 0 & 0 \\ 0 & 1 & 0 \\ 0 & 0 & \gamma_i \end{pmatrix} \begin{pmatrix} a_{it-1} \\ b_{it-1} \\ c_{it-1} \end{pmatrix} + \begin{pmatrix} 0 & 0 & 0 \\ 0 & 0 & 0 \\ 0 & 0 & (1-\gamma_i)W_i \end{pmatrix} \begin{pmatrix} 0 \\ 0 \\ C_{t-1} \end{pmatrix} + \begin{pmatrix} v_{i1t} \\ v_{i2t} \\ v_{i3t} \end{pmatrix} \tag{9}$$

where γ_i represents a scalar parameter to be estimated which is defined over a grid of values on the interval (0,1) and v_{i1t}, v_{i2t}, and v_{i3t} represent stochastic disturbances, about which more will be said later. C_{t-1} is an $n \times 1$ vector containing the previous period estimates of c_{it-1}, $i = 1, \ldots, n$ for all n institutions in the sample.

The parameter transition specification in (9) states that the intercept parameter, a_{it} and the parameter on the lagged dependent variable, b_{it}, both follow a Markov transition scheme of the type often used in specifying time-varying parameter models. The competitive influence parameter, c_{it}, follows an innovative transition pattern which reflects a linear combination of a Markov and a spatial weighting scheme. The scalar parameter γ determines the extent to which each of these influences contribute towards variation in the parameter c_{it} over time.

Expression (9) indicates that stochastic variation over time in the intercept parameter, a_{it} and lagged dependent variable parameter, b_{it}, are solely influenced by past values. This is motivated by our decision rule where the intercept parameter is a function of three factors (see the definitions after equation (7) embodied in β_1, θ_1, and ϕ. Recall, β_1 reflects the strength of the institution's own CD rate offer in historically influencing deposit levels, θ_1 contains a β_1 parameter reflecting this same influence, as well as the signal-to-noise ratio, $(\bar{s}^2 M^{11})$, and the adjustment cost parameter G. Finally, ϕ reflects the desired level of deposits. The Markov process seems best capable of modeling variation in these factors as all of them are likely to follow smooth random-walk transitions over time. By similar reasoning, the lagged dependent variable parameter b_{it}, which depends on costs of adjustment, G, and θ_1, also seems most appropriately modeled by a Markov process.

The transition specification for the parameter c_{it} was also motivated by the optimal decision rule from the previous section. The parameter c_{it}, reflects the role and influence of CD rate offers by competing institutions, and we model the transition over time as a linear combination of the Markov scheme and the spatial dimension of the sample reflected in row i of the spatial weight matrix W. Note that in the special case where γ takes on a value of unity, we have a Markov transition alone; and, in the event that γ takes on a value of zero, we have a parameter transition which relies exclusively on the spatial dimension of the sample data.

An example may help to clarify the motivation for the c_{it} parameter variation scheme shown in (9). Suppose that spatial weighting selects institutions 2 and 3 as contiguous to institution 1. In the observation equation, W_1 will produce an explanatory variable which represents the mean of the two rate offers by competing institutions 2 and 3 in time period t . The parameter c_{1t} will record the influence of this variable on the CD rate offered by institution 1 in period t. In general, the parameter c_{it} will record the spatial competitive influences on institution i, so that, the parameters c_{2t} and c_{3t} reflect similar influences by institutions competing (those contiguous) with institutions 2 and 3. The transition equation (9), uses spatial weighting to select previous period parameter estimates, $c_{2t\text{-}1}$ and $c_{3t\text{-}1}$, from the estimation equations for institutions 2 and 3 and applies a shrinkage (or biasing) factor of γ_i to the average of these two parameters.

In essence, this type of parameter transition employs the mean of spatially contiguous institutions' competitive influence parameters as Bayesian prior information. There are precedents in the literature for the use of this type of prior information. Garcia-Ferrer, Highfield et al. (1987) and Chapter 14 in this volume show that multi-country data information about linkages that exist between a sample of nine countries can be introduced by utilizing the pooled estimates as the mean of a Bayesian prior. LeSage (1990) and LeSage and Magura (1990) demonstrate that this approach also works in a multi-regional labor market setting with eight metropolitan areas from Ohio. Duncan et al. (1993) employ information from data vectors of spatially related areas as Bayesian prior information in order to improve forecasting performance.

The intuition behind this approach is that, if spatially related institutions 2 and 3 have exerted an historical influence on the deposit levels of institution 1, then the parameters $c_{2t\text{-}1}$ and $c_{3t\text{-}1}$ should contain information on the previous period forces working through space on the environment of institutions 2 and 3. The Bayesian prior forces the parameter c_{1t} to look like a linear combination of $c_{1t\text{-}1}$ and the average of $c_{2t\text{-}1}$ and $c_{3t\text{-}1}$. This parameter transition scheme will also force c_{2t} to represent a linear combination of $c_{2t\text{-}1}$ and the average of $c_{1t\text{-}1}$ and $c_{it\text{-}1}$, where the subscript i ranges over institutions contiguous to institution 2. This description applies equally to the determination of all c_{it} parameters for all other firms in the model.

From this, we conclude that the parameter transition scheme allows for diffusion effects, where competitive forces need not originate from the immediate spatial location in which institution 1 resides. To see this, consider the case where spatial contiguity takes a form such that institutions 2 or 3 are influenced by other institutions contiguous to them, but, not to institution 1. Through time these impacts on institutions 2 and 3 will work their way into the c_{1t} parameter of institution 1, creating a diffusion of competitive influences through the spatial sample which originate in other spatial locations.

To complete the specification of the model in (8) and (9) we need to specify distributions for the disturbance terms. A matrix expression for equation (9) shown below as (9′) facilitates the exposition of these ideas.

$$\beta_{it} = G b_{it-1} + H_i \Psi_{t-1} + V_{it} \tag{9'}$$

where:

$$\beta_{it} = \begin{pmatrix} a_{it} \\ b_{it} \\ c_{it} \end{pmatrix}, G = \begin{pmatrix} 1 & 0 & 0 \\ 0 & 1 & 0 \\ 0 & 0 & \gamma_i \end{pmatrix}, H_i = \begin{pmatrix} 1 & 0 & 0 \\ 0 & 1 & 0 \\ 0 & 0 & (1-\gamma_i)W_i \end{pmatrix}, \Psi_{t-1} = \begin{pmatrix} 0 \\ 0 \\ C_t \end{pmatrix}, V_{it} = \begin{pmatrix} v_{i1t} \\ v_{i2t} \\ v_{i3t} \end{pmatrix}$$

We assume means of zero for all disturbances and variances, $\text{var}(\varepsilon_t) = E_t = \sigma^2 U$, $\text{var}(V_{it}) = D_t = \sigma_2 V$ and that, $\sigma^2 = \lambda^{-1}$ is unknown. By virtue of the three parameters a_{it}, b_{it}, c_{it} we also assume that:

$$V = \begin{pmatrix} V_a & 0 & 0 \\ 0 & V_b & 0 \\ 0 & 0 & V_c \end{pmatrix} \tag{10}$$

Note that these assumptions regarding the disturbances do not contain the institution subscript i. We wish to abstract from any correlation structure across institutions, implementing separate univariate models for each institution. The motivation for this is not realism, but tractability. If we attempt to analyze all institutions in a multivariate DLM model, the observational errors become vectors so that specifying the model requires observational variance matrices which define the joint stochastic structure of the observations conditional on the time-varying parameters. The univariate DLM theory used here extends easily to this multivariate case when the observational error variance matrices are known for all time periods. However, as soon as we introduce uncertainty concerning these variances, the tractability of analysis is lost. Since it seems more realistic in this application that uncertainty concerning these variances exists, we avoid the multivariate approach in favor of the tractable Bayesian conjugate analysis, which is described in the following section for the univariate case [see West and Harrison (1989, chapter 15)].

15.3.2 Parameter Estimation

Using a procedure recommended by West and Harrison (1989) to deal with the case of unknown σ^2, we can assume that:

$$\begin{aligned} \beta_0 &\sim N(\bar{\beta}_0, \lambda^{-1}, Z_0) \\ (\beta_t \mid \theta_t, \lambda) &\sim N(\bar{\beta}_t, \lambda^{-1}, Z_t) \\ (\lambda_t \mid \phi_t) &\sim G(\frac{1}{2}n_t, \frac{1}{2}l_t) \end{aligned} \tag{11}$$

In (11) N denotes the normal and G, the gamma distribution, and the term θ_t indicates that the densities are conditional on data up to observation t. Z_t takes on the same dimension as in V equation (10) and contains the priors for the parameter variances.

$$\begin{aligned} f_t &= X_t \bar{\beta}_{t-1} \\ e_t &= r_t - f_t \\ P_t &= Z_{t-1} + D_t \\ F_t &= X_t P_t X_t' + E_t \\ S_t &= P_t X_t'(F_t)^{-1} \\ \bar{\beta}_t &= G\bar{\beta}_{t-1} + H_i \Psi_{t-1} + S_t e_t \\ Z_t &= P_t - S_t F_t S_t' \\ n_t &= n_{t-1} + 1 \\ l_t &= l_{t-1} + e_t'(F_t)^{-1} e_t \\ \mathrm{p}(r_t \mid \phi_{t-1}) &\sim F_t^{-1/2} l_{t-1}^{(1/2)n_{t-1}} l_t^{-(1/2)n_t} \end{aligned} \tag{12}$$

West and Harrison (1989) show that this set of distributional assumptions forms a natural conjugate prior for the estimation problem, allowing the posterior distribution from one observation to be used as the prior for the next in a recursive estimation procedure based on a Kalman filter shown in (12). The institution subscripts i have been dropped from (12) by virtue of our decision to proceed with a univariate analysis.

The recursions in (12) determine $\bar{\beta}_t$, Z_t, n_t and l_t along with the predictive density, $\mathrm{p}(r_t \mid \theta_{t-1})$. From these values we can construct the likelihood (or posterior) which can be used to draw inferences concerning the plausibility of various spatial weighting matrices as well as the value taken by the parameter γ_i. In order to determine the parameter γ we follow West and Harrison (1989) and define our model as dependent on the value of γ in the interval (0,1). Let the notation for this be, $M_t(\gamma) : \gamma \in (0,1)$. Now consider that, for some value $\gamma^* \in (0,1)$, $M_t(\gamma^*)$ holds for all time t, for all firms. We are however uncertain about this particular value of γ. Given any single value of γ, we can analyze the model using the Kalman recursions in (12) to produce sequences of posterior distributions for our model, $M_t(\gamma)$, that are sequentially updated over time. Starting with an initial prior density, $p(\gamma|\theta_0)$, for γ, we can examine the posterior, $p(\gamma|\theta_t)$, at time t using Bayes' theorem:

$$p(\gamma|\phi_t) \propto p(\gamma|\phi_{t-1})p(r_t|\gamma, \phi_{t-1}) \tag{13}$$

The term $p(r_t|\gamma,\theta_{t-1})$ is the one-step-ahead predictive density for $M_t(\gamma)$ provided by the Kalman recursions in (12). The posterior in (13) provides information about the relative support from the data and prior information for individual models associated with the various values of γ. West and Harrison (1989) label this a Class I, Multiprocess Mixture Model.

Taking a discrete grid of values for $\gamma_j = 0.01, 0.02, 0.03, ..., 1.0$, ($j = 1, ..., 100$), let $p_t(j)$ be the posterior mass at $\gamma = \gamma_j$, let the uniform prior probabilities at time $t = 0$ be represented by $p_0(j) = 0.01$, and the likelihood function for γ as j varies over γ_j be $L_t(j) = p(r_t|\gamma_j,\theta_{t-1})$. The discrete version of Bayes' theorem suggests updating the posterior masses as shown in (14).

$$p_t(j) = \frac{p_{t-1}(j)L_t(j)}{\sum_{j=1}^{100} p_{t-1}(j)L_t(j)} \tag{14}$$

One point to note about running the Kalman recursions for our univariate estimation procedure, we need to process all forty sample institutions at time 1 before proceeding to time 2. This is because, in order to process institution i's parameter transition equation, we require the previous time period estimates $c_{kt-1}(k{\neq}i)$ from the equations for all other institutions. In general, this creates no problems, but when attempting to iterate over values of γ_i it creates a computationally untenable situation. Assuming that a single parameter value γ holds for all institutions allows us to proceed in a computationally tractable fashion.

15.3.3 Testing Hypotheses with the Model

There are several hypotheses we might wish to investigate using the estimation results from the model. First, we can test the spatial time-variation scheme for the parameters c_{it} against an alternative model where the parameters c_{it} follow a simple Markov process. Second, we might wish to consider alternative types of spatial weighting matrices. Third, we can test for significant differences in the time-varying parameters for institutions in different spatial locations, different types of institutions, e.g., banks versus thrifts, or institutions differing in size. Fourth, we can examine the time-variation in the parameters for evidence of structural shifts overtime. There are no doubt numerous other issues that might be explored. Due to space limitations, we focus on only a few of these. One issue examined here is the appropriateness of the spatial time-variation scheme for the parameters c_{it}. This will be determined by examining the posterior distribution for the parameter γ (shown in (14)) over time. A posterior distribution for γ with a modal value less than unity suggest evidence in favor of the spatial time-variation specification. A more formal approach to examining this, as a hypothesis, would be to form two models and compute a Bayes factor or log-of-the-odds ratio for the two models. Consider any two values of γ, say, $\gamma_1 = 1.0$ and $\gamma_2 = \gamma^*$. From expression (14), we can calculate a Bayes' factor using (15).

$$\frac{p_t(2)}{p_t(1)} = \frac{p_{t-1}(2)p(r_t|\gamma_2,\theta_t)}{p_{t-1}(1)p(r_t|\gamma_1,\theta_t)} \tag{15}$$

Where we have labeled as 1 the model with a γ value of unity in (15) representing a straight Markov structure on time-variation for all parameters, including c_{it}. The competing model, labeled 2 in (15), might employ the modal value of the posterior determined value of $\gamma = \gamma^*$. At each observation, the ratio of posterior probabilities for any two values of γ, associated with any two DLM models in the multiprocess mixture is updated through multiplication by the corresponding Bayes' factor. As evidence in favor of one value for γ relative to the other values increases, we will see the posterior probability for the corresponding DLM model rise. The log-of-the-odds or Bayes' factor has been used extensively to compare alternative models having the type of structure set forth above. It should be noted that analogous comparisons of alternative models based on differing spatial weight matrices or different variable specifications can be made. That is, the procedure described here for drawing inferences concerning the parameter γ represents a general approach to inference in multiprocess mixture models, as described in West and Harrison (1989). One complication arises here: we have forty univariate models, one for each depository institution in the sample. If we treat each firm as a replicate of the underlying model, as in the case of the repeated measures treatment of multivariate autoregressive models, then the likelihood function is simply the sum of the log-likelihood function for each firm. This implies that the overall Bayes' factor should be based on a term for $L_t(j)$ in (14) equal to the sum of L_{it} where i ranges over all firms.

There are a number of advantages associated with the approach to inference set forth here. First, the Kalman filter processing of time-series data observations on a period-by-period basis allows one to identify influential data observations associated with parameter values of interest. For example, we find that observations around the Ohio depository insurance crisis period have an extremely important impact on the posterior distribution for the parameter γ. Given the propensity of regional time-series to contain outliers and influential data points, as well as regime shifts and other types of heterogeneity [see Anselin (1988)], the period-by-period inference procedures would seem advantageous in alerting the practitioner to these problems. Second, it seems likely that many regional econometric relationships exhibit inherent time-variation as regional economic growth and decline takes place. At each time period, the Bayes factor reflects a sequential test based on output from data up to the most recent time period. This approach to testing the impact of spatial variation, or other regional hypotheses, differs from previous studies where the entire data sample is usually employed to perform a single test for significance. Inferential results can be interpreted as treating each point in time as a sample end point. This is important if one conjectures that spatial and other regional effects vary considerably over time. Note that the estimation procedure produces a wealth of output information: parameter estimates and variance-covariances for each time period for all firms, likelihood values (or posterior probabilities) for the spatial influence parameter γ at each time period, and predicted and residual values for each firm's CD rate time-series. It is important to point out, all of this information is not without costs. First, the ability to handle uncertainty about parameters or model specification issues may not be straightforward. We must have a tractable, manageable expression for the posterior shown in (14) that can be easily computed. It may be the case that the parameter of interest enters the likelihood in a complicated way ruling out a conjugate analysis. (Lacking conjugate forms for the prior on the parameter of interest, one would have to perform all calculations for (13), (14) and (15) numerically.) Indeed, some simplifying assumptions concerning the covariance structure between depository institutions were invoked here in order to preserve the conjugate tractability. Second, even if the above conditions for a conjugate analysis hold, the dimension of the parameter space may be large or continuous, making it computationally impossible to carry out the calculations. This may be the case if one were interested in producing inferences regarding the spatial weighting matrix. These types of inference would most likely require that attention be restricted to a finite set of plausible spatial weighting matrices, rather than all possible configurations.

15.4 Implementing the Model

There may be concern regarding the somewhat ad hoc motivation for the estimation method proposed here. There are a number of points that can be made here. First, in order to estimate the non-operationally large number of parameters inherent in econometric models of space-time phenomena, restrictions must be imposed. These restrictions characterize all space-time econometric procedures. The approach taken here to handling these restrictions is quite different from previously employed approaches where restrictions are based on an entire vector or matrix in the model specification. The use of the Kalman filter allows stochastic restrictions to be imposed on an observation-by-observation basis, generalizing the more restrictive vector or matrix based procedures. If for example, time-variation in the parameters is inappropriate, posterior analysis will indicate this. Similarly, we will be able to draw an inference regarding the appropriateness of the spatial time-variation scheme for the parameters. The application undertaken here demonstrates how some of these issues arise in practice. A related area of concern regards the maintained (or hypothesized) model. The starting point here, as elsewhere, is to assume the standard linear hypothesis, $Y = X\beta + \varepsilon$. The proposed procedure could be viewed as introducing some uncertainty regarding this hypothesis. Perhaps, one could say that the standard linear model is likely to be the maintained model 90 percent of the time when working with space-time data samples. A small number of data samples (or observations) may deviate from this because of outliers, regime changes or spatial heterogeneity influences. Having generalized the maintained hypothesized model slightly, the method proposed here can still draw valid inferences regarding the most likely maintained model (the one in effect 90 percent of the time). In addition, we can accommodate the smaller number of cases where inferences associated with the more restrictive hypothesized model would be contaminated. As it turns out, the application introduced in this section represents a case where regime shifts and time-variation in the parameters does arise.

15.4.1 The Data

The data sample includes weekly six-month CD rates offered by 21 banks and 19 thrifts operating in or near three major cities in Ohio: Cincinnati, Cleveland, and Toledo. The classification into these three metropolitan areas reflects three regions of Ohio: Southwest, Northeast and Northwest. The time period covered by the sample was March 7, 1984 to December 31, 1987 producing a total sample of 200 weekly observations. These data were obtained by a survey. The institutions and their asset magnitudes are reported in Table 15.1. The sample of institutions does not include any of the 70 S&Ls that were closed during the Ohio crisis. The weekly observations coincide with the typical decision time frame for banks and thrifts and should be capable of reflecting timely changes in response to the Ohio deposit insurance crisis. The average weekly six-month, minimum deposit, retail CD rates were employed, since they have little implicit pricing and are the most frequently used maturity by both thrifts and banks, thus, allowing consistent data across the sample of institutions.

15.4.2 Prior Information

In order to implement the estimation procedure, prior information was used in setting initial values for parameters, their variances, observation noise and transition equation variances. In order to specify prior parameter values and variances, a sam-

Table 15.1: Ohio despository institutions in the sample.

Cincinnati			
Assets	Bank	Assets	Thrift
2197555	FNB Cincinnati	165771	Cottage Savings Assoc
499136	Park National	132852	Fidelity Federal S&L
187154	The Second Natl. Bank	100123	Mid Fed Svgs Middleton
133939	Ohio Valley	86266	Merit Savings Assoc
66761	Farmers and Traders Natl Bank	67062	Brentwood Savings Assoc
21240	First Breman Bank	57426	Harvest Home Svgs Assoc
12923	Citizens Bank of Degraff	53362	North Cin Loan and Bldg co.
Cleveland			
Assets	Bank	Assets	Thrift
7571547	Ameritrust	660655	Women's Fed Svgs Bank
5524549	National City	403745	People's Fed Svgs Wooster
873230	Dollar Savg and Trust Co	362803	First Federal Youngstown
318547	Bank One Mansfield	284286	First Federal (Lakewood)
79600	United Bank, Bucyrus	201309	First S&L Co. Masillion
22912	Hamler State Bank	180864	Horizon S&L
10844	Republic Banking Co	121188	Superior S&L
Toledo			
Assets	Bank	Assets	Thrift
6163560	Huntingon Toledo	583488	United Home Federal
2183716	Toledo Trust	402714	Peoples S&L Toledo
907144	Ohio Citizens	293847	State Home Savings
355644	Mid Am Bank	191163	First Fed Lima
180956	Metropolitan Bank Lima	83621	First Fed Bowling Green
91317	Commercial Bank at Delphos		
36952	Old Fort Banking Co.		

ple of CD rate-offers for the five largest Detroit, Michigan banks was used to estimate equation (7) using ordinary least-squares. Prior parameter values and variances were set to closely reflect the estimates from the Detroit sample. This type of procedure might be easily employed in other applications of the methods proposed here.

Given a cross-sectional sample of time-series data, one could always omit some part of the cross-sectional sample from the analysis and use it to determine prior parameter values and variances. The coefficient estimates indicated values for the intercept, a_{i0} around zero, a value of 0.75 for the lagged dependent variable, b_{i0} and a value of 0.25 for the competitive influence parameter c_{i0}. These values seem reasonable and are consistent with a prior belief that the desired level of deposits is relatively unimportant, the adjustment costs fairly important and the competitive influences important, but less so than adjustment costs. The intuition for this prior may be that depository institutions don't change the level of desired deposits on a weekly basis by very much, but rather adhere to a longer term plan embodied in the history of past CD rate-offers. Hence, the lagged dependent variable could be viewed as important and the intercept as relatively unimportant. Institutions probably react on a weekly basis to offers from competing institutions and may even need to revise their long-term plan in the face of such competition, leading to a fairly important role for the competitive influence parameter. The prior settings for the variance of the parameters a_{i0}, b_{i0} and c_{i0} were set so that a standard deviation of 0.1 existed for all three parameters. The estimated standard deviations from the Detroit sample were around 0.01, making the prior adopted here reflect relatively more uncertainty regarding these parameters. Although the prior is fairly diffuse, it is still proper. It should also be noted that the amount of sample data used before the 1985 Ohio crisis period is large (50 observations), making the impact of the prior settings within reasonable ranges fairly unimportant. Prior values for the variance of the disturbances in the parameter transition equation are a bit more difficult to determine, but again we can draw upon the Detroit sample results. We set $n_0 = 4$ and $l_0 = 1.0$ so that $E(\lambda^{-1}) = 1.0$. This reflects a prior belief that parameter variability was equal to that associated with observational noise variability. The fit of the equation to the Detroit sample was extremely good, having R^2 an value of 0.99, so that, this prior reflects a small amount of variability in the parameters over time. The prior noise estimate was set roughly equal to the noise estimate from ordinary least-squares estimation of the CD rate-setting relationship based on the Detroit sample of banks. The prior settings described above were employed for all forty depository institutions in the Ohio sample. Some experimentation was carried out with regard to the prior for parameter variability. This involved changing the prior to allow for more and less parameter variability and then examing the one-step-ahead predictive performance of the model averaged over all firms. The conclusions were that increasing or decreasing parameter variability tended to degrade the one-step-ahead predictions, suggesting that our strategy of employing the Detroit sample as the basis for the prior was quite successful. A FORTRAN program for implementing the spatial mixture model in the Appendix provides antoher approach to analyzing the impact of prior settings for the observation and parameter transition variances. The program produces a (5×5) grid around values for these two hyperparameters, showing how the posterior outcomes vary with respect to the prior.

The specification of the spatial weighting matrix involved defining all banks in a metropolitan area as exhibiting first-order contiguity. Similarly, all thrifts in a metropolitan area were specified as first-order contiguous. To make this clear, Table

15.2 shows a hypothetical spatial weighting matrix for twelve institutions, (six banks and six thrifts) based on this type of specification.

Table 15.2: A hypothetical spatial weighting matrix.

Cincinnati												
Bank 1	0	1	0	0	0	0	0	0	0	0	0	0
Bank 2	1	0	0	0	0	0	0	0	0	0	0	0
Thrift 1	0	0	0	1	0	0	0	0	0	0	0	0
Thrift 2	0	0	1	0	0	0	0	0	0	0	0	0
Cleveland												
Bank 1	0	0	0	0	0	1	0	0	0	0	0	0
Bank 2	0	0	0	0	1	0	0	0	0	0	0	0
Thrift 1	0	0	0	0	0	0	0	1	0	0	0	0
Thrift 2	0	0	0	0	0	0	1	0	0	0	0	0
Toledo												
Bank 1	0	0	0	0	0	0	0	0	0	1	0	0
Bank 2	0	0	0	0	0	0	0	0	1	0	0	0
Thrift 1	0	0	0	0	0	0	0	0	0	0	0	1
Thrift 2	0	0	0	0	0	0	0	0	0	0	1	0

The effect of this spatial weighting scheme is to specify that all banks in each metropolitan area compete with (influence) all other banks in that area, and all thrifts in the area compete with other thrifts, but do not influence banks. This weighting does not allow competition to extend beyond the limits of the metropolitan area, or between the two types of institutions. Alternative weighting schemes will be examined in future work.

A fair number of the institutions in the sample are in outlying areas, making a county based first-order contiguity scheme possible. The scheme used here illustrates that the term 'spatial weighting' can be interpreted quite broadly when it comes to specifying the weighting matrix. A Bayesian approach to this specification might involve making the weighting matrix part of the prior specification.

Assuming that the weighting matrix in Table 15.2 is standardized, (divide each element by the number of non-zero terms in each row) the impact of the weighting scheme is to shrink the competitive influence parameters of all banks in each metropolitan area towards the average parameter values of all other banks in the area. Similarly, the competitive influence parameters for thrifts will be forced to take on values similar to the mean of all other thrifts in the area.

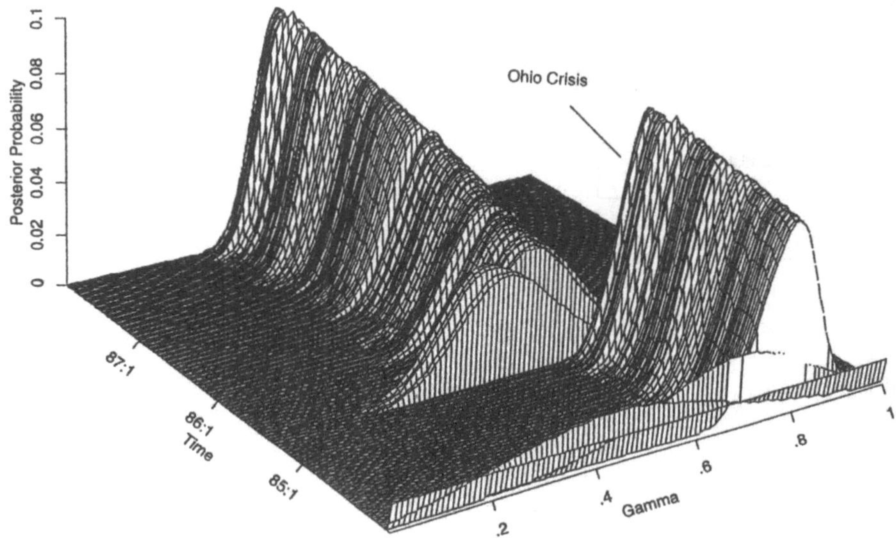

Figure 15.1: Posterior distribution of gamma parameter.

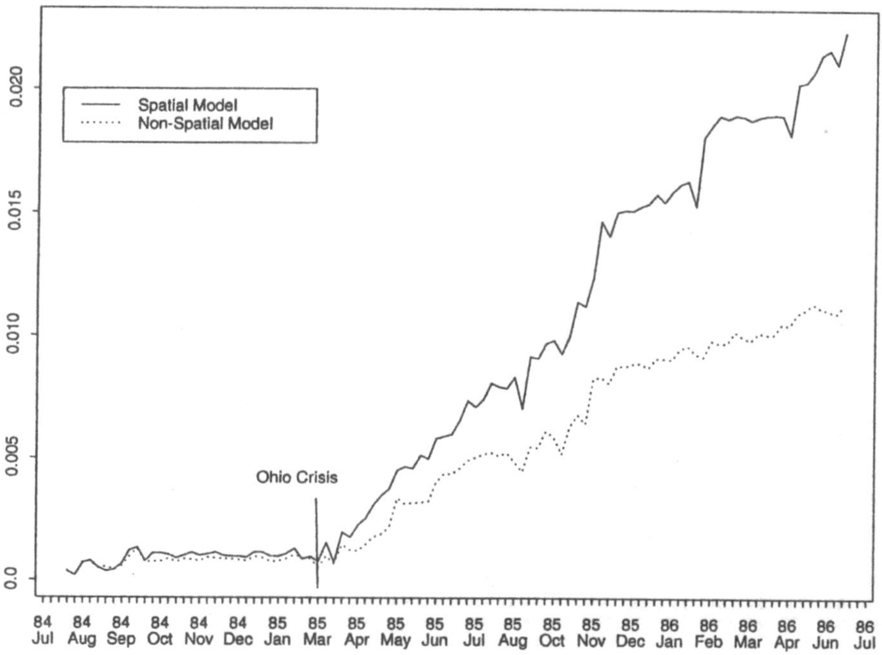

Figure 15.2a: Time-varying intercept term estimates (median of 40 depository institutions).

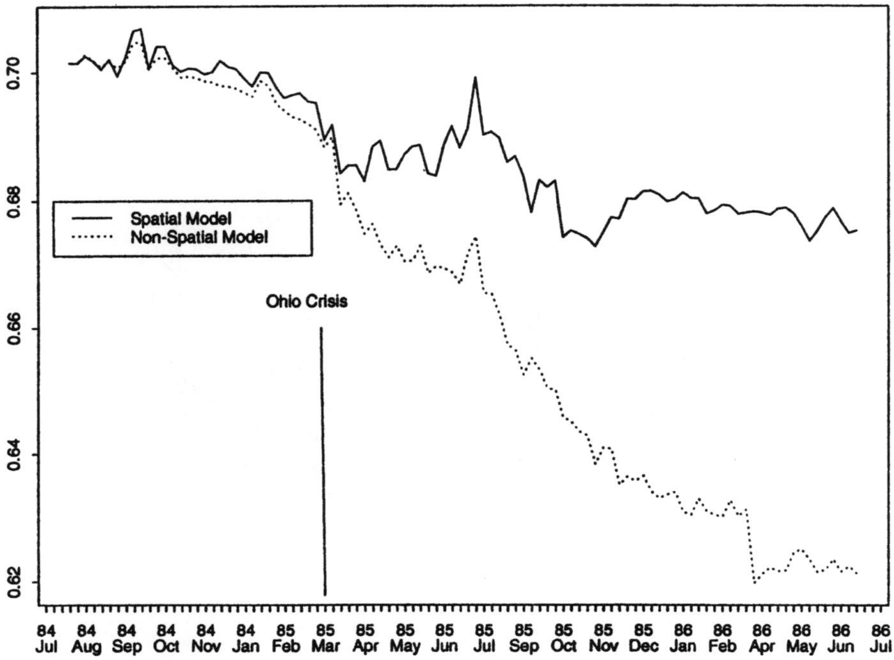

Figure 15.2b: Time-varying lagged variable estimates (median of 40 depository institutions).

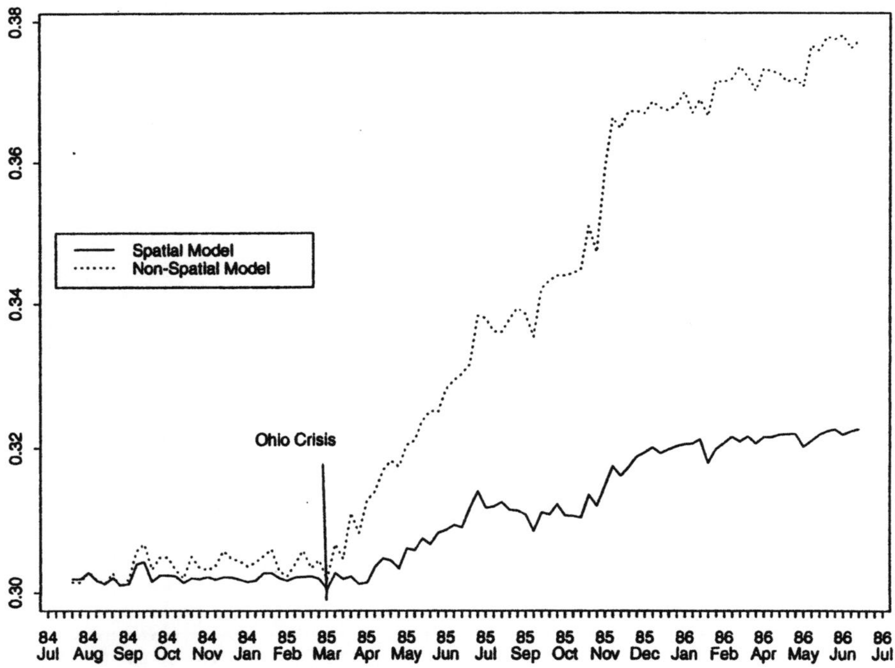

Figure 15.2c: Time-varying competitive variable estimates (median of 40 depository institutions).

15.4.3 Empirical Results

The first issue we wish to explore is the appropriateness of the spatial time-variation specification in equation (9). Figure 15.1 shows the posterior probability distribution for the γ parameter, derived using equations (13) and (14). The initial (prior) probabilities were set to uniform values of 0.01 for the one hundred point grid over γ values in the interval (0,1). These show up in the lower right of Figure 15.1 at time zero as a horizontal mass. After processing a few observations, the posterior mass moved quickly toward a distribution centered around 0.9. Recall, this indicates that spatial influences are not very important, values of unity reflecting no spatial variation in the parameters c_{it}. At the time of the Ohio depository insurance crisis in March of 1985, the posterior distribution exhibited an abrupt jump which then settled down to a modal value of $\gamma = 0.6$.

The inference we draw is that, spatial variation in the parameters became more important after the crisis event, with institutions more heavily influenced by competing institutions in their own metropolitan areas. We also conclude, that the model which allowed for spatial variation in the parameters over time dominated the Markov variation scheme for the entire data sample. One other point to note from Figure 15.1 is that the posterior distribution for γ is not all that peaked, producing a fair amount of uncertainty regarding a point estimate for this parameter. This may be the result of assuming that a single γ value holds for all firms in our sample and the accompanying averaging across all firms used to produce the posterior in Figure 15.1. Figures 15.2a, 2b, and 2c show the patterns of time-variation in the three parameters, a_{it}, b_{it}, and c_{it}, with the median value for the forty firms in the sample plotted. The two lines in Figures 15.2a, 2b and 2c compare the time paths indicated by the non-spatial model $\gamma = 1$ to that based on a model with γ set to the modal posterior values shown in Figure 15.1. The most obvious point made by Figures 15.2a, 2b and 2c is that both the spatial and non-spatial models depict the Ohio crisis event as having an impact on the time path of all three parameters. The non-spatial model assigns a relatively larger impact to the competitive influence parameter than the spatial model, and a smaller impact to the intercept parameter than the spatial model. The one clear conclusion here is that inferences drawn regarding parameter variation over time depend on the manner in which spatial influences are modeled. The posterior distribution for γ suggests that the spatial model exhibits the most appropriate inference in this regard. Intuitively, the large intercept shift indicated by the spatial model suggests that depository institutions adjusted their desired (target) deposit levels upward after the crisis. This seems consistent with the fact that there were large runs on state-insured savings and loans which resulted in more than \$ 200 million being deposited in other institutions. The parameter time paths reported by the non-spatial model seem to allocate the crisis period parameter shift to the lagged dependent and competitive influence parameters. The posterior distribution for γ suggests that this is incorrect, leading us to conclude that the Markov parameter transition scheme would inappropriately attribute routine spatial competitive influences to the crisis event. Other topics posterior analysis with the model included tests for systematic parameter differences between large, medium and small depository institutions and differences in the parameters of institutions in various regions of the state, as well as, between thrifts and banks. These results are not presented here because, further exploration using more realistic spatial weight matrices seems a first priority before proceeding with analysis of issues such as these.

15.5 Conclusions

Beginning with a theoretical model of depository institutions which viewed CD rate-offers as a control used to target desired levels of deposits, we derived an an optimal decision-rule. The rule suggested that the parameters depend on three factors: adjustment costs, the desired level of deposits, and substitution effects arising from competitive institutions. An innovative parameter transition scheme allowed the mean of parameters from spatially contiguous institutions to be incorporated during estimation as Bayesian prior information. A spatially distributed sample of forty Ohio depository institutions weekly CD rate-offers were employed to estimate the model. Bayesian posterior analysis of the model suggests that the spatial parameter transition scheme dominates a model based on the more traditional non-spatial Markov transition scheme, often used in time-varying parameter models. An implication is, we might add to the concerns of Anselin and Griffith (1988) that spatial effects need to be considered, an additional concern about modeling these effects in an appropriate way. A general methodology for posterior inference in such models, based on West and Harrison (1989), was demonstrated using the Ohio depository institution sample. The empirical results suggest that spatial influences became more important in CD rate-setting as a result of the deposit insurance crisis which took place during March of 1985.

Acknowledgements

This research was funded by the Urban Affairs Center at the University of Toledo. The author would like to thank Elizabeth S. Cooperman and Winson B. Less for kindly providing the survey data for this study. Michael Magura provided helpful comments on an earlier draft of this paper.

References

Anselin, L., *Spatial Econometrics: Methods and Models*, Dordrecht: Kluwer Academic Publishers, 1988.

Anselin, L. and D. A. Griffith, Do Spatial Effects Really Matter in Regression Analysis?, *Papers of the Regional Science Association*, 65, 11-34, 1988.

Cooperman, E.S., W.B. Lee and J.P. LeSage, Geographical Integration and the Retail CD-Pricing Decisions of Large Depository Institutions, *Review of Economics & Statistics*, 73, 546-552, 1991.

Duncan, G., W. Gorr and J. Szczypula, Bayesian Forecast for Seemingly Unrelated Time Series: Application to Local Government Revenue Forecasting, *Management Science*, 39, 275-293, 1993.

Garcia-Ferrer, A., R.A. Highfield, F. Palm and A. Zellner, Macroeconomic Forecasting using Pooled International Data, *Journal of Business & Economic Statistics*, 5, 53-68, 1987.

Harrison, P.J., and C.F. Stevens, Bayesian Forecasting, *Journal of Royal Statistical Society*, Series B, 38, 205-247, 1976.

Kane, E.J., *The S&L Insurance Mess: How Did It Happen?*, Washington, D.C.: The Urban Institute Press, 1989.

Kenen, P.B., *Capital Mobility and Financial Integration: A Survey*, Princeton: Princeton University Press, 1976.

LeSage, J.P., Forecasting Turning Points in Metropolitan Employment Growth Rates using Bayesian Techniques, *Journal of Regional Science*, 30, 533-548, 1990.

LeSage, J.P. and M. Magura, Using Bayesian Techniques for Data Pooling in Regional Payroll Forecasting, *Journal of Business & Economic Statistics*, 8, 127-135, 1990.

West, M. and J. Harrison, *Bayesian Forecasting and Dynamic Models*, New York: Springer-Verlag, 1989.

Zellner, A., *An Introduction to Bayesian Inferences in Econometrics*, New York: John Wiley & Sons, 1971.

Appendix 1: FORTRAN Program for the Spatial Mixture

This program is really three programs in one. The different programs are implemented by deleting comments labeled C1, C2 and C3. This program is set for a model with 4 explanatory variables and an intercept term. If you change the number of variables in the model, you have to change the format statements that are used to write out the results.

Program 1 performs a grid search over sige and sigw to find optimal settings for these parameters. No spatial mixing is done here. Uncomment all lines beginning with C1. The program prints out the likelihood function values for a (5×5) grid of sige and sigw values. sigel and sigwl determine the starting values for sige and sigw and asize and bsize control the movement of the grid; for example:

```
likelihood over iterations
        103.9      103.8      103.8      103.8      103.8
        100.0      100.0       99.9       99.9       99.9
         99.8       99.7       99.7       99.7       99.7
        101.8      101.8      101.8      101.8      101.8
        105.5      105.4      105.4      105.4      105.4

 bsize = 2.0
 sigwl = 15.0

 sigw0 over iterations
      17.000     19.000     21.000     23.000     25.000
      17.000     19.000     21.000     23.000     25.000
      17.000     19.000     21.000     23.000     25.000
      17.000     19.000     21.000     23.000     25.000
      17.000     19.000     21.000     23.000     25.000

 asize = 0.05
 sigel = 0.15

 sige0 over iterations
       0.200      0.200      0.200      0.200      0.200
       0.250      0.250      0.250      0.250      0.250
       0.300      0.300      0.300      0.300      0.300
       0.350      0.350      0.350      0.350      0.350
       0.400      0.400      0.400      0.400      0.400
```

Program 2 uses the optimal settings of sige and sigw in order to print out a set of mixture parameter estimates, variances and posterior probabilities for each model for each time period. To produce output of mixture parameter estimates, variances, posterior probabilities, cumulative likelihood, etcetera, after the grid search over `sige` and `sigw`, uncomment all lines beginning with `C2`. (Leave the `C1` and `C3` comments in place.) Enter the likelihood maximizing values of `sige` and `sigw` (those that produce a minimum value in the (5×5) grid) for `sige1` and `sigw1`.

Program 3 uses the parameter estimates determined by program 2 in order to form the matrix `B`. It then implements spatial mixing and grids over gamma values between 0 and 1 producing likelihood function (posterior) values for each value of gamma. To produce the spatial mixture uncomment all lines beginning with `C3`. This requires that you read in coefficient estimates produced by Program 2 in order to form the matrix `B`.

```
c--------------------------------
       implicit real*4 (a-h,o-z), integer(i-n)
C1     parameter (nk=5, nobs=70, nm=7, aiter=5)
C1     parameter (biter=5)
C2     parameter (nk=5, nobs=70, nm=7)
C3     parameter (nk=5, nobs=70, nm=7, aiter=101)
       dimension y(nobs), x(nobs,nk), bmat(nk,1)
       dimension cmat(nk,nk), xmat(1,nk), ymat(1,1)
       dimension bmix(nobs,nm,nk), sige(nm)
       dimension cmix(nobs,nm,nk,nk), rij(nm*nm)
       dimension ymix(nobs,nm), ymean(nobs)
       dimension bmean(nobs,nk), cmean(nobs,nk,nk)
       dimension blike(aiter,biter),dlike(aiter,biter)
       dimension rwtmp(nm,nk,nk), retmp(nm)
       dimension bi(nm,nk), yi(nm), ei(nm), qq(nm)
       dimension ci(nm,nk,nk), ri(nm), sigw(nm)
       dimension clike(nobs), pij(nm*nm,nk,nk)
       dimension fij(nm*nm), sij(nm*nm,nk)
       dimension bij(nm,nm,nk), cij(nm,nm,nk,nk)
       dimension zij(nm*nm), ezij(nm*nm)
       dimension prij(nm,nm), po(nobs,nm), olike(nobs)
       dimension pwght(nobs,nm*nm), pp(nobs,nm)
       dimension cijt(nm,nm,nk,nk), olike(nobs)
       dimension tmpv(1,nk), tmp(1,1), vtmp(nk,1)
       dimension vtmp2(nk,1), tmp1(nk), tmp2(nk)
       dimension tmp3(nk), tmp4(nk,nk), tmp5(nk,nk)
C3   dimension bo1(nobs,nk), bo2(nobs,nk)
C3   dimension bo3(nobs,nk)
C3   dimension alike(aiter)
C1   dimension alike(aiter,biter)

c--------------------------------
```

```
C1      asize = 0.05
C1      bsize = 2.0
C1      sigel = 0.15
C1      sigwl = 15.0
c set sige and sigw to values from Program 1
C2      sigel = 0.183
C2      sigwl = 22.1
C3      sigel = 0.183
C3      sigwl = 22.1
        dofval = 5.0
c nk   = number of parameters in regression
c nobs = number of observations
c nm   = number of models (should = nk+2)
c aiter  = points in grid over sige for Program 1
c aiter  = points in grid over gamma for Program 3
c biter  = number of points in grid over sigw
c asize  = size of steps taken for sige values
c bsize    = size of steps taken for sigw values
c dofval  = dof  for Gamma distribution
c ---------------------------
c read in the data
        do i=1,nobs
        read(5,*) y(i), x(i,2), x(i,3), x(i,4), x(i,5)
c set intercept term in the model
        x(i,1) = 1.0
        end do
c  read in the B  with parameters from other areas
C3      do i=1,nobs
C3      read(5,*) (bo1(i,k),k=1,nk)
C3      end do
C3      do i=1,nobs
C3      read(5,*) (bo2(i,k),k=1,nk)
C3      end do
C3      do i=1,nobs
C3      read(5,*) (bo3(i,k),k=1,nk)
C3      end do
C1      do i=1,aiter
C1      alike(i,1) = 0.0
C1      alike(i,2) = 0.0
C1      end do
C1      do i=1,biter
C1      blike(i,1) = 0.0
C1      blike(i,2) = 0.0
C1      end do
```

```
C3      gamma=0.0
c       sc = spatial weight matrix in the form of a scalar
C3      sc = 0.333
C3      do aaiter=0,aiter
C3      gamma = aaiter*0.01
C3      gam = 1.0-gamma
C1      do aaiter = 1,aiter
C1      sige0 = sige1 + aaiter*asize
c set sige to its optimal value
C2      sige0 = sige1
C3      sige0 = sige1
C1      do bbiter = 1,biter
C1      sigw0 = sigw1 + bbiter*bsize
c set sigw to its optimal value
C2      sigw0 = sigw1
C3      sigw0 = sigw1
        dof = dofval
c initialize the likelihood vectors
        do i=1,nobs
        olike(i) = 0.0
        clike(i) = 0.0
        end do
c-----------------------------
c priors for separate models
c 1 = steady state model,
c 2 = intercept shift,
c 3 = slope1 shift,
c 4 = slope2 shift, etc.
c nk+2 = nm = outlier model
c prior b-values (close to OLS usually works)
        bmat(1,1) =  0.0
        bmat(2,1) =  1.0
        bmat(3,1) = -1.0
        bmat(4,1) = -1.0
        bmat(5,1) =  0.3
c prior state variances
        sigw(1) = 1.0
        sigw(2) = 2.0
        sigw(3) = 2.0
        sigw(4) = 2.0
        sigw(5) = 1.0
        sigw(6) = 0.3
        sigw(7) = 0.0
c prior noise estimates
```

```
      sige(1) = 1.0
      sige(2) = 1.0
      sige(3) = 1.0
      sige(4) = 1.0
      sige(5) = 1.0
      sige(6) = 1.0
      sige(7) = 100.0
c prior probabilities over the models
      qq(1) = 0.85
      qq(2) = 0.02
      qq(3) = 0.02
      qq(4) = 0.02
      qq(5) = 0.02
      qq(6) = 0.02
      qq(7) = 0.05
      do i=1,nobs
      do k=1,nm
      pp(i,k) = 1.0/(nm*1.0)
      end do
      end do
c ----------------------
c set up prior information matrices
      do i=1,nm
      do j=1,nk
       do k=1,nk
       rwtmp(i,j,k) = 0.0
       if ((i.ne.1).or.(i.ne.nm)) then
         if((j.eq.(i-1)).and.(k.eq.(i-1)))
     1   rwtmp(i,j,j) = sigw0*sigw(i)
         endif
       end do
      end do
      ri(i) = 5.0/(dof-3.0)
      ri(i) = sige0*ri(i)
      retmp(i) = sige0*sige(i)
      end do
      retmp(nm) = sige(nm)
      do i=1,nk
      do j=1,nk
      cmat(i,j) = 0.0
      if (i .eq. j) cmat(i,i) = 1.0
      end do
      end do
      do i=1,nm
```

```
      do j=1,nk
       do k=1,nk
       ci(i,j,k) = cmat(j,k)*sige0
       end do
      bi(i,j) = bmat(j,1)
      end do
      end do
c  start looping over data
      do ii=2,nobs
c pull out the data
      ymat(1,1) = y(ii)
      do k=1,nk
      xmat(1,k) = x(ii,k)
      end do
c extended Kalman filter equations
c pij = ci(i,,) + rwtmp(j,,)
c fij = xmat(1,k) * pij * xmat(k,1) + retmp(j)
c sij = pij * xmat(k,1) * inv(fij)
c yi(i) = xmat(1,k) * btmp(i,k)
c ei(i) = ymat(1,1) - yi(i)
c rij = ei*ei/fij
c bij = bbsav(i,k) + sij * ei(i)
c cij = pij - sij * fij * sij'
c
c find pij = ci(i,,) + rwtmp(j,,)
          ij = 0
          do i=1,nm
          do j=1,nm
          ij = ij + 1
           do k=1,nk
           do l=1,nk
           pij(ij,k,l) = ci(i,k,l) + rwtmp(j,k,l)
           end do
           end do
          end do
          end do
c find fij=xmat(1,k)* pij * xmat(k,1)+retmp(j)
          ij = 0
          do i=1,nm
          do j=1,nm
          ij = ij + 1
           do k=1,nk
           do l=1,nk
           tmp4(k,l) = pij(ij,k,l)
```

```
            end do
            end do
c compute xmat(1,k) * pij
           call ab(xmat,tmp4,1,nk,nk,1,nk,tmpv,1)
c compute xmat(1,k) * pij * xmat(k,1)
           call abp(tmpv,xmat,1,nk,1,1,1,tmp,1)
c add retmp and store the result
           fij(ij) = tmp(1,1) +  retmp(j)
           end do
           end do
c find sij = pij * xmat(k,1) * inv(fij)
           ij = 0
           do i=1,nm
           do j=1,nm
           ij = ij + 1
             do k=1,nk
             do l=1,nk
             tmp4(k,l) = pij(ij,k,l)
             end do
             end do
c compute  pij * xmat(k,1)
           call abp(tmp4,xmat,nk,nk,1,nk,1,vtmp,nk)
c compute  pij * xmat(k,1) * inv(fij)
             do k=1,nk
             vtmp(k,1) = vtmp(k,1)/fij(ij)
c store the result
           sij(ij,k) = vtmp(k,1)
             end do
           end do
           end do
c find yi(i) = xmat(1,k) * bi(i,k)
           do i=1,nm
           yi(i) = 0.0
             do k=1,nk
             yi(i) = yi(i) + xmat(1,k)*bi(i,k)
             end do
           end do
c find ei(i) = ymat(1,1) - yi(i)
           do i=1,nm
           ei(i) = ymat(1,1) - yi(i)
           end do
c find rij = ei*ei/fij
           ij = 0
           do i=1,nm
```

```
          do j=1,nm
          ij = ij + 1
          rij(ij) = ri(i) + ei(i)*ei(i)/fij(ij)
          end do
          end do
c find bij = bi(i,k) + sij * ei(i)
          ij = 0
          do i=1,nm
          do j=1,nm
          ij = ij + 1
            do k=1,nk
C1         bij(i,j,k) = bi(i,k) + sij(ij,k)*ei(i)
C2         bij(i,j,k) = bi(i,k) + sij(ij,k)*ei(i)
C3         bij(i,j,k)=gamma*bi(i,k)+gam*sc*bo1(ii,k)
C3   1    +gam*sc*bo2(ii,k)+gam*sc*bo3(ii,k)
C3   1    +sij(ij,k)*ei(i)
            end do
          end do
          end do
c find cij = pij - sij * fij * sij'
          ij = 0
          do i=1,nm
          do j=1,nm
          ij = ij + 1
            do k=1,nk
            do l=1,nk
            tmp4(k,l) = pij(ij,k,l)
            end do
            vtmp(k,1) = sij(ij,k)
            end do
          tmp(1,1) = fij(ij)
c compute  sij * fij
          call ab(vtmp,tmp,nk,1,1,nk,1,vtmp2,nk)
c compute  sij * fij * sij'
          call abp(vtmp2,vtmp,nk,1,nk,nk,nk,tmp5,nk)
c compute  pij - sij * fij * sij'
            do k=1,nk
            do l=1,nk
            cij(i,j,k,l) = tmp4(k,l) - tmp5(k,l)
            end do
            end do
          end do
          end do
c update the dof for the gamma distribution
```

```
      dofs = dof
      dof = dof + 1.0
c find the posterior probabilities and do the collapsing
      ij = 0
      do i=1,nm
      do j=1,nm
      ij = ij + 1
      zij(ij)= -0.5*ALOG(fij(ij))+0.5*dofs*
     1        ALOG(ri(i))-0.5*dof*ALOG(rij(ij))
      end do
      end do
      ij = 0
      do i=1,nm
      do j=1,nm
      ij = ij + 1
      ezij(ij) = EXP(zij(ij))
      end do
      end do
c find the normalizing constant
      ct = 0.0
      k = ii-1
      ij = 0
      do i=1,nm
      do j=1,nm
      ij = ij + 1
      ct = ct + pp(k,i)*ezij(ij)*qq(j)
      end do
      end do
      ct = 1.0/ct
c compute normalized pp (posterior probabilities)
      ij = 0
c note k = ii-1 set above
      do i=1,nm
      do j=1,nm
      ij = ij + 1
      prij(i,j) = ct*ezij(ij)*qq(j)*pp(k,i)
      end do
      end do
c get one-step-back and other posterior probabilities
      do i=1,nm
      pp(ii,i) = 0.0
      po(ii,i) = 0.0
      do j=1,nm
      pp(ii,i) = pp(ii,i) + prij(j,i)
```

```
      po(ii,i) = po(ii,i) + prij(i,j)
      end do
      end do
c get weighting ratio
      ij = 0
      do j=1,nm
      do i=1,nm
      ij = ij + 1
      pwght(ii,ij) = (prij(i,j)/pp(ii,j))
      end do
      end do
c use the posterior probability collapsing weights
c to collapse the bij bhat vectors
      ij = 0
      do j=1,nm
        do k=1,nk
        bi(j,k) = 0.0
        end do
      do i=1,nm
      ij = ij + 1
        do k=1,nk
        bi(j,k) = bi(j,k) + pwght(ii,ij)*bij(i,j,k)
        end do
      end do
      end do
c use the posterior probability collapsing weights
c to collapse the rij
      ij = 0
      do j=1,nm
      icnt = j
      ri(j) = 0.0
      do i=1,nm
      ij = ij + 1
      ri(j) = ri(j) + pwght(ii,ij)*(1.0/rij(icnt))
      icnt = icnt + nm
      end do
      end do
      do i=1,nm
      ri(i) = 1.0/ri(i)
      end do
c use the posterior probability collapsing weights
c to collapse the cij var-cov matrices
      do i=1,nm
      do j=1,nm
```

```
          do k=1,nk
          tmp1(k) = bij(i,j,k)
          tmp2(k) = bi(j,k)
          tmp3(k) = tmp1(k) - tmp2(k)
          end do
        call abp(tmp3,tmp3,nk,1,nk,nk,nk,tmp4,nk)
          do k=1,nk
          do l=1,nk
          cijt(i,j,k,l) = cij(i,j,k,l)+tmp4(k,l)
          end do
          end do
        end do
        end do
c apply the probability weights
          ij = 0
          do j=1,nm
            do k=1,nk
            do l=1,nk
            ci(j,k,l) = 0.0
            end do
            end do
          do i=1,nm
          ij = ij + 1
           do k=1,nk
           do l=1,nk
           ci(j,k,l)=ci(j,k,l)+pwght(ii,ij)*cijt(i,j,k,l)
           end do
           end do
          end do
          end do
c transfer mixture results into storage
        do i=1,nm
        do k=1,nk
        bmix(ii,i,k) = bi(i,k)
        end do
        end do
        do i=1,nm
          do j=1,nk
          do k=1,nk
          cmix(ii,i,j,k) = ci(i,j,k)
          end do
          end do
        end do
        do i=1,nm
```

```
       ymix(ii,i) = yi(i)
       end do
c form mean estimates
          ymean(ii) = 0.0
          do k=1,nk
          bmean(ii,k) = 0.0
          do l=1,nk
          cmean(ii,k,l) = 0.0
          end do
          end do
          do k=1,nm
          ymean(ii) = ymean(ii) + pp(ii,k)*ymix(ii,k)
            do l=1,nk
            bmean(ii,l)=bmean(ii,l)+pp(ii,k)*bmix(ii,k,l)
            end do
          end do
c generate mean chat
          do k=1,nk
          do l=1,nk
          cmean(ii,k,l) = 0.0
          end do
          end do
          do j=1,nm
           do k=1,nk
           do l=1,nk
           cmean(ii,k,l)=cmean(ii,k,l)+pp(ii,j)*ci(j,k,l)
           end do
           end do
          end do
c calculate overall likelihood (posterior) for time t
          do j=1,nm
          olike(ii) = olike(ii) + pp(ii,j)*ri(j)
          end do
c calculate cumulative likelihood
          clike(ii) = clike(ii-1) + olike(ii)
c set to go through the observation loop again
c end of do over observations
        end do
c save this iterations likelihood values
C1       alike(aaiter,bbiter) = clike(nobs)
C1       blike(aaiter,bbiter) = sigw0
C1       dlike(aaiter,bbiter) = sige0
C3       alike(aaiter+1) = clike(nobs)
c  end of do biter
```

```
C1       end do
c  end of do aiter
C1       end do
C3       end do
c  write out grid iteration results --------------
C1       write(6,10)
C1 10    format(/,'likelihood over iterations',/)
C1       do i=1,aiter
C1       write(6,11) (alike(i,j),j=1,biter)
C1       end do
C1 11    format(5f12.3)
C1       write(6,12)
C1 12    format(/,'sigw over iterations',/)
C1       do i=1,aiter
C1       write(6,13) (blike(i,j),j=1,biter)
C1       end do
C1 13    format(5f12.3)
C1       write(6,14)
C1 14    format(/,'sige over iterations',/)
C1       do i=1,aiter
C1       write(6,15) (dlike(i,j),j=1,biter)
C1       end do
C1 15    format(5f12.3)
c  write out estimation results -------------
C2       write(6,10)
C2 10    format(/,'mean coefficients',/)
C2       do ii=1,nobs
C2       write(6,11) ii, (bmean(ii,k),k=1,nk)
C2       end do
C2 11    format('obs =',I3,5f12.4)
C2       write(6,12)
C2 12    format(/,'mixture var-cov matrices',/)
C2       do ii=1,nobs
C2       do j=1,nm
C2       do k=1,nk
C2       write(6,13) ii, j, (cmix(ii,j,k,l),l=1,nk)
C2       end do
C2       write(6,9)
C2       end do
C2       write(6,9)
C2       end do
C2 13    format('obs = ',I3,'     model = ',I2,5f16.8)
C2 9     format(/,'------------------------',/)
C2       write(6,14)
```

```
C2 14     format(/,'mean var-cov matrices',/)
C2        do ii=1,nobs
C2        write(6,15) ii, (cmean(ii,k,k),k=1,nk)
C2        write(6,9)
C2        end do
C2 15     format('obs = ',I3,5f16.8)
C2        write(6,16)
C2 16     format(/,'mixture and mean predictions',/)
C2        do ii=1,nobs
C2        write(6,17) ii, (ymix(ii,i),i=1,nm),ymean(ii)
C2        end do
C2 17     format('obs = ',I3,8f9.3)
C2        write(6,18)
C2 18     format(/,'posterior probs non-smoothed',/)
C2        do ii=1,nobs
C2        write(6,19) ii, (pp(ii,i),i=1,nm)
C2        end do
C2 19     format('obs =',I4,7f10.3)
C2        write(6,20)
C2 20     format(/,'posterior probs smoothed',/)
C2        do ii=1,nobs
C2        write(6,21) ii, (po(ii,i),i=1,nm)
C2        end do
C2 21     format('obs =',I4,7f10.3)
C2        write(6,22)
C2 22     format(/,'Cumulative likelihood',/)
C2        do ii=1,nobs
C2        write(6,23) ii, clike(ii)
C2        end do
C2 23     format('obs =',I4,f16.8)
c  write out gamma iteration results -------
C3        write(6,10)
C3 10     format(/,'likelihood over gamma
         1  iterations',/)
C3        do i=0,aiter-1
C3        gami = i/100.0
C3        write(6,11) gami, alike(i+1)
C3        end do
C3 11     format(2f10.3)
         stop
         end

c ----------------------------------
   subroutine ab(a,b,iar,iac,ibc,iadim,ibdim,abo,abr)
```

```
c multiplies a times b.
c iar is the number of rows in a.
c iac is the numberof cols in a (and rows in b).
c ibc is the number of cols in b.
c iadim is the row dimension of a in main
c ibdim is the row dimension of b in main
c ab is the result with row dimension abr in main
      integer iar, iac, ibc, iadim, ibdim, abr, i,j,k
      real*4 a(iadim,iac), b(ibdim,ibc), abo(abr,ibc)
      double precision tmp
      if (iadim .ge. iar .and. ibdim .ge. iac) goto 1
      write(6,100)
  100 format(/'* Dim Error in a or b matrix in ab'/)
      goto 2
  1   do 10 i=1,iar
        do 10 j=1,ibc
        tmp = 0.0D0
          do 20 k=1,iac
          tmp = DBLE(a(i,k))*DBLE(b(k,j))+tmp
   20     continue
        abo(i,j) = tmp
   10 continue
  2   return
      end
c ---------------------------------------------
subroutine abp(a,b,iar,iac,ibr,iadim,ibdim,abpo,abr)
c multiplies a times b-transpose
c iar is the number of rows in a (and c).
c iac is the number of cols in a (and b).
c ibr is the number of rows in b
c iadim is the row dimension of a in main
c ibdim is the row dimension of b in main
c abp is the result with row dim abr in main
       integer iar, iac, ibr, iadim, ibdim, abr,i,j,k
       real*4 a(iadim,iac),b(ibdim,iac),abpo(abr,ibr)
       double precision tmp
       if (iadim.ge.iar.and.ibdim.ge.ibr) goto 1
       write(6,100)
  100 format(/'* Dim Error in a or b matrix in abp'/)
       goto 2
  1    do 10 i=1,iar
         do 20 j=1,ibr
         tmp = 0.0D0
           do 30 k=1,iac
```

```
            tmp = tmp + DBLE(a(i,k))*DBLE(b(j,k))
   30       continue
         abpo(i,j) = tmp
   20    continue
   10 continue
   2  return
      end
```

Author Index

Subject Index

Contributors

Luc Anselin is research professor of Regional Science at the Regional Research Institute of West Virginia University, where he is also professor of economics, adjunct professor of geography and adjunct professor of agricultural and resource economics. He previously served as associate director of the U.S. National Center for Geographic Information and Analysis at the University of California, Santa Barbara, where he was professor of geography and economics. He has published widely in the areas of regional and spatial analysis and is the author of the SpaceStat software package for spatial data analysis.

Denis Bolduc is currently an associate professor of economics at Université Laval, Canada and director of the GREEN research group. His current interests are on discrete choice modeling with an emphasis on interdependent choice over space. He has published papers on multinomial probit estimations, spatial econometrics and qualitative data related econometric techniques.

Emilio Casetti received a doctorate in law and economics from the University of Rome, Italy in 1951, and a Ph.D. in geography from the Northwestern University, Chicago, Illinois in 1966. He is a professor in the Department of Geography, Ohio State University, Columbus, Ohio, and is the current editor of *Geographical Analysis*.

Robin Dubin is currently an associate professor in the Department of Economics at Case Western Reserve University. Her research has been concentrated in the areas of urban economics and spatial econometrics.

George T. Duncan is professor of statistics and director of the Decision Systems Research Institute in the H. John Heinz III School of Public Policy and Management at Carnegie Mellon University. He holds B.S. and M.S. degrees from the

University of Chicago and a Ph.D. from the University of Minnesota, all in the field of statistics. He has done research in statistical methodology and decision processes. His research focus is on privacy and information issues; he chairs the National Academy of Sciences' Panel on Confidentiality and Data Access.

Mark R. Ferguson is a Ph.D. candidate in geography at McMaster University. His interests include spatial choice theory and modeling as well as applications of discrete choice models in the fields of migration and energy conservation analysis.

Raymond J.G.M. Florax received his Ph.D. from the University of Twente, the Netherlands, in 1992. He is an assistent professor of economics in the Department of General Economics at Wageningen Agricultural University. He has published on regional economic impacts of universities, spatial econometrics, and higher education financing, in both national and international journals, including *Regional Science and Urban Economics* and the *Journal of Regional Science.*

Reinaud van Gastel is a member of the Department of Theoretical Spatial Economics at the Erasmus University in Rotterdam. He has published some articles on (hypergraph) conflict analysis.

Arthur Getis is the Birch professor of geographical studies at San Diego State University. He has written a number of articles on various aspects of spatial analysis and geographic information systems. These papers have appeared in such journals as *Geographical Analysis*, *Environment and Planning A*, *Ecology*, *Growth and Change*, and the *Annals of Regional Science*. He is the North American editor of the new journal *Geographical Systems*. In addition, he serves as secretary and full member of the Commission on Mathematical Models of the International Geographical Union.

Wilpen L. Gorr is professor of public policy and management information systems in the H. John Heinz III School of Public Policy and Management at Carnegie Mellon University. His research interests include forecasting methods, spatial adaptive filtering, neural network models, illicit drug market modeling, HIV/AIDS spatial diffusion modeling, and geographic information systems applications. His publications appear in *Geographical Analysis*, *Management Science*, *International Journal of Forecasting*, *MIS Quarterly*, and several other journals.

Daniel A. Griffith holds a Ph.D. from the University of Toronto, and is professor of geography and director of the Interdisciplinary Statistics Program at Syracuse University. To date he has published 10 books and 80 articles in geography, regional science, economics, and statistics, and has convened spatial analysis workshops in universities, government agencies, and private industry. He has been a Fulbright Research Fellow to Canada, a visiting research professor at Erasmus University Rotterdam, a visiting professor at the university of Rome, and a U.S.

E.P.A. EMAP research affiliate. Currently his research interests focus on implementing spatial statistics with commercial software packages, errors in spatial databases, the quantitative/urban/economic geography of North America, and applied statistics.

Robert P. Haining is professor of geography at the University of Sheffield, England. His major research interests are in spatial analysis. He has completed a book *Spatial Data Analysis in the Social and Environmental Sciences*, and in collaboration with colleagues at the University of Sheffield is currently developing spatial analytical software linked to GIS for application to health data. Funding is being provided by the ESRC. He is involved in several projects concerned with the application of spatial analysis techniques in the field of medical geography. He also has research interests in the application of this methodology to problems in regional science.

Pavlos S. Kanaroglou is an associate professor of geography at McMaster University. He is interested in spatial choice models, migration and integrated transportation and land use urban models with applications to environmental policy.

Harry H. Kelejian received his B.A. degree in 1962 from Hofstra University and his M.A. and Ph.D. in economics from the University of Wisconsin in 1965 and 1968. He has held academic positions at Princeton and New York University before joining the faculty at the University of Maryland in 1974. He has published a wide variety of articles relating to econometric theory and its application in a variety of journals.

Richard Laferrière holds a Ph.D. in economics from Université de Montréal. His specialisation is in transportation economics. He is now full time economist with Canadian Pacific Rail, Montréal, P.Q., Canada. He published several papers in transportation related journals.

James P. LeSage received his Ph.D. from Boston College in 1983, and taught economics at Bowling Green State University from 1982 to 1988. In 1988 he joined the University of Toledo economics department, where he is currently an associate professor and director of the department's graduate program. His most recent research involves application of Bayesian econometric methods in the area of regional science. He has published work on Bayesian analysis and development of regional leading indicators, Bayesian forecasting of turning points in regional economic activity, Bayesian error correction models for regional labor market forecasting and has recently focused on space-time applications of the Bayesian Kalman filter.

Daniel P. McMillen received his Ph.D. in economics from Northwestern University in 1987, after which he taught at the University of Oregon and Santa Clara University. He currently is professor of economics at Tulane University in New

Orleans, LA, and serves on the editorial board of the *Journal of Urban Economics*. His articles have appeared in the *Journal of Urban Economics*, the *Journal of Regional Science*, *Regional Science and Urban Economics*, *Land Economics*, *Environment and Planning A*, and the *Review of Economics and Statistics*.

Jean Paelinck is professor of theoretical spatial economics at the Rotterdam School of Economics, the Faculty of Economics of the Erasmus University Rotterdam. As an author, co-author, editor and co-editor he has published over 50 books and 300 articles. His work on theoretical spatial economics and spatial econometrics has earned him seven honorary Ph.D.'s.

Jessie Poon received her Ph.D. in 1992 from the Ohio State University, Columbus, Ohio. She is currently a visiting assistent professor at the Department of Geography, University of Georgia, Athens, Georgia.

Serge Rey received his Ph.D. in geography from the University of California, Santa Barbara and is currently on the faculty at San Diego State University. His research interests include spatial econometrics, integrated multiregional modeling, and regional economics.

Dennis P. Robinson holds a Ph.D. from George Washington University. He has been an economist for several U.S. government agencies during the past twenty years including the Bureau of Economic Analysis, the Environmental Protection Agency, and the Army Corps of Engineers. His research is in regional and interregional modeling and in spatial econometric analysis.

Gino Santarossa holds a M.A. in economics from Université Laval since 1990. He is now an economist with the Québec Ministry of Finance. His research activities are concentrated in spatial econometrics and more generally in applied econometrics.

Janusz Szczypula is a Ph.D. candidate in the H. John Heinz III School of Public Policy and Management at Carnegie Mellon University. He received Master Degrees in computer science from Jagiellonian University, Poland; public affairs from the LBJ School of Public Affairs, University of Texas at Austin; and public policy from the Heinz School. His research interests include forecasting cross-sectional time series, rule-based forecasting, neural network models in forecasting, and AI applications in decision support systems.

MIX
Papier aus verantwortungsvollen Quellen
Paper from responsible sources
FSC® C105338

If you have any concerns about our products,
you can contact us on
ProductSafety@springernature.com

In case Publisher is established outside the EU,
the EU authorized representative is:
Springer Nature Customer Service Center GmbH
Europaplatz 3, 69115 Heidelberg, Germany

Printed by Libri Plureos GmbH
in Hamburg, Germany